Marie-Angèle Grandbastien • Josep M. Casacuberta
Editors

Plant Transposable Elements

Impact on Genome Structure and Function

Springer

Editors
Marie-Angèle Grandbastien
Institut Jean Pierre Bourgin
UMR 1318 INRA/AgroParisTech
INRA-Versailles
78026 Versailles, France
e-mail: gbastien@versailles.inra.fr

Josep M. Casacuberta
Department of Molecular Genetics
Center for Research in Agricultural Genomics (CRAG)
CSIC-IRTA-UAB-UB
Campus UAB
Bellaterra - Cerdanyola del Vallés
08193 Barcelona, Spain
e-mail: josep.casacuberta@cragenomica.es

ISBN 978-3-642-31841-2 ISBN 978-3-642-31842-9 (eBook)
DOI 10.1007/978-3-642-31842-9
Springer Heidelberg New York Dordrecht London

Library of Congress Control Number: 2012956160

Printed on acid-free paper

Springer is part of Springer Science+Business Media (www.springer.com)

Series description

Topics in Current Genetics publishes review articles of wide interest in volumes that center around a specific topic in genetics; genomics; as well as cell, molecular, and developmental biology. Particular emphasis is placed on the comparison of several model organisms. Volume editors are invited by the series editor for special topics, but further suggestions for volume topics are highly welcomed. Each volume is edited by one or several acknowledged leaders in the field, who ensure the highest standard of content and presentation. All contributions are peer reviewed. All volumes of Topics in Current Genetics are part of the Springer eBook Collection. The collection includes online access to more than 3,500 newly released books, book series volumes, and reference works each year. In addition to the traditional print version, this new, state-of-the-art format of book publications gives every book a global readership and a better visibility.

Preface

Transposable elements (TEs) are ubiquitous mobile DNA sequences found in both prokaryotic and eukaryotic genomes. They are able to insert at different positions of the genome, either by excising from one position and reinserting into another or by replicating into daughter copies. TEs are particularly abundant in plant genomes, where they can represent over 80 % of the bulk of large cereal genomes. Their discovery by B. McClintock, and the subsequent introduction of the notion of genome fluidity, was a major shift in our concepts on heredity. TEs can dramatically modify the structure of host genomes, affect genome sizes and generate genetic variation, not only by transposition but also by providing the raw material for genomic rearrangements due to their repetitive nature. Until recently, and in spite of B. Mc Clintock's seminal concept of "Controlling Elements," the impact of TEs on host genome function was merely regarded as circumstantial. A rather different representation has been brought to light in the last decade, which strongly argues that TEs may also act as pivotal factors in generating genic variation and modulating cellular gene expression. This book is intended at presenting the latest advances on the importance of TEs and on their impact on plant genome dynamics and function.

The TE research scene has recently seen major advances, with new tools such as Next Generation Sequencing (NGS) technologies opening tremendous possibilities for rapid global analyses of genomes at reduced costs. This has led to an exponential increase in the amount of TE-related data and to a deeper knowledge of their impact on host genomes. As a consequence, all plant researchers engaged in genomic studies are more or less unwillingly bumping into this wealth of TEs and are now realizing that these TEs cannot be discarded as annoying junk sequences anymore. TEs are encountered in both genomic and transcriptomic data, and in a tremendous variety of elements, including highly defective and deleted versions sometimes mobilized at surprisingly high levels via related copies, making their classification a difficult task. There is therefore a need for researchers to find guidelines to recognize and classify TEs and better understand their importance and potential impact.

This book is intended both for scientists familiar with the field and for nonspecialists. It is organized in 14 chapters written by recognized researchers and is centered, on one hand on how to recognize and study plant TEs, notably using NGS technologies, and, on the other hand, on how TEs impact plant genome structure and genome function, with a few final examples of exciting TE-mediated phenotypic impacts. The first few chapters cover important aspects of what are TEs and how they can be identified and analyzed. Chapter 1 covers recent developments in TE classification and annotation and tackles the complex issue of defining consistent guidelines, while Chap. 2 summarizes and compares computational tools available for TE identification and provides a road map for efficient annotation. Chapter 3 then explores how NGS technologies can be used to study TE-mediated genome size variations and evolutionary patterns that shape the TE compartment, and Chap. 4 describes the recent development of NGS technologies to monitor TE mobility. The three following chapters provide further insights on some of the best known plant TEs. Chapter 5 describes the predominant type of TEs found in plant genomes, the LTR retrotransposons, and the subtle functional interplay between their autonomous and nonautonomous versions, while Chap. 6 explores the intriguing possibility of the existence of plant endogenous retroviruses, and Chap. 7 updates our knowledge on the highly abundant miniature elements, MITEs, and their impact on plant genomes. Chapter 8 summarizes the current state of affairs for epigenetic mechanisms developed by plant genomes to control TE mobility and highlights the plasticity of these mechanisms. The two following chapters address the important issue of TEs in polyploid contexts: Chap. 9 summarizes current knowledge on TE involvement in the drastic structural and functional changes resulting from allopolyploidy, a major speciation process in the plant kingdom, while Chap.10 compares the nature and evolution of TEs between polyploid sugarcane and other grass genomes. The four following chapters are dedicated to several striking mechanisms by which TEs have been exapted by host genomes to distil invaluable tools for modifying genome function. Chapter 11 describes how a fascinating type of TEs, Helitrons, can capture gene fragments and describes how such process can lead to new regulatory functions, and Chap. 12 reviews in detail how plant TE coding sequences have been frequently domesticated into functional cellular genes. Chapter 13 assesses current knowledge on the ubiquitous process of SINE exaptation for the production of regulatory RNAs, and Chap. 14 updates current data on plant LTR retrotransposon stress response and examines the possibility that LTRs could play a role in modulating host gene expression. Finally, the last two chapters present particularly striking examples of TE-associated phenotypic changes. Chapter 15 illustrates the role of the Rider LTR retrotransposon in several morphological and physiological changes in tomato, while Chap. 16 describes how small RNAs produced by a non-LTR retrotransposon are involved in the desiccation tolerance of resurrection plants.

The chapters were conceived and written autonomously, so that they can be read independently, even though this may have resulted in a few redundancies. Many other topics could have been covered, and many other beautiful examples of TE impact on plant genomes could have been exposed, however it was impossible to assemble all of the chapters that we would have liked to have in this volume, due to lack of space. Nevertheless, we feel that the 14 chapters presented in this book provide altogether a global overview of the most interesting current advances in the field of plant TE studies, while providing a useful reference vademecum volume for all (highly welcomed!) newcomers to the field. We hope that they will feel the urge to better understand what are these repetitive sequences that compose more than half of their data and that, after consulting this book, they will become convinced that Transposable Elements are certainly not "junk," but may actually be by far the most interesting and fun part of their data!

Finally, we wish to heartily thank all authors of this volume, that all have made substantial efforts to share our common passion with you and to provide excellent contributions. We also thank Stefan Hohmann for providing us the opportunity to compile this volume, the staff at Springer Verlag for their continuous help and support to make this book possible, and Tom Bureau for correcting this text.

September 2012
Versailles, France — Marie-Angèle Grandbastien
Barcelona, Spain — Josep M. Casacuberta

Contents

1 **So Many Repeats and So Little Time: How to Classify Transposable Elements** 1
Thomas Wicker

2 **Transposable Element Annotation in Completely Sequenced Eukaryote Genomes** 17
Timothée Flutre, Emmanuelle Permal, and Hadi Quesneville

3 **Using Nextgen Sequencing to Investigate Genome Size Variation and Transposable Element Content** 41
Concepcion Muñoz-Diez, Clémentine Vitte, Jeffrey Ross-Ibarra, Brandon S. Gaut, and Maud I. Tenaillon

4 **Genome-Wide Analysis of Transposition Using Next Generation Sequencing Technologies** 59
Moaine Elbaidouri and Olivier Panaud

5 **Hitching a Ride: Nonautonomous Retrotransposons and Parasitism as a Lifestyle** 71
Alan H. Schulman

6 **Plant Endogenous Retroviruses? A Case of Mysterious ORFs** 89
Howard M. Laten and Garen D. Gaston

7 **MITEs, Miniature Elements with a Major Role in Plant Genome Evolution** 113
Hélène Guermonprez, Elizabeth Hénaff, Marta Cifuentes, and Josep M. Casacuberta

8 Glue for Jumping Elements: Epigenetic Means for Controlling Transposable Elements in Plants 125
Thierry Pélissier and Olivier Mathieu

9 Responses of Transposable Elements to Polyploidy 147
Christian Parisod and Natacha Senerchia

10 Noise or Symphony: Comparative Evolutionary Analysis of Sugarcane Transposable Elements with Other Grasses 169
Nathalia de Setta, Cushla J. Metcalfe, Guilherme M.Q. Cruz, Edgar A. Ochoa, and Marie-Anne Van Sluys

11 *Helitron* Proliferation and Gene-Fragment Capture 193
Yubin Li and Hugo K. Dooner

12 Transposable Element Exaptation in Plants 219
Douglas R. Hoen and Thomas E. Bureau

13 SINE Exaptation as Cellular Regulators Occurred Numerous Times During Eukaryote Evolution 253
Jean-Marc Deragon

14 LTR Retrotransposons as Controlling Elements of Genome Response to Stress? 273
Quynh Trang Bui and Marie-Angèle Grandbastien

15 *Rider* Transposon Insertion and Phenotypic Change in Tomato 297
Ning Jiang, Sofia Visa, Shan Wu, and Esther van der Knaap

16 Retrotransposons and the Eternal Leaves 313
Antonella Furini

Index 325

Chapter 1
So Many Repeats and So Little Time: How to Classify Transposable Elements

Thomas Wicker

Abstract Transposable elements (TEs) are present in all genomes. Often there are hundreds to thousands of different TE families contributing the majority of the genomic DNA. Although probably only a very small portion of TEs actually contributes to the function and thereby to the survival of an organism, they still have to be analysed, annotated and classified. To filter out the scarce meaningful signals from the deluge of data produced by modern sequencing technologies, researchers need to be able to efficiently and reliably characterise TE sequences. This process requires three things: First, clear guidelines how to classify and characterise TEs. Second, high-quality databases that contain well-characterised reference sequences, and third, computational tools for efficient TE searches and annotations. This article is intended as a summary of recent developments in TE classification as well as a "little helper" for researchers burdened with the epic task of TE annotation in genomic sequences.

Keywords Transposable element • Retrotransposon • DNA transposon • Superfamily • Family • Classification

1.1 Introduction

1.1.1 Early Findings on Genome Sizes and Sequence Complexity

Even before DNA could be sequenced, researchers realised that eukaryotic genomes show an extreme variation in size (Bennett and Smith 1976). Some studies reported an over 200,000-fold variation in genome size, namely between the amoeba *Amoeba dubia* that has an estimated genome size of 670,000 Mbp (Gregory

T. Wicker (✉)
Institute of Plant Biology, University of Zurich, Ollikerstrasse 107, CH-8008 Zurich, Switzerland
e-mail: wicker@botinst.uzh.ch

M.-A. Grandbastien and J.M. Casacuberta (eds.), *Plant Transposable Elements*, Topics in Current Genetics 24, DOI 10.1007/978-3-642-31842-9_1,

2001) and the 2.9 Mbp genome of the microsporidium *Encephalitozoon cuniculi* (Biderre et al. 1995; Katinka et al. 2001). In the absence of DNA sequence information, genome sizes were measured by estimating nuclear DNA amounts through densitometric measurements (e.g. Bennett and Smith 1976). The "sequence complexity" of genomes was assessed by DNA re-association kinetics. These experiments showed that the vast differences in genome sizes are due to the presence of different amounts of "repeating DNA sequences" (Britten et al. 1974), although their nature was completely unknown at that time. Nevertheless, it was clear early on that the repetitive fraction of a genome is relatively complex and consists of many different types of repeats. Genomes could even be fractionated into highly and moderately repetitive sequences by DNA re-association kinetics (Peterson et al. 2002).

1.1.2 Definition of "Gene Space" and the "C-Value Paradox"

Only when technological advances allowed near-complete sequencing of eukaryotic genomes, actual gene numbers could finally be estimated. Here, it needs to be noted that the definition of what actually constitutes the "gene space" of a genome is still a topic of debate. It certainly includes all "typical" protein-coding genes. Additionally, many components of the gene space do not encode proteins, such as the highly repetitive ribosomal DNA clusters, tRNAs and small nucleolar and small interfering RNAs. Probably, gene space should also include conserved non-coding sequences (Freeling and Subramaniam 2009) and ultraconserved elements (Bejerano et al. 2004), although their functions are barely understood. In the following discussion of gene numbers, I will only refer to protein-coding genes.

1.1.3 The Number of Genes is Similar in All Genomes

As Table 1.1 shows, the estimates of gene numbers differ from species to species, but for all sequenced eukaryotic genomes they are in a range from 5,000 to 50,000. Thus, at a first glance, gene numbers vary only by a factor of 10 while genomes sizes, as described above, vary more than 200,000-fold. The recently finished genome of *Brachypodium distachyon* probably has the most stringent gene annotation so far and possesses 25,554 genes. This gene number is very similar to that of the most recent version of the *Arabidopsis thaliana* genome (version 9) that has 26,173 annotated genes. Even the large maize genome is estimated to contain only about 30,000 genes (Schnable et al. 2009). Interestingly, these numbers are very similar to those for vertebrate genomes, because for all sequenced vertebrate genomes, such as human, mouse, or chicken, genes numbers are now estimated in the range of 25,000–30,000 (Table 1.1). Only fungi and invertebrate animals have clearly fewer genes. Yeast, with its compact 12 Mbp genome has less than 6,000 genes while insects such as *Anopheles gambiae* or *Drosophila melanogaster* have approximately 12,000 genes

Table 1.1 Genome sizes and gene numbers in publicly available genomes

	Size [Mbp]	Genes	Reference
Animal genomes			
Anopheles gambiae	278	14,000	Holt et al. (2002)
Caenorhabditis elegans	97	19,000	CSC (1998)
Drosophila melanogaster	120	15,200	Adams et al. (2000)
Gallus gallus	1,200	20,000–23,000	ICGSC (2004)
Homo sapiens	2,850	24,000	IHGSC (2004)
Mus musculus	2,500	30,000	MGSC (2002)
Plant genomes			
Arabidopsis thaliana	120	26,200	AGI (2000)
Brachypodium distachyon	273	25,500	IBI (2010)
Fritillaria uva-vulpis	87,400	unknown	Leitch et al. (2007)
Hordeum vulgare	5,700	38,000–48,000	Mayer et al. (2009)
Oryza sativa	372	40,600	IRGSC (2005)
Physcomitrella patens	462	35,900	Rensing et al. (2008)
Populus trichocarpa	410	45,500	Tuskan et al. (2006)
Sorghum bicolor	659	34,500	Paterson et al. (2009)
Triticum aestivum	16,000	50,000	Choulet et al. (2010)
Vitis vinifera	342	30,400	Jaillon et al. (2007)
Zea mays	2,061	30,000	Schnable et al. (2009)
Fungal genomes			
Aspergillus nidulans	30	10,600	http://www.broadinstitute.org
Aspergillus flavus	36.8	12,600	http://www.broadinstitute.org
Fusarium verticilloides	41.8	14,200	http://www.broadinstitute.org
Magnaporthe grisea	42	11,100	Dean et al. (2005)
Saccharomyces cerevisiae	11.7	5,700	http://www.broadinstitute.org
Stagonospora nodurum	37	16,600	http://www.broadinstitute.org
Tuber melanosporum	125	7,500	http://www.broadinstitute.org
Botrytis cinerea	42.6	16,400	http://www.broadinstitute.org
Other genomes			
Encephalitozoon cuniculi	2.9	1,997	Katinka et al. (2001)
Amoeba dubia	670,000	unknown	Gregory et al. (2001)

AGI Arabidopsis genome initiative, *CSC C. elegans* sequencing consortium. *IBI* International Brachypodium initiative, *ICGSC* International chicken genome sequencing consortium, *IHGSC* International human genome sequencing consortium, *IRGSP* International rice genome sequencing consortium, *MGSC* Mouse genome sequencing consortium

(Table 1.1). Thus, a consensus transpires that most eukaryotes possess between 5,000 and 30,000 genes, making it obvious that only a relatively small fraction of the genomes sequenced to date actually encode functional genes.

1.1.4 The C-Value Paradox

The fact that gene numbers are very similar while genome sizes vary extremely came to be known as the “C-value Paradox”. Moreover, depending on which taxonomic group is analysed, there may be little or no correlation between genome

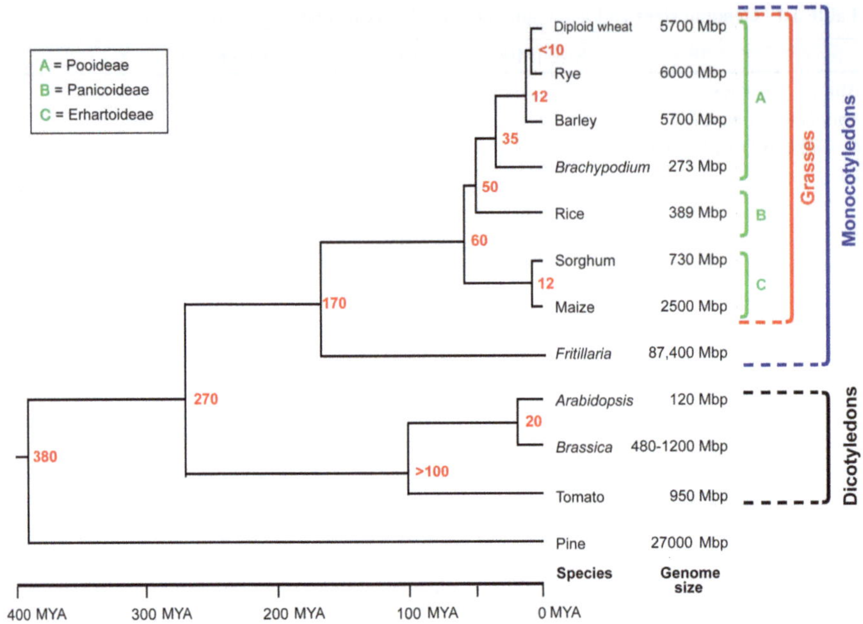

Fig. 1.1 Phylogenetic relationships and genome sizes in selected plant species. Divergence times of specific clades are indicated in red numbers next to the corresponding branching. These numbers are averages of the published values provided in Table 1.1. The scale at the bottom indicates divergence times in million years ago (MYA). Major taxonomic groups that are discussed in the text are indicated at the *left*

size and phylogenetic relationships. This effect is particularly strong on plants where even very closely related species can have very different genome sizes (Fig. 1.1). Among the dicotyledonous plants, there is *Arabidopsis thaliana*, the first plant which had its genome completely sequenced. With a size of about 120 Mbp (*Arabidopsis Genome Initiative* 2000), it is one of the smallest plant genomes known. In contrast, closely related *Brassica* species that diverged from *Arabidopsis* only 15–20 MYA (Yang et al. 1999) have five to ten times larger genomes. In monocotyledonous plants, variation is even more extreme: The grasses *Brachypodium dystachion*, rice and sorghum have genome sizes of 273 Mbp, 389 Mbp and 690 Mbp, respectively, considerably larger than the *Arabidopsis* genome but roughly an order of magnitude smaller than the genomes of some agriculturally important grass species such as wheat and maize, with haploid genome sizes of 5,700 and 2,500 Mbp, respectively. And even they are still dwarfed by the genomes of some lilies, among them *Fritillaria uva-vulpis* which has a genome size of more than 87,000 Mbp, over 700 times the size of the *Arabidopsis* genome (Leitch et al. 2007). Also among *Dicotyledons*, closely related species often differ dramatically in their genome sizes. Maize and sorghum, for example diverged only about 12 MYA (Swigonova et al. 2004), but the maize genome is more than four times the size of the sorghum genome (Table 1.1, Fig. 1.1).

1.2 Transposable Elements

1.2.1 Basics of Selfishness and Junk

As the number of genes is similar in all organisms, it became clear early on that the factor which mainly determines genome size is the amount of repetitive sequences. Nowadays we know that the vast majority of these repetitive sequences are in fact transposable elements (TEs). These elements contain no genes with apparent importance for the immediate survival of the organism. Instead they contain just enough genetic information to produce copies of themselves and/or move around in the genome. For this reason, such sequences are often referred to as "selfish" DNA (Orgel and Crick 1980). To some degree that disparaging view is justified, because TEs are small genetic units, actual "minimal genomes", which contain exactly enough information to be able to replicate, move around in the genome or both. They use the DNA replication and translation machinery of their "host" and thrive within the environment of the genome. For this reason, the term "junk DNA", is often used almost synonymously with TE sequences, reflecting the view of TEs being largely a parasitic burden to the organism.

1.2.2 TE Taxonomy and Classification

Pioneering work in TE classification was done by Hull and Covey (1986), Finnegan (1989) and Capy et al. (1996). The first publicly available database for TEs was RepBase (girinst.org/repbase/) by Jerzy Jurka and colleagues who also proposed a classification system for all TEs (Jurka et al. 2005). In 2007, a group of TE experts met at the Plant and Animal Genome Conference in San Diego (CA, USA) with the goal to define a broad consensus for the classification of all eukaryotic transposable elements. This included the definition of consistent criteria in the characterisation of the main superfamilies and families and a proposal for a naming system (Wicker et al. 2007). The proposed system is a consensus of previous TE classification systems and groups all TEs into 2 major classes, 9 orders and 29 superfamilies (Fig. 1.2). A practical aspect of the classification system is that the TE family name should be preceded by a three-letter code for class, order and superfamily (Fig. 1.2). This was intended to make working with large sets of diverse TEs easier as it enables simple text-based sorting and allows the immediate recognition of the classification when seeing the name of a TE. The proposed classification system is open to expansion as new types of TEs might still be identified in the future. A system that attempts to cover such a vast and complex biological field is by its nature reductionist and tends to oversimplify matters. Thus, there is still an ongoing scientific debate about various aspects of the system (Kapitonov and Jurka 2008; Seberg and Petersen 2009), some of which will be discussed in more detail below.

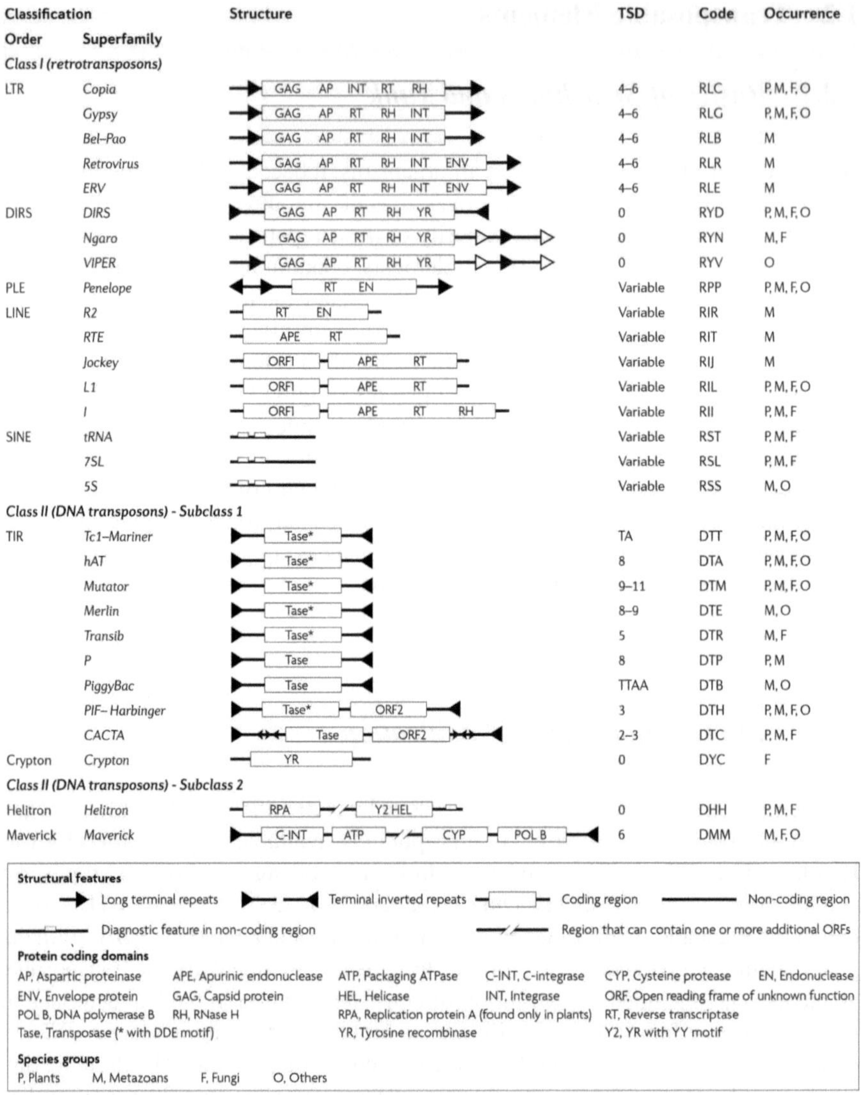

Fig. 1.2 Classification system for transposable elements (Wicker et al. 2007a). The classification divides TEs into two main classes on the basis of the presence or absence of RNA as a transposition intermediate. They are further subdivided into subclasses, orders and superfamilies. The size of the target site duplication (TSD), which is characteristic for most superfamilies, can be used as a diagnostic feature. A three-letter code describes all major groups and is added to the family name of each TE

1.2.3 Class and Subclass: The Highest Levels of TE Classification

At the highest taxonomic level, TEs are divided into two classes. Class 1 contains all TEs that replicate via an RNA intermediate in a "copy-and-paste" process. This class includes both LTR as well as non-LTR retrotransposons. In Class 2 elements, the DNA itself is moved analogous to a "cut-and-paste" process. Class 2 elements are further subdivided into subclass 1 and 2. Subclass 1 are the classic cut-and-paste elements where the DNA is moved with the help of a transposase enzyme. Subclass 2 includes TEs whose transposition process entails replication without double-stranded cleavage and the displacement of only one strand. The Order Helitron from Subclass 2 seems to replicate via a rolling-circle mechanism (Kapitonov and Jurka 2001). Their placement within class 2 reflects the common lack of an RNA intermediate, but not necessarily common ancestry.

1.2.4 TE Superfamilies Represent Ancient Evolutionary Lineages

The most commonly used level of classification is the assignment of a TE to a particular superfamily. Superfamilies are ancient evolutionary lineages that arose during the very early evolution of eukaryotes, some even before the divergence of prokaryotes and eukaryotes. Superfamilies are mainly defined by homology at the protein level. That means that two TEs belong to the same superfamily if their predicted protein sequences show clear homology and can be aligned over most of their length. Terms like "clear homology" and "most of their length" reflect a plea to common sense and should not be tightly bound to arbitrary cut-offs based on E-Values or percent sequence similarity. The fact is that TEs belonging to the same superfamily (even if they come from very distantly related species) usually share many conserved amino acid motifs along the length of their predicted proteins which, importantly for practical work, is usually picked up in a blastx or blastp search. In contrast, TEs from different superfamilies usually show hardly any sequence similarity in their encoded proteins. Protein similarity between members of different superfamilies is reduced to very ancient sequence motifs such as the DDE or Zn-finger motifs (Capy et al. 1997). Here it has to be noted that sequence similarity within the same superfamily can only be expected in the "core" enzymes of the TE elements such as the transposase, reverse transcriptase or integrase, while fast-evolving proteins such as gag (in LTR retrotransposon) and ORF2 (in many DNA transposons) often cannot be aligned between members of the same superfamily. The superfamily of SINEs (small interspersed nuclear elements) has a special status. These small elements do not encode any proteins but are derived from RNA Polymerase promoters and can therefore only be classified based on specific DNA motifs.

1.2.5 TEs Show Most Diversity at the Family Level

It is at the family level is where things get really complicated. While the 29 superfamilies are relatively clearly defined, the exact definition of a TE family is still topic of debate (Kapitonov and Jurka 2008; Seberg and Petersen 2009).

It is clear that within superfamilies TEs have diverged in to an almost incomprehensibly large number of sub-groups and clades. Here, researchers usually introduce the family as the next lower level (after Superfamily). Early on, it became clear that there must be hundreds or even thousands of different types of TEs populating genomes (SanMiguel et al. 1998; Wicker et al. 2001). However, the challenge has been to define criteria for a family that, on one hand, make at least some biological sense and on the other hand are reasonably simple to apply. Of course, the most biologically meaningful TE classification would be based on phylogenetic analysis (Seberg and Petersen 2009). Construction of phylogenetic trees deduced from DNA or predicted protein sequences allows the identification of specific clades, and is therefore a classification scheme based on biological criteria. Such analyses are essential for our understanding of how TEs and genomes evolve. However, phylogenetic analyses are complex and very labour intensive and require a thorough knowledge of TEs, but they are relatively irrelevant when it comes to the initial task of TE identification and annotation, especially in large-scale genome projects.

1.2.6 The 80–80–80 Rule Revisited

In 2007, several colleagues and I proposed the "80–80–80" rule (Wicker et al. 2007) which became both famous and infamous among researchers working on TE annotation. The rule says that two TEs belong to the same family if they share at least 80 % sequence identity at the DNA level over at least 80 % of their total size. The third criterion simply refers to the minimal size of a putative TE sequence that should be analysed in order to avoid that unspecific signals are over-interpreted. The rule was mainly based on practical criteria. We assumed that most researchers on task to annotate TE sequences would need a simple guideline to classify TE sequences. In most cases, blastn (DNA against DNA) searches would be performed as a first step for TE identification. The BLAST algorithm is not able to align DNAs which are significantly less than 80 % identical. Thus, a given TE sequence will produce no strong BLASTN alignments if its sequence is significantly less than 80 % identical to sequences in the reference database. The second criterion (80 % of the entire length of the TE) was introduced to address the problem that different parts show different levels of sequence conservation within the same TE family. Most TEs are comprised of protein-coding sequences and regulatory regions. Good examples illustrating that problem are the long terminal repeat (LTR) retrotransposon superfamilies. The two LTRs contain promoter and downstream regions while the internal domain contains mainly protein-coding regions. Comparisons between many different TE families shows that the regulatory regions evolve much faster than the coding sequences. Thus, often the DNA sequences of the coding region might be alignable while up- and downstream regions (e.g. LTRs) are completely diverged and cannot be aligned. The second criterion of the 80–80–80 rule requires that at least some of the regulatory sequences can be aligned at the

DNA level. There is at least some biological justification for the 80/80 rule, as elements which are similar at the DNA level must have originated from a common "mother" copy in evolutionary recent times.

1.2.7 Biological Meaning vs. Pragmatism in TE Classification

It is clear that a classification rule based simply on the fact that DNA sequences can be aligned is arbitrary, and it was justifiably criticised (Kapitonov and Jurka 2008; Seberg and Petersen 2009). Indeed, TE families (we shall stick to the term "family" for this discussion) sometimes form a continuum, where a sequence from one end of the spectrum might not be properly alignable with one from the other end. But within the continuum, it is possible to move from one end to the other by continuously aligning the most similar sequences. Thus, the simple criterion of whether the DNA sequence of two TEs can be aligned over most of their length can lead to unclear situations. Nevertheless, in most cases, the criterion works quite well. Indeed, usually it is not possible to cross the boundary from one TE family to the other simply by continuously aligning the most similar sequences. For example the *Copia* families *BARE1* and *Maximus* from barley show practically no DNA sequence identity, not even in the most conserved parts of the CDS (Wicker and Keller 2007). It is, therefore, not possible to cross the boundary from one family to the other based on alignments of the DNA sequences. If nothing else, the strategy of defining TE families based on sequence homology is at least pragmatic and allows classification without complex phylogenetic analyses. Nevertheless, it does not replace phylogenetic analyses when it comes to the study of evolution.

1.2.8 How Many Different TE Families Are There?

Recently, the classification system of Wicker et al. (2007) was put to the test in the framework of the International Brachypodium Initiative (2010). The stated goal was to obtain a TE annotation that is comparable in quality to gene annotation. Thus, Brachypodium became the first plant genome where a special group, the Brachypodium repeat annotation consortium (BRAC), was responsible solely for TE annotation. Great care was taken to isolate and characterise as many TE families as possible. As shown in Table 1.2, a total of 499 TE families were characterised. The largest variety was found in LTR retrotransposons which contribute over two-thirds of all families. They are also the class of elements that contributes most to the total genome sequence due to their large size. Most abundant in numbers of copies were small Miniature Inverted-Repeat Transposable Elements (MITEs; Bureau and Wessler 1994), small non-autonomous DNA transposons. Over 20,000 Stowaway MITEs of 23 different families were identified. Despite the large effort invested in TE annotation in the Brachypodium genome, TE annotation is still not complete.

Table 1.2 Numbers of TE families in the genome of the model grass *Brachypodium distachyon*

Superfamily	Code	Families
Gypsy	RLG	147
Copia	RLC	133
LTR unknown	RLX	56
Non-LTR	RIX	3
CACTA	DTC	13
Harbinger	DTH	44
Mariner	DTT	36
Mutator	DTM	62
Helitron	DHH	5
Total		499

TE are categorised into superfamilies. These numbers refer to TE families that were characterised in detail in the framework of the Brachypodium repeat annotation consortium. The actual number of TE families is known to be higher

When sequences were annotated carefully in comparative analyses, dozens of additional TE families could be identified (Jan Buchmann, pers. comm). Many of them are low-copy elements which have weak or no homology to previously described TE families. Thus, the 499 TE families identified in the framework of the genome project are certainly a minimal number. The Brachypodium genome is relatively small compared to other plant genomes. However, there is evidence that the size of larger genomes is mainly due to the excessive expansion of relatively few TE families, rather than the diversification of countless small families. Especially in plants, single or a few LTR retrotransposon families can contribute large parts to the genome (Paterson et al. 2009; Schnable et al. 2009; Wicker et al. 2009). In fungi, the situation is similar: in the very repetitive genome of barley powdery mildew, a few dozen TEs completely dominate the repetitive fraction (Spanu et al. 2010). In summary, in most genomes one has to expect hundreds of different TE families, in some probably thousands. However, fears that there might more TE families in a single genome than words in the English language (SanMiguel et al. 2002), and thus naming of all individual families would be impossible, seem to be unfounded.

1.2.9 The Necessity of TE Databases

For the researcher confronted with the epic task to annotate TEs in a genome, it is essential to have a good reference database of TE sequences. In the best case, this is a dataset of well-characterised TE sequences. In the worst case, it is a collection of sequences that are simply known to be repetitive and which were assembled automatically into contigs. Often the reality lies somewhere between the two. The most abundant TEs are usually well characterised with respect to their precise termini and proteins they encode. But for many sequences, one only knows that they are repetitive, but the exact size or classification is not known. Repeat classification and characterisation is still done very much on a species by species. This is mainly because TEs from different species (if they diverged more than a dozen million years ago) share very little sequence identity at the DNA level. Thus, only protein-coding TEs can usually be identified across species boundaries. If one also wants to precisely annotate non-coding regions and non-autonomous TEs, one usually needs to generate a TE database for the respective species. There are too many TE databases for different species available to describe here. The most inclusive product available today is probably RepBase (girinst.org/repbase/), which includes TE sequences from many different species. However, the task of compiling an all-inclusive TE database which adheres to consistent rules is a monumental one, and it is growing literally by the day.

References

Adams MD, Celniker SE, Holt RA, Evans CA, Gocayne JD, Amanatides PG, Scherer SE, Li PW, Hoskins RA, Galle RF, George RA, Lewis SE, Richards S, Ashburner M, Henderson SN, Sutton GG, Wortman JR, Yandell MD, Zhang Q, Chen LX, Brandon RC, Rogers YH, Blazej RG, Champe M, Pfeiffer BD, Wan KH, Doyle C, Baxter EG, Helt G, Nelson CR, Gabor GL, Abril JF, Agbayani A, An HJ, Andrews-Pfannkoch C, Baldwin D, Ballew RM, Basu A, Baxendale J, Bayraktaroglu L, Beasley EM, Beeson KY, Benos PV, Berman BP, Bhandari D, Bolshakov S, Borkova D, Botchan MR, Bouck J, Brokstein P, Brottier P, Burtis KC, Busam DA, Butler H, Cadieu E, Center A, Chandra I, Cherry JM, Cawley S, Dahlke C, Davenport LB, Davies P, de Pablos B, Delcher A, Deng Z, Mays AD, Dew I, Dietz SM, Dodson K, Doup LE, Downes M, Dugan-Rocha S, Dunkov BC, Dunn P, Durbin KJ, Evangelista CC, Ferraz C, Ferriera S, Fleischmann W, Fosler C, Gabrielian AE, Garg NS, Gelbart WM, Glasser K, Glodek A, Gong F, Gorrell JH, Gu Z, Guan P, Harris M, Harris NL, Harvey D, Heiman TJ, Hernandez JR, Houck J, Hostin D, Houston KA, Howland TJ, Wei MH, Ibegwam C, Jalali M, Kalush F, Karpen GH, Ke Z, Kennison JA, Ketchum KA, Kimmel BE, Kodira CD, Kraft C, Kravitz S, Kulp D, Lai Z, Lasko P, Lei Y, Levitsky AA, Li J, Li Z, Liang Y, Lin X, Liu X, Mattei B, McIntosh TC, McLeod MP, McPherson D, Merkulov G, Milshina NV, Mobarry C, Morris J, Moshrefi A, Mount SM, Moy M, Murphy B, Murphy L, Muzny DM, Nelson DL, Nelson DR, Nelson KA, Nixon K, Nusskern DR, Pacleb JM, Palazzolo M, Pittman GS, Pan S, Pollard J, Puri V, Reese MG, Reinert K, Remington K, Saunders RD, Scheeler F, Shen H, Shue BC, Sidén-Kiamos I, Simpson M, Skupski MP, Smith T, Spier E, Spradling AC, Stapleton M, Strong R, Sun E, Svirskas R, Tector C, Turner R, Venter E, Wang AH, Wang X, Wang ZY, Wassarman DA, Weinstock GM, Weissenbach J, Williams SM, WoodageT WKC, Wu D, Yang S, Yao QA, Ye J, Yeh RF, Zaveri JS, Zhan M, Zhang G, Zhao Q, Zheng L, Zheng XH, Zhong FN, Zhong W, Zhou X, Zhu S, Zhu X, Smith HO, Gibbs RA, Myers EW, Rubin GM, Venter JC (2000) The genome sequence of *Drosophila melanogaster*. Science 287:2185–2195

Arabidopsis Genome Initiative (2000) Analysis of the genome sequence of the flowering plant *Arabidopsis thaliana*. Nature 408:796–815

Bejerano G, Pheasant M, Makunin I, Stephen S, Kent WJ, Mattick JS, Haussler D (2004) Ultraconserved elements in the human genome. Science 304:1321–1325

Bennett MD, Smith JB (1976) Nuclear DNA amounts in angiosperms. Philos Trans R Soc Lond B Biol Sci 274:227–274

Biderre C, Pages M, Metenier G, Canning EU, Vivaras CP (1995) Evidence for the smallest nuclear genome (2.9 Mb) in the microsporidium *Encephalitozoon cuniculi*. Mol Biochem Parasitol 74:229–231

Britten RJ, Graham DE, Neufeld BR (1974) Analysis of repeating DNA sequences by reassociation methods. Enzymology 29:363–418

Bureau TE, Wessler SR (1994) Stowaway: a new family of inverted repeat elements associated with the genes of both monocotyledonous and dicotyledonous plants. Plant Cell 6:907–916

C. elegans Sequencing Consortium (1998) Genome sequence of the nematode *C. elegans*: a platform for investigating biology. Science 282:2012–2018

Capy P, Vitalis R, Langin T, Higuet D, Bazin C (1996) Relationships between transposable elements based upon the integrase-transposase domains: is there a common ancestor? J Mol Evol 42:359–368

Capy P, Langin T, Higuet D, Maurer P, Bazin C (1997) Do the integrases of LTR-retrotransposons and class II element transposases have a common ancestor? Genetica 100:63–72

Choulet F, Wicker T, Rustenholz C, Paux E, Salse J, Leroy P, Schlub S, Le Paslier MC, Magdelenat G, Gonthier C, Couloux A, Budak H, Breen J, Pumphrey M, Liu S, Kong X, Jia J, Gut M, Brunel D, Anderson JA, Gill BS, Appels R, Keller B, Feuillet C (2010) Megabase level sequencing reveals contrasted organization and evolution patterns of the wheat gene and transposable element spaces. Plant Cell 22:1686–1701

Dean RA, Talbot NJ, Ebbole DJ, Farman ML, Mitchell TK, Orbach MJ, Thon M, Kulkarni R, Xu JR, Pan H, Read ND, Lee YH, Carbone I, Brown D, Oh YY, Donofrio N, Jeong JS, Soanes DM, Djonovic S, Kolomiets E, Rehmeyer C, Li W, Harding M, Kim S, Lebrun MH, Bohnert H, Coughlan S, Butler J, Calvo S, Ma LJ, Nicol R, Purcell S, Nusbaum C, Galagan JE, Birren BW (2005) The genome sequence of the rice blast fungus *Magnaporthe grisea*. Nature 434:980–986

Finnegan DJ (1989) Eukaryotic transposable elements and genome evolution. Trends Genet 5:103–107

Freeling M, Subramaniam S (2009) Conserved noncoding sequences (CNSs) in higher plants. Curr Opin Plant Biol 12:126–132

Gregory TR (2001) Coincidence, coevolution, or causation? DNA content, cell size, and the C-value enigma. Biol Rev Camb Philos Soc 76:65–101

Holt RA, Subramanian GM, Halpern A, Sutton GG, Charlab R, Nusskern DR, Wincker P, Clark AG, Ribeiro JM, Wides R, Salzberg SL, Loftus B, Yandell M, Majoros WH, Rusch DB, Lai Z, Kraft CL, Abril JF, Anthouard V, Arensburger P, Atkinson PW, Baden H, de Berardinis V, Baldwin D, Benes V, Biedler J, Blass C, Bolanos R, Boscus D, Barnstead M, Cai S, Center A, Chaturverdi K, Christophides GK, Chrystal MA, Clamp M, Cravchik A, Curwen V, Dana A, Delcher A, Dew I, Evans CA, Flanigan M, Grundschober-Freimoser A, Friedli L, Gu Z, Guan P, Guigo R, Hillenmeyer ME, Hladun SL, Hogan JR, Hong YS, Hoover J, Jaillon O, Ke Z, Kodira C, Kokoza E, Koutsos A, Letunic I, Levitsky A, Liang Y, Lin JJ, Lobo NF, Lopez JR, Malek JA, McIntosh TC, Meister S, Miller J, Mobarry C, Mongin E, Murphy SD, O'Brochta DA, Pfannkoch C, Qi R, Regier MA, Remington K, Shao H, Sharakhova MV, Sitter CD, Shetty J, Smith TJ, Strong R, Sun J, Thomasova D, Ton LQ, Topalis P, Tu Z, Unger MF, Walenz B, Wang A, Wang J, Wang M, Wang X, Woodford KJ, Wortman JR, Wu M, Yao A, Zdobnov EM, Zhang H, Zhao Q, Zhao S, Zhu SC, Zhimulev I, Coluzzi M, della Torre A, Roth CW, Louis C, Kalush F, Mural RJ, Myers EW, Adams MD, Smith HO, Broder S, Gardner MJ, Fraser CM, Birney E, Bork P, Brey PT, Venter JC, Weissenbach J, Kafatos FC, Collins FH, Hoffman SL (2002) The genome sequence of the malaria mosquito *Anopheles gambiae*. Science 298:129–149

Hull R, Covey SN (1986) Genome organization and expression of reverse transcribing elements: variations and a theme. J Gen Virol 67:1751–1758

International Brachypodium Initiative (2010) Genome sequencing and analysis of the model grass *Brachypodium distachyon*. Nature 463:763–768

International Chicken Genome Sequencing Consortium (2004) Sequence and comparative analysis of the chicken genome provide unique perspectives on vertebrate evolution. Nature 432:695–716

International Human Genome Sequencing Consortium (2004) Finishing the euchromatic sequence of the human genome. Nature 431:931–945

International Rice Genome Sequencing Project (2005) The map-based sequence of the rice genome. Nature 436:793–800

Jaillon O, Aury JM, Noel B, Policriti A, Clepet C, Casagrande A, Choisne N, Aubourg S, Vitulo N, Jubin C, Vezzi A, Legeai F, Hugueney P, Dasilva C, Horner D, Mica E, Jublot D, Poulain J, Bruyère C, Billault A, Segurens B, Gouyvenoux M, Ugarte E, Cattonaro F, Anthouard V, Vico V, Del Fabbro C, Alaux M, Di Gaspero G, Dumas V, Felice N, Paillard S, Juman I, Moroldo M, Scalabrin S, Canaguier A, Le Clainche I, Malacrida G, Durand E, Pesole G, Laucou V, Chatelet P, Merdinoglu D, Delledonne M, Pezzotti M, Lecharny A, Scarpelli C, Artiguenave F, Pè ME, Valle G, Morgante M, Caboche M, Adam-Blondon AF, Weissenbach J, Quétier F, Wincker P (2007) The grapevine genome sequence suggests ancestral hexaploidization in major angiosperm phyla. Nature 449:463–467

Jurka J, Kapitonov VV, Pavlicek A, Klonowski P, Kohany O, Walichiewicz J (2005) Repbase update, a database of eukaryotic repetitive elements. Cytogenet Genome Res 110:462–467

Kapitonov V, Jurka J (2001) Rolling-circle transposons in eukaryotes. Proc Natl Acad Sci USA 98:8714–8719

Kapitonov V, Jurka J (2008) A universal classification of eukaryotic transposable elements implemented in Repbase. Nat Rev Genet 9:411–412

Katinka MD, Duprat S, Cornillot E, Metenier G, Thomarat F, Prensier G, Barbe V, Peyretaillade E, Brottier P, Wincker P, Delbac F, El Alaoui H, Peyret P, Saurin W, Gouy M, Weissenbach J, Vivares CP (2001) Genome sequence and gene compaction of the eukaryote parasite *Encephalitozoon cuniculi*. Nature 414:450–453

Leitch IJ, Beaulieu JM, Cheung K, Hanson L, Lysak MA, Fay MF (2007) Punctuated genome size evolution in *Liliaceae*. J Evol Biol 20:2296–2308

Mayer KF, Taudien S, Martis M, Simková H, Suchánková P, Gundlach H, Wicker T, Petzold A, Felder M, Steuernagel B, Scholz U, Graner A, Platzer M, Dolezel J, Stein N (2009) Gene content and virtual gene order of barley chromosome 1H. Plant Physiol 151:496–505

Mouse Genome Sequencing Consortium (2002) Initial sequencing and comparative analysis of the mouse genome. Nature 420:520–562

Orgel LE, Crick FHC (1980) Selfish DNA: the ultimate parasite. Nature 284:604–607

Paterson AH, Bowers JE, Bruggmann R, Dubchak I, Grimwood J, Gundlach H, Haberer G, Hellsten U, Mitros T, Poliakov A, Schmutz J, Spannagl M, Tang H, Wang X, Wicker T, Bharti AK, Chapman J, Feltus FA, Gowik U, Grigoriev IV, Lyons E, Maher CA, Martis M, Narechania A, Otillar RP, Penning BW, Salamov AA, Wang Y, Zhang L, Carpita NC, Freeling M, Gingle AR, Hash CT, Keller B, Klein P, Kresovich S, McCann MC, Ming R, Peterson DG, Mehboob-ur-Rahman WD, Westhoff P, Mayer KF, Messing J, Rokhsar DS (2009) The *Sorghum bicolor* genome and the diversification of grasses. Nature 457:551–556

Peterson DG, Schulze SR, Sciara EB, Lee SA, Nagel A, Jiang N, Tibbetts DC, Wessler SR, Paterson AH (2002) Integration of Cot analysis, DNA cloning, and high-throughput sequencing facilitates genome characterization and gene discovery. Genome Res 12:795–807

Rensing SA, Lang D, Zimmer AD, Terry A, Salamov A, Shapiro H, Nishiyama T, Perroud PF, Lindquist EA, Kamisugi Y, Tanahashi T, Sakakibara K, Fujita T, Oishi K, Shin-I T, Kuroki Y, Toyoda A, Suzuki Y, Hashimoto S, Yamaguchi K, Sugano S, Kohara Y, Fujiyama A, Anterola A, Aoki S, Ashton N, Barbazuk WB, Barker E, Bennetzen JL, Blankenship R, Cho SH, Dutcher SK, Estelle M, Fawcett JA, Gundlach H, Hanada K, Heyl A, Hicks KA, Hughes J, Lohr M, Mayer K, Melkozernov A, Murata T, Nelson DR, Pils B, Prigge M, Reiss B, Renner T, Rombauts S, Rushton PJ, Sanderfoot A, Schween G, Shiu SH, Stueber K, Theodoulou FL, Tu H, Van de Peer Y, Verrier PJ, Waters E, Wood A, Yang L, Cove D, Cuming AC, Hasebe M, Lucas S, Mishler BD, Reski R, Grigoriev IV, Quatrano RS, Boore JL (2008) The *Physcomitrella* genome reveals evolutionary insights into the conquest of land by plants. Science 319:64–69

SanMiguel P, Gaut BS, Tikhonov A, Nakajima Y, Bennetzen JL (1998) The paleontology of intergene retrotransposons of maize. Nat Genet 20:43–45

SanMiguel PJ, Ramakrishna W, Bennetzen JL, Busso CS, Dubcovsky J (2002) Transposable elements, genes and recombination in a 215-kb contig from wheat chromosome 5A(m). Funct Integr Genomics 2:70–80

Schnable PS, Ware D, Fulton RS, Stein JC, Wei F, Pasternak S, Liang C, Zhang J, Fulton L, Graves TA, Minx P, Reily AD, Courtney L, Kruchowski SS, Tomlinson C, Strong C, Delehaunty K, Fronick C, Courtney B, Rock SM, Belter E, Du F, Kim K, Abbott RM, Cotton M, Levy A, Marchetto P, Ochoa K, Jackson SM, Gillam B, Chen W, Yan L, Higginbotham J, Cardenas M, Waligorski J, Applebaum E, Phelps L, Falcone J, Kanchi K, Thane T, Scimone A, Thane N, Henke J, Wang T, Ruppert J, Shah N, Rotter K, Hodges J, Ingenthron E, Cordes M, Kohlberg S, Sgro J, Delgado B, Mead K, Chinwalla A, Leonard S, Crouse K, Collura K, Kudrna D, Currie J, He R, Angelova A, Rajasekar S, Mueller T, Lomeli R, Scara G, Ko A, Delaney K, Wissotski M, Lopez G, Campos D, Braidotti M, Ashley E, Golser W, Kim H, Lee S, Lin J, Dujmic Z, Kim W, Talag J, Zuccolo A, Fan C, Sebastian A, Kramer M, Spiegel L, Nascimento L, Zutavern T, Miller B, Ambroise C, Muller S, Spooner W, Narechania A, Ren L, Wei S, Kumari S, Faga B, Levy MJ, McMahan L, Van Buren P, Vaughn MW, Ying K, Yeh CT, Emrich SJ, Jia Y, Kalyanaraman A, Hsia AP, Barbazuk WB, Baucom RS, Brutnell TP, Carpita NC, Chaparro C, Chia JM, Deragon JM, Estill JC, Fu Y, Jeddeloh JA, Han Y, Lee H, Li P, Lisch DR, Liu S, Liu Z, Nagel DH, McCann MC, SanMiguel P, Myers AM, Nettleton D, Nguyen J, Penning BW, Ponnala L, Schneider KL, Schwartz DC, Sharma A, Soderlund C, Springer NM, Sun Q, Wang H, Waterman M, Westerman R, Wolfgruber TK, Yang L, Yu Y, Zhang L, Zhou S, Zhu Q, Bennetzen JL, Dawe RK, Jiang J, Jiang N, Presting GG, Wessler SR, Aluru S, Martienssen RA, Clifton SW, McCombie WR, Wing RA, Wilson RK (2009) The B73 maize genome: complexity, diversity, and dynamics. Science 326:1112–1115

Seberg O, Petersen G (2009) A unified classification system for eukaryotic transposable elements should reflect their phylogeny. Nat Rev Genet 10:276

Spanu PD, Abbott JC, Amselem J, Burgis TA, Soanes DM, Stüber K, Loren V, van Themaat E, Brown JK, Butcher SA, Gurr SJ, Lebrun MH, Ridout CJ, Schulze-Lefert P, Talbot NJ, Ahmadinejad N, Ametz C, Barton GR, Benjdia M, Bidzinski P, Bindschedler LV, Both M, Brewer MT, Cadle-Davidson L, Cadle-Davidson MM, Collemare J, Cramer R, Frenkel O, Godfrey D, Harriman J, Hoede C, King BC, Klages S, Kleemann J, Knoll D, Koti PS, Kreplak J, López-Ruiz FJ, Lu X, Maekawa T, Mahanil S, Micali C, Milgroom MG, Montana G, Noir S, O'Connell RJ, Oberhaensli S, Parlange F, Pedersen C, Quesneville H, Reinhardt R, Rott M, Sacristán S, Schmidt SM, Schön M, Skamnioti P, Sommer H, Stephens A, Takahara H, Thordal-Christensen H, Vigouroux M, Wessling R, Wicker T, Panstruga R (2010) Genome expansion and gene loss in powdery mildew fungi reveal functional tradeoffs in extreme parasitism. Science 330:1543–1546

Swigonova Z, Lai J, Ma J, Ramakrishna W, Llaca V, Bennetzen JL, Messing J (2004) On the tetraploid origin of the maize genome. Comp Funct Genomics 5:281–284

Tuskan GA, Difazio S, Jansson S, Bohlmann J, Grigoriev I, Hellsten U, Putnam N, Ralph S, Rombauts S, Salamov A, Schein J, Sterck L, Aerts A, Bhalerao RR, Bhalerao RP, Blaudez D, Boerjan W, Brun A, Brunner A, Busov V, Campbell M, Carlson J, Chalot M, Chapman J, Chen GL, Cooper D, Coutinho PM, Couturier J, Covert S, Cronk Q, Cunningham R, Davis J, Degroeve S, Déjardin A, Depamphilis C, Detter J, Dirks B, Dubchak I, Duplessis S, Ehlting J, Ellis B, Gendler K, Goodstein D, Gribskov M, Grimwood J, Groover A, Gunter L, Hamberger B, Heinze B, Helariutta Y, Henrissat B, Holligan D, Holt R, Huang W, Islam-Faridi N, Jones S, Jones-Rhoades M, Jorgensen R, Joshi C, Kangasjärvi J, Karlsson J, Kelleher C, Kirkpatrick R, Kirst M, Kohler A, Kalluri U, Larimer F, Leebens-Mack J, Leplé JC, Locascio P, Lou Y, Lucas S, Martin F, Montanini B, Napoli C, Nelson DR, Nelson C, Nieminen K, Nilsson O, Pereda V, Peter G, Philippe R, Pilate G, Poliakov A, Razumovskaya J, Richardson P, Rinaldi C, Ritland K, Rouzé P, Ryaboy D, Schmutz J, Schrader J, Segerman B, Shin H, Siddiqui A, Sterky F, Terry A, Tsai CJ, Uberbacher E, Unneberg P, Vahala J, Wall K, Wessler S, Yang G, Yin T, Douglas C, Marra M, Sandberg G, Van de Peer Y, Rokhsar D (2006) The genome of black cottonwood, *Populus trichocarpa* (Torr. & Gray). Science 313:1596–1604

Wicker T, Keller B (2007) Genome-wide comparative analysis of copia retrotransposons in Triticeae, rice, and Arabidopsis reveals conserved ancient evolutionary lineages and distinct dynamics of individual *copia* families. Genome Res 17:1072–1081

Wicker T, Stein N, Albar L, Feuillet C, Schlagenhauf E, Keller B (2001) Analysis of a contiguous 211 kb sequence in diploid wheat (*Triticum monococcum* L.) reveals multiple mechanisms of genome evolution. Plant J 26:307–316

Wicker T, Sabot F, Hua-Van A, Bennetzen JL, Capy P, Chalhoub B, Flavell A, Leroy P, Morgante M, Panaud O, Paux E, SanMiguel P, Schulman AH (2007) A unified classification system for eukaryotic transposable elements. Nat Rev Genet 8:973–982

Wicker T, Taudien S, Houben A, Keller B, Graner A, Platzer M, Stein N (2009) A hole-genome snapshot of 454 sequences exposes the composition of the barley genome and provides evidence for parallel evolution of genome size in wheat and barley. Plant J 59:712–722

Yang YW, Lai KN, Tai PY, Li WH (1999) Rates of nucleotide substitution in angiosperm mitochondrial DNA sequences and dates of divergence between Brassica and other angiosperm lineages. J Mol Evol 48:597–604

Chapter 2
Transposable Element Annotation in Completely Sequenced Eukaryote Genomes

Timothée Flutre, Emmanuelle Permal, and Hadi Quesneville

Abstract With the development of new sequencing techniques, the number of sequenced plant genomes is increasing. However, accurate annotation of these sequences remains a major challenge, in particular with regard to transposable elements (TEs). The aim of this chapter is to provide a roadmap for researchers involved in genome projects to address this issue. We list several widely used tools for each step of the TE annotation process, from the identification of TE families to the annotation of TE copies. We assess the complementarities of these tools and suggest that combined approaches, using both *de novo* and knowledge-based TE detection methods, are likely to produce reasonably comprehensive and sensitive results. Nevertheless, existing approaches still need to be supplemented by expert manual curation. Hence, we describe good practice required for manual curation of TE consensus sequences.

Keywords Annotation • Bioinformatics • Classification • Curation • Identification • Pipeline

2.1 Introduction

Transposable elements (TEs) are mobile genetic elements that shape the eukaryotic genomes in which they are present. They are virtually ubiquitous and make up, for instance, 20% of a typical *D. melanogaster* genome (Bergman et al. 2006), 50% of a *H. sapiens* genome (Lander et al. 2001), and 85% of a *Z. mays* genome (Schnable et al. 2009). They are classified into two classes depending on their transposition mode: via RNA for class I retrotransposons and via DNA for class II transposons

T. Flutre • E. Permal • H. Quesneville (✉)
INRA, UR 1164, URGI, Unité de Recherche en Génomique-Info, 78026 Versailles cedex, France
e-mail: hadi.quesneville@versailles.inra.fr

M.-A. Grandbastien and J.M. Casacuberta (eds.), *Plant Transposable Elements*, Topics in Current Genetics 24, DOI 10.1007/978-3-642-31842-9_2,

(Finnegan 1989). Each class is also subdivided into several orders, superfamilies, and families (Wicker et al. 2007). Due to their unique ability to transpose and because they frequently amplify, TEs are major determinants of genome size (Petrov 2001; Piegu et al. 2006) and cause genome rearrangements (Gray 2000; Fiston-Lavier et al. 2007). Once described as the "ultimate parasites" (Orgel and Crick 1980), TEs are commonly found to regulate the expression of neighboring genes (Feschotte 2008; Bourque 2009) or even to have been domesticated so as to provide a specific host function (Zhou et al. 2004; Bundock and Hooykaas 2005; Santangelo et al. 2007; Kapitonov and Jurka 2005).

As a consequence of the development of new rapid sequencing techniques, the number of available sequenced eukaryotic genomes is constantly increasing. However, the first step of the analysis, i.e., accurate annotation, remains a major challenge, particularly concerning TEs. Correct genome annotation of genes and TEs is an indispensable part of thorough genome-wide studies. Consequently, efficient computational methods have been proposed for TE annotation (Bergman and Quesneville 2007; Lerat 2010; Janicki et al. 2011). Given that the pace at which genomes are sequenced is unlikely to decrease in the coming years; the process of TE annotation needs to be made widely accessible.

This chapter lays down a clear road map detailing the order in which computational tools (or combinations of such tools) should be used to annotate TEs in a whole genome. We distinguish three steps (1) identifying TEs by searching for reference sequences (e.g., full-length TE sequences) and building consensuses from similar sequences, (2) manual curation to define and classify TE families, and (3) annotation of every TE copy. We also provide some hints on manual curation, a step that is still necessary.

2.2 *De Novo* Detection of Transposable Elements

Various efficient computational methods are available to identify unknown TEs in genomic sequences. Each method is based on specific assumptions that have to be understood to optimize selection and combination of the methods to ensure they are appropriate for any particular analytic goal.

2.2.1 Computing Highly-Repeated Words

TEs, due to their capacity to transpose, are often present in a large number of copies within the same genome. Although TE sequences degenerate with time, words (i.e., short subsequences of few nucleotides) that compose them are consequently repeated throughout the genome. Software, such as the TALLYMER (Kurtz et al. 2008) and P-CLOUDS (Gu et al. 2008), has been designed to find repeats rapidly in genome sequences by counting highly frequent words of a given length *k*, called *k*-mers. These programs are very useful for quickly providing a view of the repeated fraction in a given set of genomic sequences, including especially unassembled sequences. However, they do not provide much detail about the TEs present in these sequences. Their output only identifies highly repeated regions without indicating precise TE fragment boundaries or TE family assignments. These methods are quick and simple to use but allow only limited biological interpretations and no real TE annotation.

Other methods also start by counting frequent *k*-mers but then go on to try to define consensuses. ReAS (Li et al. 2005) applies this approach directly to shotgun reads. For each frequent *k*-mer, a multiple alignment of all short reads containing it is built and then extended iteratively. REPEATSCOUT (Price et al. 2005) has a similar approach but works on assembled sequences. These tools return a library of consensus sequences. Although their results are more biologically relevant than those of previous methods, the consensuses are usually too short and correspond to truncated versions of ancestral TEs (Flutre et al. 2011). Substantial manual inspection and editing is therefore needed to obtain a meaningful list of consensus sequences.

2.2.2 All-by-All Alignment and Clustering of Interspersed Repeats

Repeats can also be identified by self-alignment of genomic sequences, starting with an all-by-all alignment of the assembled sequences.

Several tools can be used for this. Some, such as BLAST (Altschul et al. 1997) and BLAST-like algorithms, use heuristics. For instance, BLASTER (Quesneville et al. 2003) performs this search by launching BLAST repeatedly over the genome sequences. Others are exact algorithms. Hence, PALS uses "q-Gram filters" that unlike a heuristic (e.g., BLAST), it rapidly and stringently eliminates a large part of the search space from consideration before the alignment search but nevertheless guarantees not to eliminate a region containing a match (Rasmussen et al. 2005). As the amount of input data is usually large, the computations are intensive. Consequently, stringent parameters are applied: good results are obtained with BLAST-like tools when matches shorter than 100 bp or with identity below 90% or with an E-value above 1e-300 are dismissed (Flutre et al. 2011). As most TEs are shorter than 25 kb, segmental duplications can also be filtered out by removing longer matches. To speed up the computations, such alignment tools can be launched in parallel on a computer cluster.

With these parameters, only closely related TE copies will be found. Note that the aim of this step is not to recover all TE copies of a family but to use those that are well conserved to build a robust consensus (see below). Stringent alignment parameters are crucial for successful reconstruction of a valid consensus. Interestingly, even with these stringent criteria, this approach is still more sensitive than other methods for identifying repeats. However, it is also the most computer intensive. It also misses single-copy TE families because at least two copies are required for detection by self-alignment.

Once the matches corresponding to repeats have been obtained, they need to be clustered into groups of similar sequences. The aim is for each cluster to correspond to copies of a single TE family. However, TEs may include divergent interspersed repeats, often nested within each other, making the task difficult. Algorithms have been designed to cluster identified sequences appropriately, limiting the artifacts induced by nested and deleted TE copies and non-TE repeats such as segmental duplications. The various tools that are available are based on different assumptions about (1) the sequence diversity within a TE family, (2) the evolutionary dynamics of TE sequences, (3) nested patterns, and (4) repeat numbers.

GROUPER (Quesneville et al. 2003; Flutre et al. 2011) starts by connecting fragments belonging to the same copy by dynamic programming, and then applies a single link clustering algorithm with (1) a 95% coverage constraint between copies of the same cluster and (2) cluster selection based on the number of copies not included in larger copies of other clusters. The rational here is to detect copies that have the same length as they most probably correspond to mobile entities. Indeed, copies can diverge rapidly by accumulating deletions leading to copies with different sizes. Copies that are almost intact can transpose conserving their original, presumably functional, size. RECON (Bao and Eddy 2002) also starts with a single link-clustering step. If a cluster includes nested repeats and is thus chimerical, it can be subdivided according to the distribution of its all-by-all genome alignment ends. Indeed, nested repeats exhibit a specific pattern in alignments of sequences obtained in an all-by-all genome comparison: the alignment ends of any one inner repeat are all in the relative same position.

PILER-DF (Edgar and Myers 2005) identifies lists of matches covering a maximal contiguous region, defines them as piles, and then builds clusters of globally alignable piles. The rational here is identical to that used by GROUPER where copies of identical length are sought; however, PILER-DF has no specific attitude to indels.

The three clustering programs behave differently according to the sequence diversity of TE families. For instance, GROUPER better distinguishes groups of mobile elements differing by their sizes inside a TE family. It also better recovers fragmented copies due to its dynamic programming joining algorithm. But, it produces more redundant results and only correctly recovers TE families if there are at least three complete copies. RECON is better for TE families with fewer than three complete copies, being able to reconstruct the complete TE from fragments. PILER is fast and very specific. It is a useful option for large genomes when time is an issue, or if a non-exhaustive search is sufficient.

Once clusters are defined, a filter is usually applied to retain only those with at least three members, thereby eliminating the vast majority of segmental duplications. Finally, for each remaining cluster, a multiple alignment is built from which a consensus sequence is derived. Numerous algorithms are available for this but only those complying with the following criteria should be used (1) speed, because the number of clusters is usually very large and (2) ability to handle appropriately sequences of different lengths, which is the case for the clusters generated by RECON. MAP (Huang 1994) and MAFFT (Katoh et al. 2002) comply

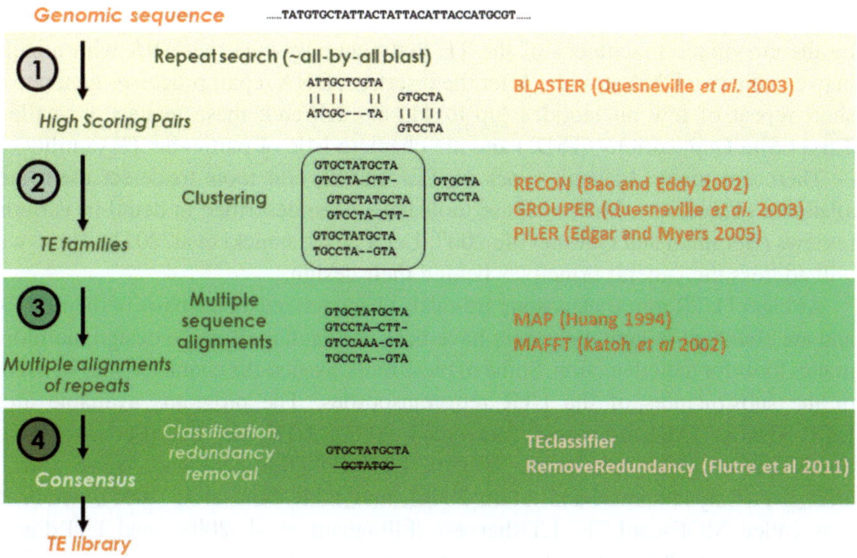

Fig. 2.1 Workflow of the 4-step *de novo* TE detection pipeline (Flutre et al. 2011)

with these criteria and give good results (Flutre et al. 2011). Taking the 20 longest sequences is generally sufficient to build the consensus. The set of consensus sequences obtained represents a condensed view of all TE families present in the genome being studied.

For easy identification of TE families, i.e., those for which there are full-length copies that are very similar to each other, all clustering methods will find roughly the same consensus. However, for other families, which may be numerous, different methods generate different clusters, because they rely on different assumptions. Therefore, manual curation is required to identify an appropriate set of representative sequences (see below).

This all-by-all genome comparison strategy has been implemented in a pipeline called TEdenovo (Fig. 2.1). The TEdenovo pipeline is part of the REPET package (Flutre et al. 2011) and was designed to be used on a computer cluster for fast calculations. It allows the use of different software at each step to exploit the best strategy according to the genome size and the TE identification goal.

2.2.3 Features-Based Methods

Alternatively, TEs can be detected using prior knowledge about TE features. For example, class I LTR retrotransposons characteristically have LTR at both ends of the element, and this can be used for their detection. Numerous class II TEs encompass TIR structures that can be used as markers. Many TE families generate a double-strand break when they insert into the DNA sequence. The break is caused by the enzymatic machinery of the TE that generally cuts the DNA with a shift between the two DNA strands. After the insertion, DNA repair processes generate a short repeat of few nucleotides (up to 11) at each end; these repeats are called Target Site Duplications (TSDs) and are characteristic of particular TE families.

There are many different types of TEs and several tools to detect them are available (Table 2.1). Most of these tools have been described in detail in various reviews (Bergman and Quesneville 2007; Lerat 2010; Janicki et al. 2011). Here, we will address the general principles behind their design.

As class I LTR retrotransposons are easily characterized on the basis of their LTRs and are abundant in genomes, there have been substantial efforts to design bioinformatics tools for their detection. Some of these tools also use the characteristics of some of the substructures of the LTR retrotransposons. The programs available are: LTR_STRUC (McCarthy and McDonald 2003), LTR_MINER (Pereira 2004), SmaRTFinder (Morgante et al. 2005b), LTR_FINDER (Xu and Wang 2007), LTR_par (Kalyanaraman and Aluru 2006), find_LTR (Rho et al. 2007), which is now called MGEscanLTR, LTRharvest (Ellinghaus et al. 2008), and LTRdigest (Steinbiss et al. 2009) that also identifies protein-coding regions within the LTR element. The algorithms of these tools are generally divided into two parts: they first build a data structure to speed up searches for repeats, and then use this structure to search for repeats in the genomic sequences. For example, LTRharvest builds suffix-array using the "suffixerator" tool from GenomeTools package (Lee and Chen 2002). Some of these tools add a third step to refine the search by looking for additional substructures, such as Primer Binding Sites (PBS) and Poly-Purine Tracks (PPT) that are important signals for LTR retrotransposon transposition. These programs also allow searching for TSD and coding regions, including those encoding protein domains, specific to these TEs.

There are also tools aimed at detecting class I non-LTR retrotransposons, e.g., *Long Interspersed Nuclear Elements* (*LINE*) and *Short Interspersed Nuclear Elements* (*SINE*). TSDfinder (Szak et al. 2002) is based on the L1 TE insertion signature which is constituted in part by two Target Site Duplications (TSDs) and a polyA tail. RTAnalyzer (Lucier et al. 2007) is a Web server that follows the same approach as TSDFinder. SINEDR (Tu et al. 2004) is designed to look for SINE elements, a group of non-LTR retrotransposons, in sequence databases. MGEScan-non-LTR (Rho and Tang 2009) identifies and classifies non-LTR TEs in genomic sequences using probabilistic models. It is based on the structure of the 12 TE clades that are non-LTR TEs. It uses two separate Hidden Markov Model (HMM) profiles, one for the Reverse Transcriptase (RT) gene and one for the endonuclease (APE) gene, both of which are well conserved among non-LTR TEs.

Class II TEs, but not Helitrons and Cryptons, are structurally characterized by TIRs. Some class II-specific bioinformatics tools, for example, FindMite (Tu 2001), Transpo (Santiago et al. 2002), and MAK (Yang and Hall 2003), search for defined TIR features in sequences. Must (Chen et al. 2009) is designed to search for TEs containing two TIRs and two direct repeats (i.e., TSD) to identify MITE candidates. Two new tools were published recently: MITE-Hunter (Han and Wessler 2010)

Table 2.1 Availability of feature-based detection programs for TE *de novo* identification

TEclass	Program	URL	Web server or program
I LTR	LTR_STRUC	http://www.mcdonaldlab.biology.gatech.edu/finalLTR.htm	Software available upon request
	LTR_MINER	http://genomebiology.com/2004/5/10/R79/additional	Script available as additional file
	SmartFinder	http://services.appliedgenomics.org/software/smartfinder/	Downloadable software
	LTR_FINDER	http://tlife.fudan.edu.cn/ltr_finder/	Web server and software available upon request
	LTR_par	http://www.eecs.wsu.edu/~ananth/software.htm	Software available upon request
	Find_LTR (MGEscanLTR)	http://darwin.informatics.indiana.edu/cgi-bin/evolution/ltr.pl	Downloadable software
	LTRharvest	http://www.zbh.uni-hamburg.de/forschung/genominformatik/software/ltrharvest.html	Downloadable as part of the genomeTool package (http://genometools.org/pub/)
	LTRdigest	http://www.zbh.uni-hamburg.de/forschung/genominformatik/software/ltrdigest.html	Downloadable as part of the genomeTool package (http://genometools.org/pub/)
I Non-LTR	TSDfinder	http://www.ncbi.nlm.nih.gov/CBBresearch/Landsman/TSDfinder/	Script available as additional file
	RTAanalyzer	http://www.riboclub.org/cgi-bin/RTAnalyzer/index.pl?page = rt_find	Web server
	SINEDR	Not available	Software available upon request
	MGEScan-non-LTR	http://darwin.informatics.indiana.edu/cgi-bin/evolution/nonltr/nonltr.pl	Downloadable software
II TIR	Transpo	http://alggen.lsi.upc.es/recerca/search/transpo/transpo.html	Web server and downloadable software
	FindMITE	No longer available	No longer available
	MAK	No longer available	No longer available
	MUST	http://csbl1.bmb.uga.edu/ffzhou/MUST/	Web server
	MITE-Hunter	http://target.iplantcollaborative.org/mite_hunter.html	Downloadable software
	TS clustering	Not available	Software available upon request
II Helitron	HelSearch	http://sourceforge.net/projects/helsearch/files/	Downloadable software
	HelitronFinder	http://limei.montclair.edu/HF.html	Web server and software available upon request

Feature-based *de novo* TE identification is generally fast and efficient. Unfortunately, only well-described TEs that also have a strong signature can be found. Some TEs do not have such characteristics and thus cannot be found by this type of approach. Consequently, feature-based *de novo* TE identification cannot be used alone to provide an exhaustive inventory of TEs in a genome. Nevertheless, this approach can be used to supplement the findings of all-by-all genome comparison TE searches, in particular for low copy TE families that are otherwise difficult to detect. Surprisingly, these feature-based tools also suffer from high false-positive detection rates such that careful curation is required (data not shown)

which is a five-step pipeline, with the first step involving a TIR-like structure search and TS clustering (Hikosaka and Kawahara 2010), which is dedicated to finding T2-MITEs.

Despite there being no TIR structures in Helitrons, programs have also been designed for their detection: HelitronFinder (Du et al. 2008) is based on known consensus sequences and HelSearch (Yang and Bennetzen 2009) looks for a Helend structure constituted by a six base-pair hairpin and CTRR nucleotide motif.

2.2.4 *Evidence for TE Mobility*

The identification of a long indel by sequence alignments between two closely related species is suggestive of the presence of a TE. The rest of the genome can then be searched for this sequence to assess its repetitive nature. This approach has been used (Caspi and Pachter 2006) and appears to work well for recent TE insertions: indeed, it will only detect insertions that occurred after speciation. Using several alignments with species diverging at different times may lead to more TEs being identified (Caspi and Pachter 2006), as each alignment allows detection of TEs inserted at different times. However, one limitation is the difficulty of correctly aligning long genomic sequences from increasingly divergent species.

This idea could be also used within a genomic sequence by considering segmental duplications. A long indel apparent in sequence alignments of genomic duplications may similarly be an indication of the presence of a TE (Le et al. 2000). Various controls are needed, however, to confirm the TE status of the sequence. For example, TE features such as terminal repeats (e.g., LTR, TIR) or similarity to other TE sequences could be used. This approach only detects TE insertions that occur after the duplication event and may thus be limited to rare events.

TSDs are hallmarks of a transposition event, but they can be difficult to find in old insertions because they are short, and they can be altered by mutations or deletions. In addition, the size of the TSD depends on the family and not all TEs generate a TSD upon insertion.

2.3 Classification and Curation of Transposable Element Sequences

When they amplify, TE copies may nest within each other in complex patterns (Bergman et al. 2006), thereby fragmenting the elements. With time, the sequences accumulate (1) point substitutions, (2) deletions that truncate copies, and (3) insertions that interrupt their sequences (Blumenstiel et al. 2002). These events generate complex remnants of TEs. Various *de novo* tools use these remnants to try to infer the ancestral sequence that actually transposed.

When starting with a self-alignment (i.e., all-by-all genome comparison) of genomic sequences, the optimal strategy is to use several tools and even combine them. However, all the relevant tools and every *de novo* approach can encounter difficulties when trying to distinguish true TEs from segmental duplications, multi-member gene families, tandem repeats, and satellites. It is, therefore, strongly recommended to confirm that the predicted sequences can be classified as being TEs. Computerized analysis therefore still needs to be complemented by manual curation.

2.3.1 Classification

Sequences believed to correspond to TEs can be classified according to their similarity to known TEs, for example, those recorded in databases like Repbase Update (Jurka et al. 2005). A tool called TEclass (Abrusan et al. 2009) implements a support vector machine, using oligomer frequencies, to classify TE candidates.

However, for most previously unknown TE sequences obtained via *de novo* approaches from nonmodel organisms, classification requires the specific identification of several TE features [see (Wicker et al. 2007) for complete description]. By searching for structural features, such as terminal repeats, features characteristic of various TE types can be identified: long terminal repeats specific to class I LTR retrotransposons, terminal inverted repeats specific to the class II DNA transposons, and poly-A or SSR-like tails specific to class I non-LTR retrotransposons. In addition, using BLASTN, BLASTX, and TBLASTX to compare TE candidates with a reference data bank, can provide hints for classification, as long as the reference data bank contains elements similar to the TE candidate. Therefore, it is also recommended to search for matches for sequences encoding TE-specific protein profiles in TE sequences. For example, the presence of a transposase gene is strongly indicative of a class II DNA transposon. Such protein profiles can be obtained from the Pfam database which includes protein families represented by multiple sequence alignments and hidden Markov models (HMM) (Finn et al. 2010). These profiles can be used by programs such as HMMER to find matches within the candidate TE sequences.

Some tools classify TE sequences according to their features, usually via a decision tree. The TEclassifier in the REPET package (Flutre et al. 2011) and REPCLASS (Feschotte et al. 2009) searches for all the features listed above. In addition, REPCLASS allows TE candidates to be filtered on the basis of the number of copies they have in the genome. TEclassifier interestingly allows the removal of redundancy from among potential TE sequences. It uses the classification to eliminate redundant copies (a sequence contained within a longer one) and retains well-classified TE candidate sequences preferentially over less well-classified TE candidate sequences. This tool is particularly useful for reconciling different TE reference libraries obtained independently, as it guarantees to retain well-classified TE candidate sequences.

2.3.2 Identification of Families

Once the newly identified TE sequences have been classified, manual curation is required as some consensus sequences may not have been classified previously and there may still be some redundant consensus sequences. Manual curation is crucial because the annotation of TE copies, as described in the next section, depends on the quality of the TE library. One way to curate a library of TE consensus sequences is to gather these sequences into clusters that may constitute TE families. A tool like BLASTCLUST in the NCBI-BLAST suite can quickly build such clusters via simple link clustering based on sequence alignment coverage and identity. Eighty percent identity and coverage, as proposed by (Wicker et al. 2007), gives good results. Typical clusters will contain well-classified consensuses (e.g., class I—LTR—Gypsy element) as well as unclassified consensuses (without structural features and little sequence similarity either with known TEs or any TE domain).

Then, computing a multiple sequence alignment (MSA) for each cluster gives a useful view of the relationships between the consensus sequences such that it is possible to assess whether they belong to the same TE family. One of the programs detailed above, MAP or MAFFT, can be used. It can also be informative to build a MSA with the consensus and with the genomic sequences from which these consensuses were derived and/or the genomic copies that each of these consensuses can detect. In such cases, we advise first building a single MSA for each consensus with the genomic sequences it detects, and then building a global MSA by aligning these multiple alignments together, for example, using the "profile" option of the MUSCLE program (Edgar 2004). Finally, after a visual check of the MSA with the evidence used to assign a classification to the consensus, it is then possible to tag all consensus sequences in the same cluster with the most frequent TE class, order, superfamily, and family, if one has been assigned (Fig. 2.2). The MSA can be also edited by splitting it or deleting sequences to obtain a MSA corresponding to a single TE family. Indeed, in some cases, consensuses are only similar along a small segment or display substantial sequence divergence. In these cases, the MSA can be split into as many MSA as there are candidate TE families. In other cases, an insertion appears to be specific to one consensus sequences and may sometimes show evidence (e.g., BLAST hits) for a different TE order. This may indicate a chimeric consensus that can be either removed from the library, if artifactual according to the sequences used to build the consensus (also visible in the MSA), or used to build a new TE family (if several copies support it). In all these cases, finding a genomic copy that aligns along almost all the length of a consensus (e.g., 95% coverage) appears to be a reasonable criterion for retaining the consensus. Those that fail generally appear to be artifacts or at least could be considered to be of no value.

Phylogenies of TE family copies and/or consensus sequences provide another view of the members in a TE family. This can serve as an aid to curation if the cluster has many members or if two or more subfamilies are present. In such cases, sub-families can be hard to detect by examination of the MSA alone, but may

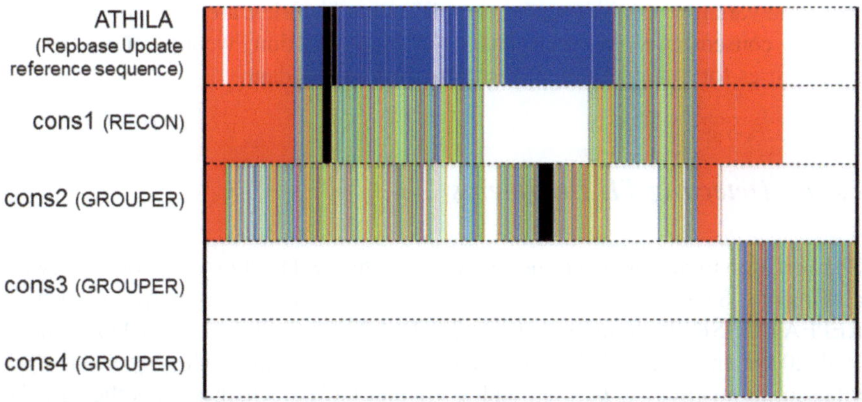

Fig. 2.2 Alignment (Jalview (Clamp et al. 2004) screenshot) of *de novo* TE consensus sequences with Athila, the best-matching known TEs in the Repbase Update. They are represented with some of the features shown: LTRs (*red zones*), ORFs (*blue zone*), and matches with HMM profiles (*black*). The differences between the consensuses obtained by different methods, here RECON (cons1) and GROUPER (cons2, cons3, cons4), are indicated. Manual curation would remove cons3 as it corresponds to a single LTR with short sequences not present in the Athila family and cons4 as it corresponds to a LTR probably formed from the Athila solo-LTRs of the genome. A good consensus for the family would be a combination of cons1 and cons2

become evident in a phylogeny if distinct sub-trees emerge. Such phylogenies can be constructed from the MSA with currently available software, including the PhyML program (Guindon and Gascuel 2003). Note, however, as most phylogeny programs do not consider gaps, branch length may be biased when sequences are of very different lengths. Divergence between the sequences can also be a criterion. Some authors (Wicker et al. 2007) have suggested a 80–80–80 rule: two sequences can be considered to belong to same TE family if they can be aligned along more than 80 bp, over more than 80% of their length, with more than 80% of identity. This rule is empirical but appears to be useful for classifying TE sequences into families that are consistent for the following annotation step, the annotation of their copies. These authors also suggest a nomenclature system for naming new TEs.

2.4 Annotation of Transposable Element Copies

This third phase annotates all TE copies in the genome, resolving the most complex degenerate or nested structures. This requires a library of reference sequences representing the TE families. In the best case, the library is both exhaustive and non-redundant, i.e., each ancestral TE, autonomous or not, is represented by a single consensus sequence. We usually use the manually curated library built as described in the previous section, as well as known TE sequences present in the public data banks. Note that some TE families, particularly those including structural variants with independent amplification histories, are best represented by several consensuses. In such cases, manual curation would retain several consensuses for a family, considered here as nonredundant.

2.4.1 Detecting TE Fragments

The first step mines the genomic sequences with the TE library via local pairwise alignments. Several tools were designed specifically for this purpose, such as REPEATMASKER (Smit et al. 1996–2004), CENSOR (Jurka et al. 1996; Kohany et al. 2006), and BLASTER (Quesneville et al. 2003). Some of these tools incorporate scoring matrices to be used with particular GC percentages, as is the case for isochores in the human genome. All these tools propose a small set of parameter combinations depending on the level of sensitivity required by the user.

Although similar, these tools are complementary. We have shown previously that combining these three programs is the best strategy (Quesneville et al. 2005). The MATCHER program (Quesneville et al. 2003) can then be used to assess the multiple results and keep only the best for each location.

Whatever parameters are used for the pairwise alignments, some of the matches will be false positives, i.e., a TE reference sequence will match a locus although no TE is present. For protein-coding genes, full-length cDNAs can be used for confirmation; unfortunately, there is no equivalent way of checking for TE annotation. An empirical statistical filter, such as implemented in the TEannot pipeline (REPET package) (Flutre et al. 2011), can be used to assess the false positive risk. The genomic sequences are shuffled and screened with the TE library. The alignments obtained on a shuffled sequence can be considered as false positives, then the 95-percentile alignment score is used to filter out spurious alignments obtained with the true genome. Only the matches with the true genomic sequences having a higher score are kept. This procedure guarantees that no observed match scores used for the annotation can be obtained for random sequences with a probably greater than 5%.

2.4.2 Filtering Satellites

Short simple repeats (SSRs) are short motifs repeated in tandem. Many TE sequences contain SSRs but SSRs are also present in the genome independently. It is therefore necessary to filter out TE matches if they are restricted to SSR that the TE consensus may contain. This can be done by annotating SSRs and then removing TE matches included in SSR annotations. Several efficient programs, for example, TRF (Benson 1999), MREPS (Kolpakov et al. 2003), and REPEATMASKER, are available for SSR annotation. In TEannot from the REPET package, these three programs are launched in parallel, and their results are subsequently combined to be used to eliminate hits due to only SSRs in TE consensuses.

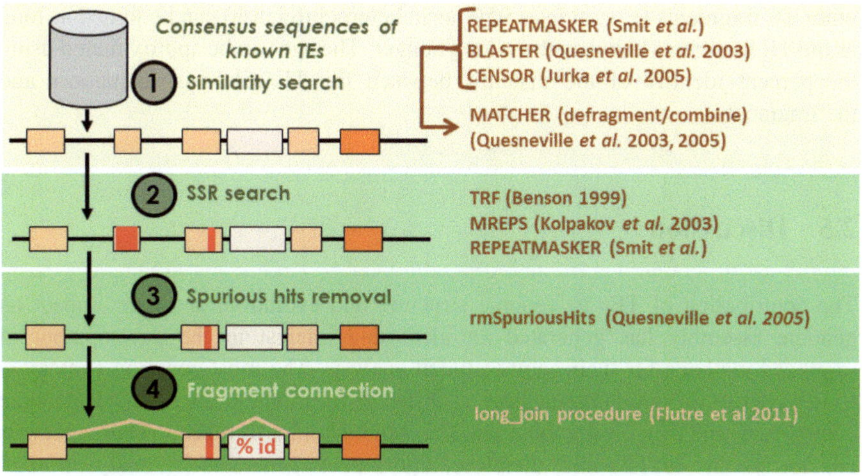

Fig. 2.3 The four steps of the TEannot pipeline (Quesneville et al. 2005)

Satellites are longer motifs, around 100 bp long, also repeated in tandem. Although they are not TEs, they are sometimes difficult to distinguish because they may contain parts of TEs. PILER-TA (Edgar and Myers 2005) detects pyramids in a self-alignment of the genomic sequences. These pyramids can be used to make a consensus of the satellite unit motif. These consensuses can then be aligned on the whole genome to find all their occurrences and to distinguish them from TEs.

2.4.3 *Connecting TE Fragments to Recover TE Copies*

Even when TE fragments have been mapped in the genome, the work is only half-finished. Indeed, TE copies can be disrupted into several fragments. A complete TE annotation requires retrieving all copies and thus linking fragments belonging to the same copy when it has transposed.

The first, historical method was manual curation using dot plots. However, this is laborious and curator dependent, and is impractical for large genomes. It requires the curator having detailed knowledge of transposable elements. Moreover, it ignores the age of nested fragments, potentially leading to incongruities. Therefore several computational approaches have been proposed. Many of them are reviewed in the article by Pereira (Pereira 2008).

Joining TE fragments to reconstruct a TE copy is known as a "chain problem" as it corresponds to finding the best chain of local pairwise alignments. The optimal solution is found via dynamic programming as implemented in MATCHER. Subsequently, an additional procedure implemented in the TEannot pipeline (Fig. 2.3) called "long join," can be used to take into account additional considerations related to TE biology. Two TE fragments distant from each other but mostly separated by other TE fragments (e.g., at least 95% as in heterochromatin) can be joined as long as the TE fragments between them are younger. The age can be approximated using the percent identity of the matches between the TE reference sequences and the fragments.

2.5 Discussion

The contribution of TEs to genome structure and evolution, and their impact on genome assembly has generated an increasing interest in the development of improved methods for their computational analysis. The most common strategy is to detect pairs of similar sequences at different locations in an all-by-all genome comparison, and then cluster these pairs to obtain families of repeats. These methods are not specific to TEs and, therefore, find repeats generated by many different processes, including tandem repeats, segmental duplications, and satellites. Moreover, TE copies can be highly degenerated, deleted, or nested. So repeat detection methods can make errors in the detection of individual TE copies and consequently in defining TE families. We believe that existing automatic approaches still need to be supplemented by expert manual curation. At this step, careful examination is required because some identified families that may appear to be artifactual can in fact be unusual TE families. Indeed, well documented cases illustrate how TEs families can appear confusing as they may (1) include cellular genes or parts of genes [e.g., pack-MULEs (Jiang et al. 2004) or *Helitrons* (Morgante et al. 2005a)], (2) be restricted to rDNA genes [e.g., the R2 Non-LTR retroelement superfamily (Eickbush et al. 1997)], or (3) form telomeres [in *Drosophila* (Clark et al. 2007)]. Close examination of noncanonical cases may also reveal new and interesting TE families or particular transposition events [e.g., macrotranspositions (Gray 2000)].

Knowledge-based TE detection methods (i.e., based on structure or similarity to distant TEs) have distinct advantages over *de novo* repeat discovery methods. They capitalize on prior knowledge established from the large number of previously reported TE sequences. Thus, they are more likely to detect *bona fide* TEs, including even those present as only a single copy in the genome. However, these methods are not well suited to the discovery of new TEs (especially of new types). Moreover, these methods have intrinsic ascertainment biases. For example, miniature inverted repeat transposable elements (MITEs) and short interspersed nuclear elements (SINEs) will be under-identified if only similarity-based methods are used because these TEs are composed entirely of noncoding sequences.

For some species, only parts of the genomic sequences are available as BAC sequences assembly. Working on a genome subset could be difficult for all-by-all genome comparison approaches as a TE might appear not repeated if other copies are not yet sequenced. Detection sensitivity of such approaches increase on both the sequenced fraction of the genome and its repeat density. Consequently, according to the sequence size and the repeat density, all-by-all genome comparison approaches may be used with more or less success. Interestingly, detection

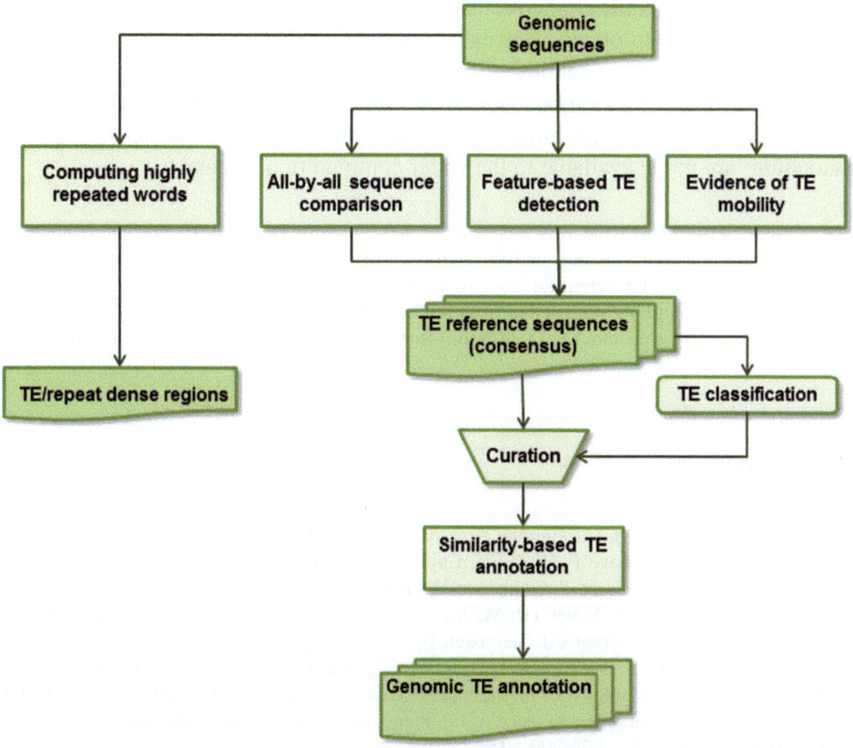

Fig. 2.4 Workflow for annotating TEs in genomic sequences

sensitivity of knowledge-based approaches (i.e., based on structure or similarity to distant TEs) is independent of the sequenced fraction, making them highly recommended here.

Through our experience with many genome projects (Cock et al. 2010; Abad et al. 2008; Amselem et al. 2011; Cuomo et al. 2007; Duplessis et al. 2011; Martin et al. 2008, 2010; Nene et al. 2007; Quesneville et al. 2003, 2005; Rouxel et al. 2011; Spanu et al. 2010), we have assessed the relative benefits of using different programs for TE detection, clustering, and multiple alignments. Our investigations suggest that only combined approaches, using both *de novo* and knowledge-based TE detection methods, are likely to produce reasonably comprehensive and sensitive results. Figure 2.4 shows the general workflow to follow for annotating TEs. In view of this, the REPET package (Flutre et al. 2011) has been developed. It is composed of two pipelines, TEdenovo and TEannot. These pipelines launch several different prediction programs in parallel and then combine their results to optimize the accuracy and exhaustiveness of TE detection. Even with this sophisticated pipeline, manual curation is still needed. Hence, in addition to the automation of all the steps required for the TE annotation, it computes data that are useful for the manual curation, including TE sequence multiple alignments, TE sequence phylogenies, and TE evidence. Sequencing costs have dropped dramatically and sequences have thus become easier to obtain. However, sequence analysis remains a major bottleneck. Efficient analysis pipelines are required. They need to be quick and robust to accelerate the pace of data production; they should also exploit the knowledge of the few specialists able to perform genome analysis on a large scale so that TE annotations are made available to the wider community of scientists.

Acknowledgments This work was supported in part by grants from the Agence Nationale de la Recherche (Holocentrism project, to HQ [grant number ANR-07-BLAN-0057]) and the Centre National de la Recherche Scientifique—Groupement de Recherche "Elements Transposables." TF was supported by a PhD studentship from the Institut National de la Recherche Agronomique. EP was supported by a postdoctoral fellowship from the Agence Nationale de la Recherche.

References

Abad P, Gouzy J, Aury JM, Castagnone-Sereno P, Danchin EG, Deleury E, Perfus-Barbeoch L, Anthouard V, Artiguenave F, Blok VC, Caillaud MC, Coutinho PM, Dasilva C, De Luca F, Deau F, Esquibet M, Flutre T, Goldstone JV, Hamamouch N, Hewezi T, Jaillon O, Jubin C, Leonetti P, Magliano M, Maier TR, Markov GV, McVeigh P, Pesole G, Poulain J, Robinson-Rechavi M, Sallet E, Segurens B, Steinbach D, Tytgat T, Ugarte E, van Ghelder C, Veronico P, Baum TJ, Blaxter M, Bleve-Zacheo T, Davis EL, Ewbank JJ, Favery B, Grenier E, Henrissat B, Jones JT, Laudet V, Maule AG, Quesneville H, Rosso MN, Schiex T, Smant G, Weissenbach J, Wincker P (2008) Genome sequence of the metazoan plant-parasitic nematode Meloidogyne incognita. Nat Biotechnol 26:909–915

Abrusan G, Grundmann N, DeMester L, Makalowski W (2009) TEclass–a tool for automated classification of unknown eukaryotic transposable elements. Bioinformatics 25:1329–1330

Altschul SF, Madden TL, Schaffer AA, Zhang J, Zhang Z, Miller W, Lipman DJ (1997) Gapped BLAST and PSI-BLAST: a new generation of protein database search programs. Nucleic Acids Res 25:3389–3402

Amselem J, Cuomo CA, van Kan JA, Viaud M, Benito EP, Couloux A, Coutinho PM, de Vries RP, Dyer PS, Fillinger S, Fournier E, Gout L, Hahn M, Kohn L, Lapalu N, Plummer KM, Pradier JM, Quevillon E, Sharon A, Simon A, ten Have A, Tudzynski B, Tudzynski P, Wincker P, Andrew M, Anthouard V, Beever RE, Beffa R, Benoit I, Bouzid O, Brault B, Chen Z, Choquer M, Collemare J, Cotton P, Danchin EG, Da Silva C, Gautier A, Giraud C, Giraud T, Gonzalez C, Grossetete S, Guldener U, Henrissat B, Howlett BJ, Kodira C, Kretschmer M, Lappartient A, Leroch M, Levis C, Mauceli E, Neuveglise C, Oeser B, Pearson M, Poulain J, Poussereau N, Quesneville H, Rascle C, Schumacher J, Segurens B, Sexton A, Silva E, Sirven C, Soanes DM, Talbot NJ, Templeton M, Yandava C, Yarden O, Zeng Q, Rollins JA, Lebrun MH, Dickman M (2011) Genomic analysis of the necrotrophic fungal pathogens Sclerotinia sclerotiorum and Botrytis cinerea. PLoS Genet 7:e1002230

Bao Z, Eddy SR (2002) Automated *de novo* identification of repeat sequence families in sequenced genomes. Genome Res 12:1269–1276

Benson G (1999) Tandem repeats finder: a program to analyze DNA sequences. Nucleic Acids Res 27:573–580

Bergman CM, Quesneville H (2007) Discovering and detecting transposable elements in genome sequences. Brief Bioinform 8:382–392

Bergman CM, Quesneville H, Anxolabehere D, Ashburner M (2006) Recurrent insertion and duplication generate networks of transposable element sequences in the Drosophila melanogaster genome. Genome Biol 7:R112

Blumenstiel JP, Hartl DL, Lozovsky ER (2002) Patterns of insertion and deletion in contrasting chromatin domains. Mol Biol Evol 19:2211–2225

Bourque G (2009) Transposable elements in gene regulation and in the evolution of vertebrate genomes. Curr Opin Genet Dev 19:607–612

Bundock P, Hooykaas P (2005) An Arabidopsis hAT-like transposase is essential for plant development. Nature 436:282–284

Caspi A, Pachter L (2006) Identification of transposable elements using multiple alignments of related genomes. Genome Res 16:260–270

Chen Y, Zhou F, Li G, Xu Y (2009) MUST: a system for identification of miniature inverted-repeat transposable elements and applications to Anabaena variabilis and Haloquadratum walsbyi. Gene 436:1–7

Clamp M, Cuff J, Searle SM, Barton GJ (2004) The Jalview Java alignment editor. Bioinformatics 20:426–427

Clark AG, Eisen MB, Smith DR, Bergman CM, Oliver B, Markow TA, Kaufman TC, Kellis M, Gelbart W, Iyer VN, Pollard DA, Sackton TB, Larracuente AM, Singh ND, Abad JP, Abt DN, Adryan B, Aguade M, Akashi H, Anderson WW, Aquadro CF, Ardell DH, Arguello R, Artieri CG, Barbash DA, Barker D, Barsanti P, Batterham P, Batzoglou S, Begun D, Bhutkar A, Blanco E, Bosak SA, Bradley RK, Brand AD, Brent MR, Brooks AN, Brown RH, Butlin RK, Caggese C, Calvi BR, Bernardo de Carvalho A, Caspi A, Castrezana S, Celniker SE, Chang JL, Chapple C, Chatterji S, Chinwalla A, Civetta A, Clifton SW, Comeron JM, Costello JC, Coyne JA, Daub J, David RG, Delcher AL, Delehaunty K, Do CB, Ebling H, Edwards K, Eickbush T, Evans JD, Filipski A, Findeiss S, Freyhult E, Fulton L, Fulton R, Garcia AC, Gardiner A, Garfield DA, Garvin BE, Gibson G, Gilbert D, Gnerre S, Godfrey J, Good R, Gotea V, Gravely B, Greenberg AJ, Griffiths-Jones S, Gross S, Guigo R, Gustafson EA, Haerty W, Hahn MW, Halligan DL, Halpern AL, Halter GM, Han MV, Heger A, Hillier L, Hinrichs AS, Holmes I, Hoskins RA, Hubisz MJ, Hultmark D, Huntley MA, Jaffe DB, Jagadeeshan S, Jeck WR, Johnson J, Jones CD, Jordan WC, Karpen GH, Kataoka E, Keightley PD, Kheradpour P, Kirkness EF, Koerich LB, Kristiansen K, Kudrna D, Kulathinal RJ, Kumar S, Kwok R, Lander E, Langley CH, Lapoint R, Lazzaro BP, Lee SJ, Levesque L, Li R, Lin CF, Lin MF, Lindblad-Toh K, Llopart A, Long M, Low L, Lozovsky E, Lu J, Luo M, Machado CA, Makalowski W, Marzo M, Matsuda M, Matzkin L, McAllister B, McBride CS, McKernan B, McKernan K, Mendez Lago M, Minx P, Mollenhauer MU, Montooth K, Mount SM, Mu X, Myers E, Negre B, Newfeld S, Nielsen R, Noor MA, O'Grady P, Pachter L, Papaceit M, Parisi MJ, Parisi M, Parts L, Pedersen JS, Pesole G, Phillippy AM, Ponting CP, Pop M, Porcelli D, Powell JR, Prohaska S, Pruitt K, Puig M, Quesneville H, Ram KR, Rand D, Rasmussen MD, Reed LK, Reenan R, Reily A, Remington KA, Rieger TT, Ritchie MG, Robin C, Rogers YH, Rohde C, Rozas J, Rubenfield MJ, Ruiz A, Russo S, Salzberg SL, Sanchez-Gracia A, Saranga DJ, Sato H, Schaeffer SW, Schatz MC, Schlenke T, Schwartz R, Segarra C, Singh RS, Sirot L, Sirota M, Sisneros NB, Smith CD, Smith TF, Spieth J, Stage DE, Stark A, Stephan W, Strausberg RL, Strempel S, Sturgill D, Sutton G, Sutton GG, Tao W, Teichmann S, Tobari YN, Tomimura Y, Tsolas JM, Valente VL, Venter E, Venter JC, Vicario S, Vieira FG, Vilella AJ, Villasante A, Walenz B, Wang J, Wasserman M, Watts T, Wilson D, Wilson RK, Wing RA, Wolfner MF, Wong A, Wong GK, Wu CI, Wu G, Yamamoto D, Yang HP, Yang SP, Yorke JA, Yoshida K, Zdobnov E, Zhang P, Zhang Y, Zimin AV, Baldwin J, Abdouelleil A, Abdulkadir J, Abebe A, Abera B, Abreu J, Acer SC, Aftuck L, Alexander A, An P, Anderson E, Anderson S, Arachi H, Azer M, Bachantsang P, Barry A, Bayul T, Berlin A, Bessette D, Bloom T, Blye J, Boguslavskiy L, Bonnet C, Boukhgalter B, Bourzgui I, Brown A, Cahill P, Channer S, Cheshatsang Y, Chuda L, Citroen M, Collymore A, Cooke P, Costello M, D'Aco K, Daza R, De Haan G, DeGray S, DeMaso C, Dhargay N, Dooley K, Dooley E, Doricent M, Dorje P, Dorjee K, Dupes A, Elong R, Falk J, Farina A, Faro S, Ferguson D, Fisher S, Foley CD, Franke A, Friedrich D, Gadbois L, Gearin G, Gearin CR, Giannoukos G, Goode T, Graham J, Grandbois E, Grewal S, Gyaltsen K, Hafez N, Hagos B, Hall J, Henson C, Hollinger A, Honan T, Huard MD,

Hughes L, Hurhula B, Husby ME, Kamat A, Kanga B, Kashin S, Khazanovich D, Kisner P, Lance K, Lara M, Lee W, Lennon N, Letendre F, LeVine R, Lipovsky A, Liu X, Liu J, Liu S, Lokyitsang T, Lokyitsang Y, Lubonja R, Lui A, MacDonald P, Magnisalis V, Maru K, Matthews C, McCusker W, McDonough S, Mehta T, Meldrim J, Meneus L, Mihai O, Mihalev A, Mihova T, Mittelman R, Mlenga V, Montmayeur A, Mulrain L, Navidi A, Naylor J, Negash T, Nguyen T, Nguyen N, Nicol R, Norbu C, Norbu N, Novod N, O'Neill B, Osman S, Markiewicz E, Oyono OL, Patti C, Phunkhang P, Pierre F, Priest M, Raghuraman S, Rege F, Reyes R, Rise C, Rogov P, Ross K, Ryan E, Settipalli S, Shea T, Sherpa N, Shi L, Shih D, Sparrow T, Spaulding J, Stalker J, Stange-Thomann N, Stavropoulos S, Stone C, Strader C, Tesfaye S, Thomson T, Thoulutsang Y, Thoulutsang D, Topham K, Topping I, Tsamla T, Vassiliev H, Vo A, Wangchuk T, Wangdi T, Weiand M, Wilkinson J, Wilson A, Yadav S, Young G, Yu Q, Zembek L, Zhong D, Zimmer A, Zwirko Z, Jaffe DB, Alvarez P, Brockman W, Butler J, Chin C, Gnerre S, Grabherr M, Kleber M, Mauceli E, MacCallum I (2007) Evolution of genes and genomes on the Drosophila phylogeny. Nature 450:203–218

Cock JM, Sterck L, Rouze P, Scornet D, Allen AE, Amoutzias G, Anthouard V, Artiguenave F, Aury JM, Badger JH, Beszteri B, Billiau K, Bonnet E, Bothwell JH, Bowler C, Boyen C, Brownlee C, Carrano CJ, Charrier B, Cho GY, Coelho SM, Collen J, Corre E, Da Silva C, Delage L, Delaroque N, Dittami SM, Doulbeau S, Elias M, Farnham G, Gachon CM, Gschloessl B, Heesch S, Jabbari K, Jubin C, Kawai H, Kimura K, Kloareg B, Kupper FC, Lang D, Le Bail A, Leblanc C, Lerouge P, Lohr M, Lopez PJ, Martens C, Maumus F, Michel G, Miranda-Saavedra D, Morales J, Moreau H, Motomura T, Nagasato C, Napoli CA, Nelson DR, Nyvall-Collen P, Peters AF, Pommier C, Potin P, Poulain J, Quesneville H, Read B, Rensing SA, Ritter A, Rousvoal S, Samanta M, Samson G, Schroeder DC, Segurens B, Strittmatter M, Tonon T, Tregear JW, Valentin K, von Dassow P, Yamagishi T, Van de Peer Y, Wincker P (2010) The Ectocarpus genome and the independent evolution of multicellularity in brown algae. Nature 465:617–621

Cuomo CA, Guldener U, Xu JR, Trail F, Turgeon BG, Di Pietro A, Walton JD, Ma LJ, Baker SE, Rep M, Adam G, Antoniw J, Baldwin T, Calvo S, Chang YL, Decaprio D, Gale LR, Gnerre S, Goswami RS, Hammond-Kosack K, Harris LJ, Hilburn K, Kennell JC, Kroken S, Magnuson JK, Mannhaupt G, Mauceli E, Mewes HW, Mitterbauer R, Muehlbauer G, Munsterkotter M, Nelson D, O'Donnell K, Ouellet T, Qi W, Quesneville H, Roncero MI, Seong KY, Tetko IV, Urban M, Waalwijk C, Ward TJ, Yao J, Birren BW, Kistler HC (2007) The Fusarium graminearum genome reveals a link between localized polymorphism and pathogen specialization. Science 317:1400–1402

Du C, Caronna J, He L, Dooner HK (2008) Computational prediction and molecular confirmation of Helitron transposons in the maize genome. BMC Genomics 9:51

Duplessis S, Cuomo CA, Lin YC, Aerts A, Tisserant E, Veneault-Fourrey C, Joly DL, Hacquard S, Amselem J, Cantarel BL, Chiu R, Coutinho PM, Feau N, Field M, Frey P, Gelhaye E, Goldberg J, Grabherr MG, Kodira CD, Kohler A, Kues U, Lindquist EA, Lucas SM, Mago R, Mauceli E, Morin E, Murat C, Pangilinan JL, Park R, Pearson M, Quesneville H, Rouhier N, Sakthikumar S, Salamov AA, Schmutz J, Selles B, Shapiro H, Tanguay P, Tuskan GA, Henrissat B, Van de Peer Y, Rouze P, Ellis JG, Dodds PN, Schein JE, Zhong S, Hamelin RC, Grigoriev IV, Szabo LJ, Martin F (2011) Obligate biotrophy features unraveled by the genomic analysis of rust fungi. Proc Natl Acad Sci USA 108:9166–9171

Edgar RC (2004) MUSCLE: a multiple sequence alignment method with reduced time and space complexity. BMC Bioinformatics 5:113

Edgar RC, Myers EW (2005) PILER: identification and classification of genomic repeats. Bioinformatics 21(Suppl 1):i152–i158

Eickbush TH, Burke WD, Eickbush DG, Lathe WC 3rd (1997) Evolution of R1 and R2 in the rDNA units of the genus Drosophila. Genetica 100:49–61

Ellinghaus D, Kurtz S, Willhoeft U (2008) LTRharvest, an efficient and flexible software for de novo detection of LTR retrotransposons. BMC Bioinformatics 9:18

Feschotte C (2008) Transposable elements and the evolution of regulatory networks. Nat Rev Genet 9:397–405

Feschotte C, Keswani U, Ranganathan N, Guibotsy ML, Levine D (2009) Exploring repetitive DNA landscapes using REPCLASS, a tool that automates the classification of transposable elements in eukaryotic genomes. Genome Biol Evol 2009:205–220

Finn RD, Mistry J, Tate J, Coggill P, Heger A, Pollington JE, Gavin OL, Gunasekaran P, Ceric G, Forslund K, Holm L, Sonnhammer EL, Eddy SR, Bateman A (2010) The Pfam protein families database. Nucleic Acids Res 38:D211–D222

Finnegan DJ (1989) Eukaryotic transposable elements and genome evolution. Trends Genet 5:103–107

Fiston-Lavier AS, Anxolabehere D, Quesneville H (2007) A model of segmental duplication formation in Drosophila melanogaster. Genome Res 17:1458–1470

Flutre T, Duprat E, Feuillet C, Quesneville H (2011) Considering transposable element diversification in *de novo* annotation approaches. PLoS One 6:e16526

Gray YH (2000) It takes two transposons to tango: transposable-element-mediated chromosomal rearrangements. Trends Genet 16:461–468

Gu W, Castoe TA, Hedges DJ, Batzer MA, Pollock DD (2008) Identification of repeat structure in large genomes using repeat probability clouds. Anal Biochem 380(1):77–83

Guindon S, Gascuel O (2003) A simple, fast, and accurate algorithm to estimate large phylogenies by maximum likelihood. Syst Biol 52:696–704

Han Y, Wessler SR (2010) MITE-Hunter: a program for discovering miniature inverted-repeat transposable elements from genomic sequences. Nucleic Acids Res 38:e199

Hikosaka A, Kawahara A (2010) A systematic search and classification of T2 family miniature inverted-repeat transposable elements (MITEs) in Xenopus tropicalis suggests the existence of recently active MITE subfamilies. Mol Genet Genomics 283:49–62

Huang X (1994) On global sequence alignment. Comput Appl Biosci 10:227–235

Janicki M, Rooke R, Yang G (2011) Bioinformatics and genomic analysis of transposable elements in eukaryotic genomes. Chromosome Res 19:787–808

Jiang N, Bao Z, Zhang X, Eddy SR, Wessler SR (2004) Pack-MULE transposable elements mediate gene evolution in plants. Nature 431:569–573

Jurka J, Klonowski P, Dagman V, Pelton P (1996) CENSOR–a program for identification and elimination of repetitive elements from DNA sequences. Comput Chem 20:119–121

Jurka J, Kapitonov VV, Pavlicek A, Klonowski P, Kohany O, Walichiewicz J (2005) Repbase Update, a database of eukaryotic repetitive elements. Cytogenet Genome Res 110:462–467

Kalyanaraman A, Aluru S (2006) Efficient algorithms and software for detection of full-length LTR retrotransposons. J Bioinform Comput Biol 4:197–216

Kapitonov VV, Jurka J (2005) RAG1 core and V(D)J recombination signal sequences were derived from Transib transposons. PLoS Biol 3:e181

Katoh K, Misawa K, Kuma K, Miyata T (2002) MAFFT: a novel method for rapid multiple sequence alignment based on fast Fourier transform. Nucleic Acids Res 30:3059–3066

Kohany O, Gentles AJ, Hankus L, Jurka J (2006) Annotation, submission and screening of repetitive elements in Repbase: RepbaseSubmitter and Censor. BMC Bioinformatics 7:474

Kolpakov R, Bana G, Kucherov G (2003) mreps: efficient and flexible detection of tandem repeats in DNA. Nucleic Acids Res 31:3672–3678

Kurtz S, Narechania A, Stein JC, Ware D (2008) A new method to compute K-mer frequencies and its application to annotate large repetitive plant genomes. BMC Genomics 9:517

Lander ES, Linton LM, Birren B, Nusbaum C, Zody MC, Baldwin J, Devon K, Dewar K, Doyle M, FitzHugh W, Funke R, Gage D, Harris K, Heaford A, Howland J, Kann L, Lehoczky J, LeVine R, McEwan P, McKernan K, Meldrim J, Mesirov JP, Miranda C, Morris W, Naylor J, Raymond C, Rosetti M, Santos R, Sheridan A, Sougnez C, Stange-Thomann N, Stojanovic N, Subramanian A, Wyman D, Rogers J, Sulston J, Ainscough R, Beck S, Bentley D, Burton J, Clee C, Carter N, Coulson A, Deadman R, Deloukas P, Dunham A, Dunham I, Durbin R, French L, Grafham D, Gregory S, Hubbard T, Humphray S, Hunt A,

Jones M, Lloyd C, McMurray A, Matthews L, Mercer S, Milne S, Mullikin JC, Mungall A, Plumb R, Ross M, Shownkeen R, Sims S, Waterston RH, Wilson RK, Hillier LW, McPherson JD, Marra MA, Mardis ER, Fulton LA, Chinwalla AT, Pepin KH, Gish WR, Chissoe SL, Wendl MC, Delehaunty KD, Miner TL, Delehaunty A, Kramer JB, Cook LL, Fulton RS, Johnson DL, Minx PJ, Clifton SW, Hawkins T, Branscomb E, Predki P, Richardson P, Wenning S, Slezak T, Doggett N, Cheng JF, Olsen A, Lucas S, Elkin C, Uberbacher E, Frazier M, Gibbs RA, Muzny DM, Scherer SE, Bouck JB, Sodergren EJ, Worley KC, Rives CM, Gorrell JH, Metzker ML, Naylor SL, Kucherlapati RS, Nelson DL, Weinstock GM, Sakaki Y, Fujiyama A, Hattori M, Yada T, Toyoda A, Itoh T, Kawagoe C, Watanabe H, Totoki Y, Taylor T, Weissenbach J, Heilig R, Saurin W, Artiguenave F, Brottier P, Bruls T, Pelletier E, Robert C, Wincker P, Smith DR, Doucette-Stamm L, Rubenfield M, Weinstock K, Lee HM, Dubois J, Rosenthal A, Platzer M, Nyakatura G, Taudien S, Rump A, Yang H, Yu J, Wang J, Huang G, Gu J, Hood L, Rowen L, Madan A, Qin S, Davis RW, Federspiel NA, Abola AP, Proctor MJ, Myers RM, Schmutz J, Dickson M, Grimwood J, Cox DR, Olson MV, Kaul R, Raymond C, Shimizu N, Kawasaki K, Minoshima S, Evans GA, Athanasiou M, Schultz R, Roe BA, Chen F, Pan H, Ramser J, Lehrach H, Reinhardt R, McCombie WR, de la Bastide M, Dedhia N, Blocker H, Hornischer K, Nordsiek G, Agarwala R, Aravind L, Bailey JA, Bateman A, Batzoglou S, Birney E, Bork P, Brown DG, Burge CB, Cerutti L, Chen HC, Church D, Clamp M, Copley RR, Doerks T, Eddy SR, Eichler EE, Furey TS, Galagan J, Gilbert JG, Harmon C, Hayashizaki Y, Haussler D, Hermjakob H, Hokamp K, Jang W, Johnson LS, Jones TA, Kasif S, Kaspryzk A, Kennedy S, Kent WJ, Kitts P, Koonin EV, Korf I, Kulp D, Lancet D, Lowe TM, McLysaght A, Mikkelsen T, Moran JV, Mulder N, Pollara VJ, Ponting CP, Schuler G, Schultz J, Slater G, Smit AF, Stupka E, Szustakowski J, Thierry-Mieg D, Thierry-Mieg J, Wagner L, Wallis J, Wheeler R, Williams A, Wolf YI, Wolfe KH, Yang SP, Yeh RF, Collins F, Guyer MS, Peterson J, Felsenfeld A, Wetterstrand KA, Patrinos A, Morgan MJ, de Jong P, Catanese JJ, Osoegawa K, Shizuya H, Choi S, Chen YJ (2001) Initial sequencing and analysis of the human genome. Nature 409:860–921

Le QH, Wright S, Yu Z, Bureau T (2000) Transposon diversity in Arabidopsis thaliana. Proc Natl Acad Sci USA 97:7376–7381

Lee W, Chen SL (2002) Genome-tools: a flexible package for genome sequence analysis. Biotechniques 33:1334–1341

Lerat E (2010) Identifying repeats and transposable elements in sequenced genomes: how to find your way through the dense forest of programs. Heredity 104:520–533

Li R, Ye J, Li S, Wang J, Han Y, Ye C, Wang J, Yang H, Yu J, Wong GK, Wang J (2005) ReAS: Recovery of ancestral sequences for transposable elements from the unassembled reads of a whole genome shotgun. PLoS Comput Biol 1:e43

Lucier JF, Perreault J, Noel JF, Boire G, Perreault JP (2007) RTAnalyzer: a web application for finding new retrotransposons and detecting L1 retrotransposition signatures. Nucleic Acids Res 35:W269–W274

Martin F, Aerts A, Ahren D, Brun A, Danchin EG, Duchaussoy F, Gibon J, Kohler A, Lindquist E, Pereda V, Salamov A, Shapiro HJ, Wuyts J, Blaudez D, Buee M, Brokstein P, Canback B, Cohen D, Courty PE, Coutinho PM, Delaruelle C, Detter JC, Deveau A, DiFazio S, Duplessis S, Fraissinet-Tachet L, Lucic E, Frey-Klett P, Fourrey C, Feussner I, Gay G, Grimwood J, Hoegger PJ, Jain P, Kilaru S, Labbe J, Lin YC, Legue V, Le Tacon F, Marmeisse R, Melayah D, Montanini B, Muratet M, Nehls U, Niculita-Hirzel H, Oudot-Le Secq MP, Peter M, Quesneville H, Rajashekar B, Reich M, Rouhier N, Schmutz J, Yin T, Chalot M, Henrissat B, Kues U, Lucas S, Van de Peer Y, Podila GK, Polle A, Pukkila PJ, Richardson PM, Rouze P, Sanders IR, Stajich JE, Tunlid A, Tuskan G, Grigoriev IV (2008) The genome of Laccaria bicolor provides insights into mycorrhizal symbiosis. Nature 452:88–92

Martin F, Kohler A, Murat C, Balestrini R, Coutinho PM, Jaillon O, Montanini B, Morin E, Noel B, Percudani R, Porcel B, Rubini A, Amicucci A, Amselem J, Anthouard V, Arcioni S,

Artiguenave F, Aury JM, Ballario P, Bolchi A, Brenna A, Brun A, Buee M, Cantarel B, Chevalier G, Couloux A, Da Silva C, Denoeud F, Duplessis S, Ghignone S, Hilselberger B, Iotti M, Marcais B, Mello A, Miranda M, Pacioni G, Quesneville H, Riccioni C, Ruotolo R, Splivallo R, Stocchi V, Tisserant E, Viscomi AR, Zambonelli A, Zampieri E, Henrissat B, Lebrun MH, Paolocci F, Bonfante P, Ottonello S, Wincker P (2010) Perigord black truffle genome uncovers evolutionary origins and mechanisms of symbiosis. Nature 464:1033–1038

McCarthy EM, McDonald JF (2003) LTR_STRUC: a novel search and identification program for LTR retrotransposons. Bioinformatics 19:362–367

Morgante M, Brunner S, Pea G, Fengler K, Zuccolo A, Rafalski A (2005a) Gene duplication and exon shuffling by helitron-like transposons generate intraspecies diversity in maize. Nat Genet 37:997–1002

Morgante M, Policriti A, Vitacolonna N, Zuccolo A (2005b) Structured motifs search. J Comput Biol 12:1065–1082

Nene V, Wortman JR, Lawson D, Haas B, Kodira C, Tu ZJ, Loftus B, Xi Z, Megy K, Grabherr M, Ren Q, Zdobnov EM, Lobo NF, Campbell KS, Brown SE, Bonaldo MF, Zhu J, Sinkins SP, Hogenkamp DG, Amedeo P, Arensburger P, Atkinson PW, Bidwell S, Biedler J, Birney E, Bruggner RV, Costas J, Coy MR, Crabtree J, Crawford M, Debruyn B, Decaprio D, Eiglmeier K, Eisenstadt E, El-Dorry H, Gelbart WM, Gomes SL, Hammond M, Hannick LI, Hogan JR, Holmes MH, Jaffe D, Johnston JS, Kennedy RC, Koo H, Kravitz S, Kriventseva EV, Kulp D, Labutti K, Lee E, Li S, Lovin DD, Mao C, Mauceli E, Menck CF, Miller JR, Montgomery P, Mori A, Nascimento AL, Naveira HF, Nusbaum C, O'Leary S, Orvis J, Pertea M, Quesneville H, Reidenbach KR, Rogers YH, Roth CW, Schneider JR, Schatz M, Shumway M, Stanke M, Stinson EO, Tubio JM, Vanzee JP, Verjovski-Almeida S, Werner D, White O, Wyder S, Zeng Q, Zhao Q, Zhao Y, Hill CA, Raikhel AS, Soares MB, Knudson DL, Lee NH, Galagan J, Salzberg SL, Paulsen IT, Dimopoulos G, Collins FH, Birren B, Fraser-Liggett CM, Severson DW (2007) Genome sequence of Aedes aegypti, a major arbovirus vector. Science 316:1718–1723

Orgel LE, Crick FH (1980) Selfish DNA: the ultimate parasite. Nature 284:604–607

Pereira V (2004) Insertion bias and purifying selection of retrotransposons in the Arabidopsis thaliana genome. Genome Biol 5:R79

Pereira V (2008) Automated paleontology of repetitive DNA with REANNOTATE. BMC Genomics 9:614

Petrov DA (2001) Evolution of genome size: new approaches to an old problem. Trends Genet 17:23–28

Piegu B, Guyot R, Picault N, Roulin A, Saniyal A, Kim H, Collura K, Brar DS, Jackson S, Wing RA, Panaud O (2006) Doubling genome size without polyploidization: dynamics of retrotransposition-driven genomic expansions in Oryza australiensis, a wild relative of rice. Genome Res 16:1262–1269

Price AL, Jones NC, Pevzner PA (2005) *De novo* identification of repeat families in large genomes. Bioinformatics 21(Suppl 1):i351–i358

Quesneville H, Nouaud D, Anxolabehere D (2003) Detection of new transposable element families in Drosophila melanogaster and Anopheles gambiae genomes. J Mol Evol 57(Suppl 1): S50–S59

Quesneville H, Bergman CM, Andrieu O, Autard D, Nouaud D, Ashburner M, Anxolabehere D (2005) Combined evidence annotation of transposable elements in genome sequences. PLoS Comput Biol 1:166–175

Rasmussen K, Stoye J, Myers EW (2005) Efficient q-gram filters for finding all e-matches over a given length. In: Heidelberg SB (ed) RECOMB, pp 189–203

Rho M, Tang H (2009) MGEScan-non-LTR: computational identification and classification of autonomous non-LTR retrotransposons in eukaryotic genomes. Nucleic Acids Res 37:e143

Rho M, Choi JH, Kim S, Lynch M, Tang H (2007) *De novo* identification of LTR retrotransposons in eukaryotic genomes. BMC Genomics 8:90

Rouxel T, Grandaubert J, Hane JK, Hoede C, van de Wouw AP, Couloux A, Dominguez V, Anthouard V, Bally P, Bourras S, Cozijnsen AJ, Ciuffetti LM, Degrave A, Dilmaghani A, Duret L, Fudal I, Goodwin SB, Gout L, Glaser N, Linglin J, Kema GH, Lapalu N, Lawrence CB, May K, Meyer M, Ollivier B, Poulain J, Schoch CL, Simon A, Spatafora JW, Stachowiak A, Turgeon BG, Tyler BM, Vincent D, Weissenbach J, Amselem J, Quesneville H, Oliver RP, Wincker P, Balesdent MH, Howlett BJ (2011) Effector diversification within compartments of the Leptosphaeria maculans genome affected by repeat-induced point mutations. Nat Commun 2:202

Santangelo AM, de Souza FS, Franchini LF, Bumaschny VF, Low MJ, Rubinstein M (2007) Ancient exaptation of a CORE-SINE retroposon into a highly conserved mammalian neuronal enhancer of the proopiomelanocortin gene. PLoS Genet 3:1813–1826

Santiago N, Herraiz C, Goni JR, Messeguer X, Casacuberta JM (2002) Genome-wide analysis of the Emigrant family of MITEs of Arabidopsis thaliana. Mol Biol Evol 19:2285–2293

Schnable PS, Ware D, Fulton RS, Stein JC, Wei F, Pasternak S, Liang C, Zhang J, Fulton L, Graves TA, Minx P, Reily AD, Courtney L, Kruchowski SS, Tomlinson C, Strong C, Delehaunty K, Fronick C, Courtney B, Rock SM, Belter E, Du F, Kim K, Abbott RM, Cotton M, Levy A, Marchetto P, Ochoa K, Jackson SM, Gillam B, Chen W, Yan L, Higginbotham J, Cardenas M, Waligorski J, Applebaum E, Phelps L, Falcone J, Kanchi K, Thane T, Scimone A, Thane N, Henke J, Wang T, Ruppert J, Shah N, Rotter K, Hodges J, Ingenthron E, Cordes M, Kohlberg S, Sgro J, Delgado B, Mead K, Chinwalla A, Leonard S, Crouse K, Collura K, Kudrna D, Currie J, He R, Angelova A, Rajasekar S, Mueller T, Lomeli R, Scara G, Ko A, Delaney K, Wissotski M, Lopez G, Campos D, Braidotti M, Ashley E, Golser W, Kim H, Lee S, Lin J, Dujmic Z, Kim W, Talag J, Zuccolo A, Fan C, Sebastian A, Kramer M, Spiegel L, Nascimento L, Zutavern T, Miller B, Ambroise C, Muller S, Spooner W, Narechania A, Ren L, Wei S, Kumari S, Faga B, Levy MJ, McMahan L, Van Buren P, Vaughn MW, Ying K, Yeh CT, Emrich SJ, Jia Y, Kalyanaraman A, Hsia AP, Barbazuk WB, Baucom RS, Brutnell TP, Carpita NC, Chaparro C, Chia JM, Deragon JM, Estill JC, Fu Y, Jeddeloh JA, Han Y, Lee H, Li P, Lisch DR, Liu S, Liu Z, Nagel DH, McCann MC, SanMiguel P, Myers AM, Nettleton D, Nguyen J, Penning BW, Ponnala L, Schneider KL, Schwartz DC, Sharma A, Soderlund C, Springer NM, Sun Q, Wang H, Waterman M, Westerman R, Wolfgruber TK, Yang L, Yu Y, Zhang L, Zhou S, Zhu Q, Bennetzen JL, Dawe RK, Jiang J, Jiang N, Presting GG, Wessler SR, Aluru S, Martienssen RA, Clifton SW, McCombie WR, Wing RA, Wilson RK (2009) The B73 maize genome: complexity, diversity, and dynamics. Science 326:1112–1115

Smit AFA, Hubley R, Green P (1996–2004) RepeatMasker open-3.0. Institute for Systems Biology

Spanu PD, Abbott JC, Amselem J, Burgis TA, Soanes DM, Stuber K, Loren V, van Themaat E, Brown JK, Butcher SA, Gurr SJ, Lebrun MH, Ridout CJ, Schulze-Lefert P, Talbot NJ, Ahmadinejad N, Ametz C, Barton GR, Benjdia M, Bidzinski P, Bindschedler LV, Both M, Brewer MT, Cadle-Davidson L, Cadle-Davidson MM, Collemare J, Cramer R, Frenkel O, Godfrey D, Harriman J, Hoede C, King BC, Klages S, Kleemann J, Knoll D, Koti PS, Kreplak J, Lopez-Ruiz FJ, Lu X, Maekawa T, Mahanil S, Micali C, Milgroom MG, Montana G, Noir S, O'Connell RJ, Oberhaensli S, Parlange F, Pedersen C, Quesneville H, Reinhardt R, Rott M, Sacristan S, Schmidt SM, Schon M, Skamnioti P, Sommer H, Stephens A, Takahara H, Thordal-Christensen H, Vigouroux M, Wessling R, Wicker T, Panstruga R (2010) Genome expansion and gene loss in powdery mildew fungi reveal tradeoffs in extreme parasitism. Science 330:1543–1546

Steinbiss S, Willhoeft U, Gremme G, Kurtz S (2009) Fine-grained annotation and classification of *de novo* predicted LTR retrotransposons. Nucleic Acids Res 37:7002–7013

Szak ST, Pickeral OK, Makalowski W, Boguski MS, Landsman D, Boeke JD (2002) Molecular archeology of L1 insertions in the human genome. Genome Biol 3:research0052

Tu Z (2001) Eight novel families of miniature inverted repeat transposable elements in the African malaria mosquito, Anopheles gambiae. Proc Natl Acad Sci USA 98:1699–1704

Tu Z, Li S, Mao C (2004) The changing tails of a novel short interspersed element in Aedes aegypti: genomic evidence for slippage retrotransposition and the relationship between 3' tandem repeats and the poly(dA) tail. Genetics 168:2037–2047

Wicker T, Sabot F, Hua-Van A, Bennetzen JL, Capy P, Chalhoub B, Flavell A, Leroy P, Morgante M, Panaud O, Paux E, SanMiguel P, Schulman AH (2007) A unified classification system for eukaryotic transposable elements. Nat Rev Genet 8:973–982

Xu Z, Wang H (2007) LTR_FINDER: an efficient tool for the prediction of full-length LTR retrotransposons. Nucleic Acids Res 35(Web Server issue):W265–W268

Yang L, Bennetzen JL (2009) Structure-based discovery and description of plant and animal Helitrons. Proc Natl Acad Sci USA 106:12832–12837

Yang G, Hall TC (2003) MAK, a computational tool kit for automated MITE analysis. Nucleic Acids Res 31:3659–3665

Zhou L, Mitra R, Atkinson PW, Hickman AB, Dyda F, Craig NL (2004) Transposition of hAT elements links transposable elements and V(D)J recombination. Nature 432:995–1001

Chapter 3
Using Nextgen Sequencing to Investigate Genome Size Variation and Transposable Element Content

Concepcion Muñoz-Diez, Clémentine Vitte, Jeffrey Ross-Ibarra, Brandon S. Gaut, and Maud I. Tenaillon

Abstract Transposable element (TE) content explains a large part of Eukaryotic genome size variation. TE content is determined by transposition, removal and host responses, but the efficiency of these forces is ultimately governed by genetic drift and natural selection. Contribution of TE families to genome size variation has been recently quantified using next generation sequencing (NGS) in two species pairs: *Zea mays* ssp. *mays* and *Zea luxurians*, *Arabidopsis lyrata* and *A. thaliana*. In both interspecific comparisons, genome-wide differences in TE content rather than the proliferation of a small subset of TE families was observed. We discuss three nonexclusive hypotheses to explain this pattern: selection for genome shrinkage, differential efficiency of epigenetic control, and a purely stochastic process of genome size evolution. Additional genome-wide assessments are needed to assess the extent to which selection shapes TE genomic content. To facilitate such studies, we discuss the use of NGS in "orphan" species.

Keywords Repetitive DNA • Selection • Genome shrinkage • Effective population size • Epigenetic control • Maize • Arabidopsis

C. Muñoz-Diez • B.S. Gaut
Department of Ecology and Evolutionary Biology, UC Irvine, 321 Steinhaus Hall, Irvine, CA 92617, USA

C. Vitte • M.I. Tenaillon (✉)
CNRS, UMR 0320 / UMR 8120 Génétique Végétale, INRA/CNRS/Univ Paris-Sud/ AgroParisTech, Ferme du Moulon, F-91190 Gif-sur-Yvette, France
e-mail: tenaillon@moulon.inra.fr

J. Ross-Ibarra
The Department of Plant Sciences and The Genome Center and Center for Population Biology, UC Davis, 262 Robbins Hall, Davis, CA 95616-5294, USA

M.-A. Grandbastien and J.M. Casacuberta (eds.), *Plant Transposable Elements*, Topics in Current Genetics 24, DOI 10.1007/978-3-642-31842-9_3,

3.1 Introduction

Eukaryotes vary widely in genome size both within and among species. Genome sizes were first compared among species based on flow cytometry; subsequently CoT analyses revealed that most genome size variation is attributable to repetitive DNA. However, it is only with the development of DNA sequencing that we have been able to determine both the basis of this variation and to identify the mechanisms underlying it. In plants, for example, the comparison of large orthologous regions through BAC sequencing has lead to two important observations: first, the intergenic fraction of genomes is primarily comprised of transposable elements (TEs) and second, much of the genomic variation observed between species is due to the rapid turnover of TE sequences in intergenic regions (Ramakrishna et al. 2002; Ma and Bennetzen 2004; Wang and Dooner 2006).

Further analyses based on complete genome sequences has enabled precise quantification of the TE fraction for several taxa, revealing that the genomic fraction of TEs is positively correlated with genome size [Fig. 3.1, see Gaut and Ross-Ibarra (2008) for a review]. Moreover, analysis of full genomes has allowed characterization of the molecular bases of sequence turnover in intergenic regions: TE proliferation and elimination of TE sequences through homologous recombination and illegitimate recombination (Devos et al. 2002; reviewed in Vitte and Panaud 2003). Comparison of the extent and timing of the counteracting forces of proliferation and removal have revealed that large genomes harbor at least a few highly repetitive TE families in their genome, suggesting that some of the differences observed may be due to the capacity of some TEs to escape epigenetic control by the host genome (Vitte and Bennetzen 2006).

Genome size may therefore be determined by (1) the genome's intrinsic capacity to suppress TE activity by epigenetic mechanisms, and (2) the ability of TEs to escape this suppression system. In recent years, this idea has been strengthened by characterization of the molecular bases underlying this suppression system: the transcriptional and posttranscriptional silencing of TE sequences through pathways involving small interfering RNAs (siRNAs) (Lisch 2009). This characterization has revealed that siRNAs serve as molecular guides for silencing protein complexes to target TE sequences. Their presence is, therefore, an indicator of the deployment of a genomic defence mechanism toward silencing TEs and is correlated with the DNA methylation status of targeted sequences (Lister et al. 2008; Schmitz et al. 2011).

Beyond its structural impact on the genomic landscape, variation in TE content and genome size may have an evolutionary significance (Biemont 2008). For example, genome size correlates with rates of plant development, because smaller genomes presumably facilitate faster cell division and therefore a higher growth rate. In addition, a few studies have reported within-species correlations between genome size and ecological variables such as altitude, latitude, and temperature (see Knight et al. 2005 and references therein) and between genome size and phenotypes such as flowering time, flower size, leaf size, and photosynthetic rate (for a review, Knight et al. 2005; Meagher and Vassiliadis 2005). Species with smaller genomes

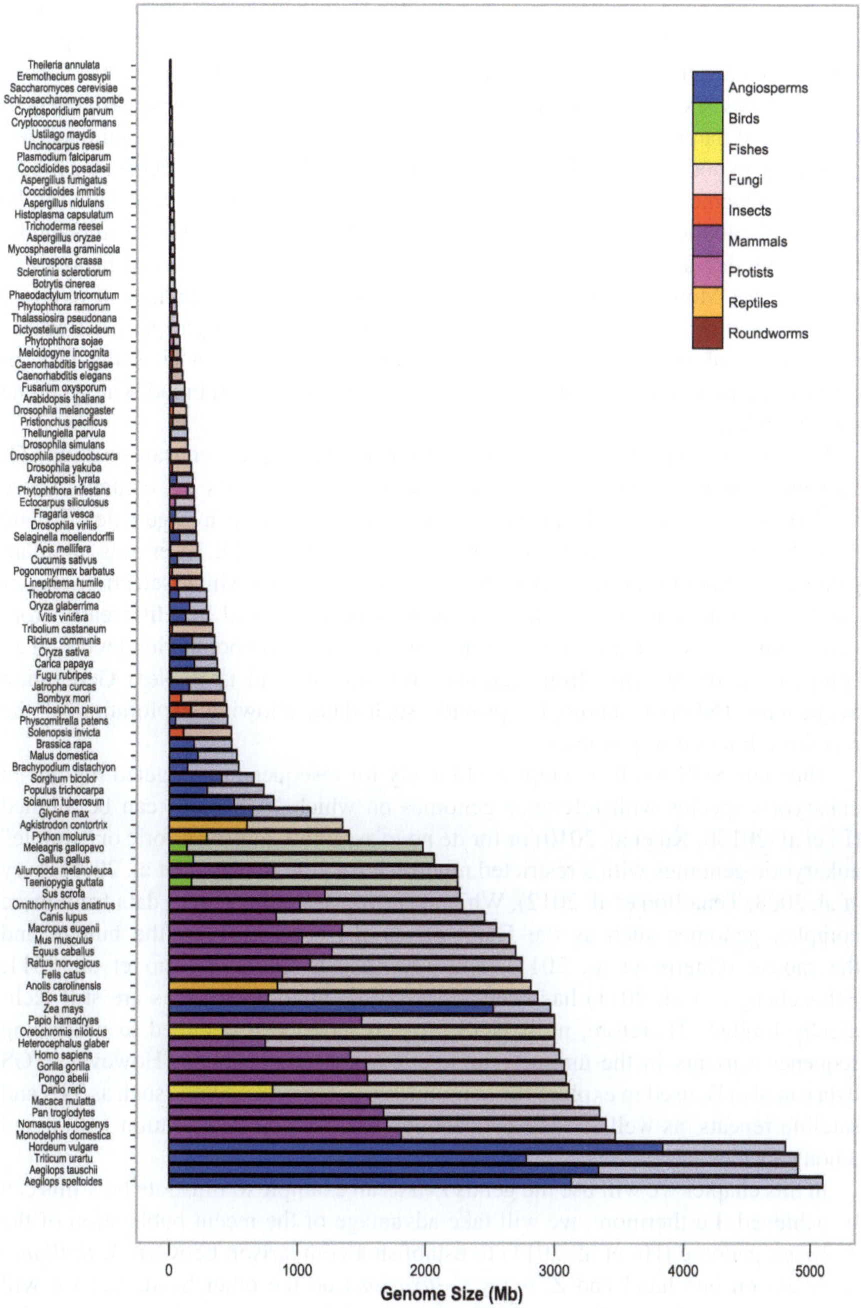

Fig. 3.1 Genome size (GS) and transposable element (TE) content of 98 eukaryote species, whose genomes have been sequenced. The total length of the *bars* indicates GS while the *darker portion* indicates TE content

also have enhanced colonization potential, due to an increase in seed mass, growth related traits, and decrease in generation time (Bennett et al. 1998; Grotkopp et al. 2004) that may altogether translate into a greater invasiveness (Lavergne et al. 2010).

While these examples appear to offer convincing evidence of the pervasiveness of the action of natural selection on genome size variation between and within-species, alternative hypotheses have been proposed. For example, a purely mechanistic model in which genome size evolves stochastically at a proportional rate can account for the skewed distributions of eukaryotic genome size (Oliver et al. 2007), but this model fails to provide a compelling reason for correlates between ecological factors and genome size. More recently, Whitney et al. (2010) have reported a lack of relationship between effective population size and genome size in angiosperms. Because the efficacy of selection is expected to scale with population size, the lack of relationship may indicate that selection has had little impact on broad-scale genome size evolution.

In summary, it is now well established that a balance between transposition, TE sequence removal, and host response determines a genome's TE content. These mechanisms are, in turn, affected by population processes, such as genetic drift and natural selection that ultimately determine the fate of TE insertions in plant genomes (Tenaillon et al. 2010). However, the extent to which selection shapes the TE genomic content is still debated. This debate would benefit greatly from genome-wide assessments that integrate across species and population levels—i.e., comparisons of genomes from various environments and taxa. Next Generation Sequencing (NGS) technologies provide such data, allowing exploration of the repetitive fraction of genomes.

Thus far, NGS has been employed largely for resequencing targeted regions in eukaryotic species with reference genomes on which NGS reads can be aligned (Li et al. 2010b; Xu et al. 2010) or for de novo assembly of prokaryotic or "simple" eukaryotic genomes with a restricted repetitive fraction (Galagan et al. 2005; Aury et al. 2008; Tenaillon et al. 2012). While de novo assembly of NGS data from more complex genomes such as the Giant panda (Li et al. 2010a), the human and the mouse (Gnerre et al. 2011), and *Arabidopsis thaliana* (Cao et al. 2011; Schneeberger et al. 2011) has been achieved, de novo approaches are still technically limited. Therefore, most NGS projects have been confined to describing sequence variants in the unique (single-copy) genomic fraction. However, NGS data can also be used to explore the components of repetitive DNA, such as TEs and satellite repeats, as well as their contribution to genome size variation within and among species.

In this chapter we will use the genus *Zea* as an example to illustrate how this can be achieved. Furthermore, we will take advantage of the recent publication of the *A. lyrata* genome (Hu et al. 2011) to establish a comparison between *A. thaliana*/*A. lyrata* on one hand and *Z. mays*/*Z. luxurians* on the other hand, and we will use these examples to discuss the factors that have contributed to genome size difference between closely related species. Finally, we will also provide some guidelines to determine TE content from NGS data in non-model species.

3.2 Exploring the Repetitive Fraction Within and Among Species Using NGS: An Example from the Genus *Zea*

3.2.1 Genome Size Variation in the Genus Zea

The genus *Zea* is traditionally divided into two sections (Fig. 3.2): *Luxuriantes* and *Zea*. The former encompasses several species, including the annual diploids *Z. luxurians and Z. diploperennis*. Section *Zea* includes a single diploid annual species (*Zea mays*), which consists of the cultivated maize (*Z. mays* ssp. *mays*) and its closest wild relatives (ssp. *parviglumis* and ssp. *mexicana*). The divergence between *Zea luxurians* and *Zea mays* is estimated to have occurred ~140,000 years ago (Hanson et al. 1996; Ross-Ibarra et al. 2009).

The genus encompasses extensive variation in genome size both within and between species. For example, within *Zea mays* genome size varies 30 % among cultivated accessions (i.e., landraces and inbred lines) and up to 32 % and 10 % in ssp. *mexicana* and ssp. *parviglumis*, respectively (Fig. 3.2 and included references). Between species, the average genome size of the diploid *Z. luxurians*, $2C = 9.07$ pg, is nearly 30 % larger than that of the average *Zea mays* ssp. *mays* genome (Fig. 3.2 and included references).

Differences in genome size may have multiple, potentially nonexclusive sources including whole genome duplication (polyploidy), segmental duplications, an increase of repetitive DNA (i.e., satellite sequences or TEs), or differential loss of TEs associated with recombination (Petrov et al. 2000). While *Z. luxurians* and *Z. mays* are both ancient polyploids (Gaut et al. 2000), extensive chromosomal rearrangements associated with the loss of some homeologs have resulted in the diploidization of *Zea* species, with $2n = 10$ chromosomes (Table 3.1). Therefore, variation between and within-species may arise from differences in the retention and the rate of production of segmental duplications as well as differential proliferation/elimination of repeated DNA.

In *Zea*, most repetitive DNA consists of interspersed TEs and heterochromatic blocks (knobs) which harbor 180- and 360-bp tandem repeats interspersed with retrotransposons (Peacock et al. 1981; Ananiev et al. 1998). Knob content varies among individuals of *Z. mays*, and knobs may be more abundant in *Z. luxurians* than in *Z. mays* (Tito et al. 1991; Gonzalez and Poggio 2011). Fully 85 % of the maize reference genome sequence consists of TEs, but the 20 most common TE families comprise ~70 % of the total (Baucom et al. 2009). These 20 families are all LTR retrotransposons (RNA elements). Amplification of LTR retrotransposons in the maize genome has been particularly dramatic in the last few million years, leading to a doubling of genome size (San Miguel and Bennetzen 1998; Brunner et al. 2005). Investigation of variation in TE copy number between *Z. luxurians* and *Z. mays* for six retrotransposon families using dotblots revealed little evidence of variation between species (Meyers et al. 2001), suggesting that these TEs may not have played a major role in genome size differentiation.

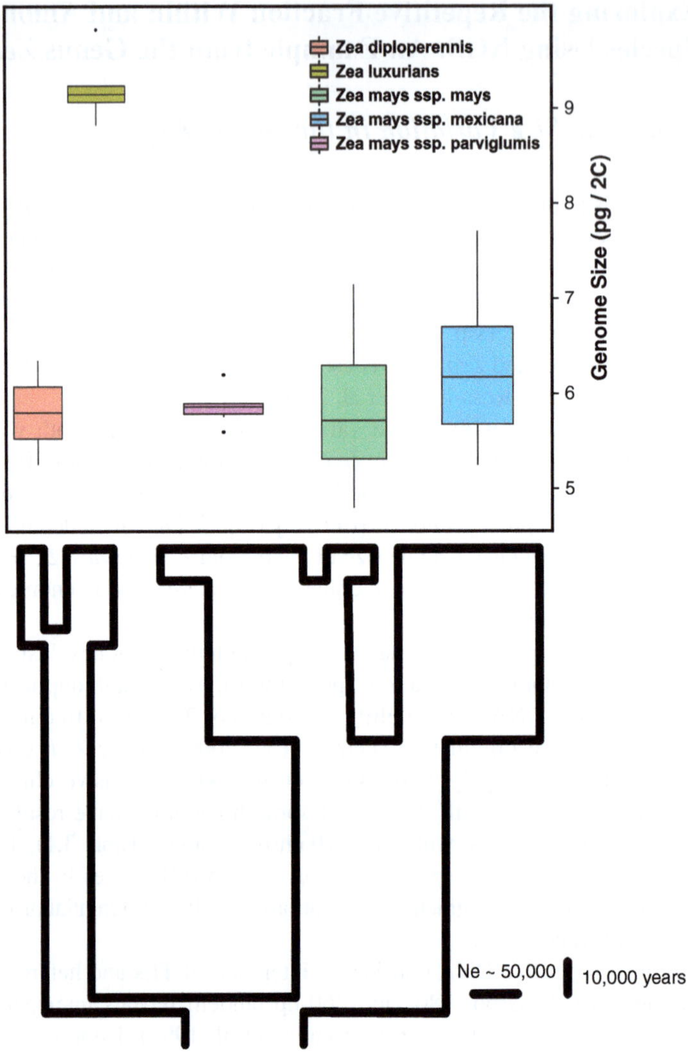

Fig. 3.2 Dendrogram and box plots showing demographic history and genome size variation in *Zea*. The branch width and length of the dendrogram are proportional to population size (N_e) and time, respectively, with scale bars shown (Ross-Ibarra et al. 2009). Divergence between *Z. mays* ssp. *parviglumis* and ssp. *mays*, and between *Z. mays* and *Z. luxurians*, was estimated to be 9,000 years (Piperno et al. 2009) and 140,000 years, respectively (Hanson et al. 1996; Ross-Ibarra et al. 2009). The *boxes* indicate the first quartile (*lower line*), the second quartile or median (*central line*), and the third quartile (*upper line*). Additionally the *whiskers* represent the standard deviation with the *dots* as the outliers. Genome size data were obtained from Laurie and Bennett (1985), Rayburn et al. (1985), Rayburn and Auger (1990), Tito et al. (1991), Guillin et al. (1992), Rayburn et al. (1993), Poggio et al. (1998), and Tenaillon et al. (2011) for a total of 2, 5, 8, 10, and 80 measures in *Z. diploperennis*, *Z. luxurians*, *Z. mays* ssp. *parviglumis*, *Z. mays* ssp. *mexicana* and Z *mays* ssp, *mays*, respectively

Table 3.1 Comparison of life-history traits, population parameters, and genomic content of *Arabidopsis* and *Zea* species

	A. lyrata	*A. thaliana*	*Z. mays*	*Z. luxurians*
Divergence time (Myr)	10[a]		0.140[b,c]	
Effective population size (N_e)	75,000[d]	250,000–300,000[e]	600,000[f]	50,000[b]
Mating system	Outcrosser	Selfer	Outcrosser +Recent inbreeding	Outcrosser
Genome size (Mb/C)	207[a]	125[a]	2,914[g]	4,435[g]
Chromosome number	$2n = 2x = 16$	$2n = 2x = 10$	$2n = 2x = 20$	$2n = 2x = 20$
Genes	32,670[a]	27,025[a]	39,656[h]	NA
TE content (% genome)	29.7[a]	23.7[a]	85[i]	NA
Ratio gene/TE	0.96[a]	1.78[a]	0.18[g]	0.18[g]

[a]Hu et al. (2011)
[b]Ross-Ibarra et al. (2009)
[c]Hanson et al. (1996)
[d]Ross-Ibarra et al. (2008). N_e value was calculated as the average among five subdivided populations
[e]Cao et al. (2011)
[f]Gossmann et al. (2010)
[g]Tenaillon et al. (2011)
[h]http://www.maizesequence.org
[i]Schnable et al. (2009)
NA not available

3.2.2 *Assessing the Contribution of TE Families to Genome Size Variation Between Maize and* Z. luxurians *Using NGS*

Recently, Tenaillon et al. (2011) performed a detailed analysis of TE content in one maize and one *Z. luxurians* genome using NGS. The approach was bolstered by the availability of a maize Filtered Gene Set (FGS) consisting of >32,000 high-quality annotated genes and a maize database of 1,526 exemplar (consensus) sequences representing distinct TE families and subfamilies (Baucom et al. 2009; Schnable et al. 2009). The method consisted of three discrete steps (Fig. 3.3). The first was creating a unique TE database (UTE) from the curated maize exemplar TE database (Baucom et al. 2009). The purpose of the UTE was to represent each of the 1,526 TE families of maize by their unique sequence signatures in order to minimize NGS reads that map ambiguously to more than one TE exemplar. In order to do so, each element of the exemplar TE database was cut into 104 bp fragments that were mapped against the exemplar TE database using the short read assembler SSAHA2 version 0.1 (Ning et al. 2001) with 80 % identity. Mapping results were used to determine the per base pair coverage of all 1,526 elements by the other elements contained in the exemplar TE database. This procedure allowed identification of portions of TEs not overlapping other elements in the exemplar database and to

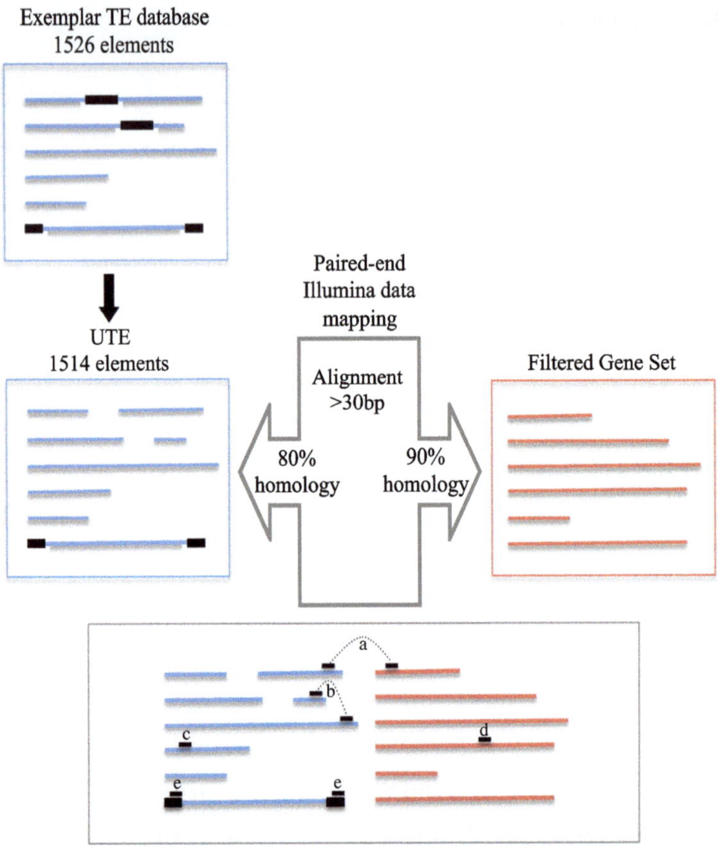

Fig. 3.3 Flowchart of the strategy used to analyze the TE content of maize and *Z. luxurians* genome using NGS data. The original exemplar TE database, represented here by six elements (*blue bars*), is filtered against the repeated regions among elements (*black boxes*). The resulting UTE contains the unique portion of each element, sometimes cut into several segments. Paired-end Illumina data are mapped against the UTE and the FGS, represented here by six genes (*red bars*). TE-gene pairs (**a**) and TE-nested pairs (**b**) are used to infer the proportions of TEs inserted into genes versus TEs inserted into other TEs. Read mapping against TEs (**c**) and genes (**d**) are used respectively to count the number of hits against a given element and estimate the coverage of the Illumina data. Note that because the UTE was not filtered against repeated regions within element (*black boxes*), two hits against a single element are counted only once

restrict the UTE to the sequences found in only a single TE in the exemplar database. Ultimately, the UTE consisted of 83 % of the original exemplar database, with 1,514 elements represented for read mapping (Tenaillon et al. 2011).

The second step was to generate high-throughput paired-end Illumina sequencing of the B73 maize inbred line and the *Z. luxurians* accession PI441933 (hereafter, *luxurians*). The paired-end libraries produced for each sample (B73 and *luxurians*) were each sequenced on a single lane of a flow cell with an Illumina Genome

Analyzer II, generating ~19 million paired-end reads of 84 and 104 bp in length. Tenaillon et al. (2011) also determined the genome size of the two accessions sequenced by flow cytometry: 5.96 pg/2 C for B73 and 9.07 pg/2 C for *luxurians*.

The third step was mapping the sequencing reads to the B73 reference genome, the UTE and the FGS, the latter providing an internal control for coverage. Using SSAHA2 version 0.1 (Ning et al. 2001) reads were mapped against the 1,514 elements of the UTE with 80 % identity, considering alignment length $\geq$30 bp. Reads aligning to a TE under these criteria were counted as single hit to the TE. One obvious caveat of the UTE is that the method as implemented is only as good as the annotated TE set, i.e., reads can only be mapped to annotated TEs. Median values of the distribution of per bp coverage from mapping of B73 and *luxurians* against each gene in the FGS were used to determine the genomic coverage of the Illumina data. In addition, by combining information about mapping against the UTE and FGS, it was possible to differentiate TEs inserted into other TEs (i.e., the two paired-ends mapped to two different TEs), from TEs inserted near genes (i.e., one paired-end mapped to a TE and the other to a gene).

Using the UTE and FGS from the maize reference genome, Tenaillon et al. (2011) were able to map 76.4 % and 75.5 % of reads to B73 and *luxurians*, respectively. They also verified reliability of their method via comparison between the Illumina data for B73 and the reference B73 genome. They observed >fivefold more TE-nested pairs than TE-gene pairs in both B73 and *luxurians*, indicating that TEs insert much more often in other TEs than genes. Assuming that gene content was similar between species, Tenaillon et al. (2011) found that at least 70 % of the 50 % genome size difference between maize and *Z. luxurians* (as determined by flow cytometry) was due to variability in TE copy number.

But the difference in genome size may have multiple origins. For example, it is possible that the *luxurians* genome encompasses genes and TEs that are absent from the B73 maize genome. These differences may occur as a consequence of differential genomic loss since species divergence. However, that similar proportions of reads were observed to map both to the UTE and FGS in both B73 and *luxurians* gives little support to this hypothesis, i.e., we would expect to observe significantly less mapping if TEs or genes present in *luxurians* were absent from B73. Alternatively, *luxurians* may exhibit a higher rate of retention of duplicated segments. If these duplicated segments offer a fair representation of the genome, encompassing both unique and repetitive DNA, one would expect to conserve similar proportions of gene to TEs and also TE families between species. Consistently, the proportion of mapped reads against FGS and UTE was similar in B73 (15.4:84.6) and in *luxurians* (14.8:85.2) and the number of hits to TE families was highly correlated between B73 and *Z. luxurians* (Fig. 3.4a, $r = 0.94$).

These observations are consistent with both TEs and genes being involved in genome size difference. They also reveal that differences in TE content between species are not due to the proliferation of a handful of TE families, as has been observed in other genera (Hawkins et al. 2006; Piegu et al. 2006), but rather due to a shift toward higher copy numbers in *Z. luxurians* for several hundred different TE families. Note, however, that *Gossypium* (Hawkins et al. 2006) and *Oryza* (Piegu

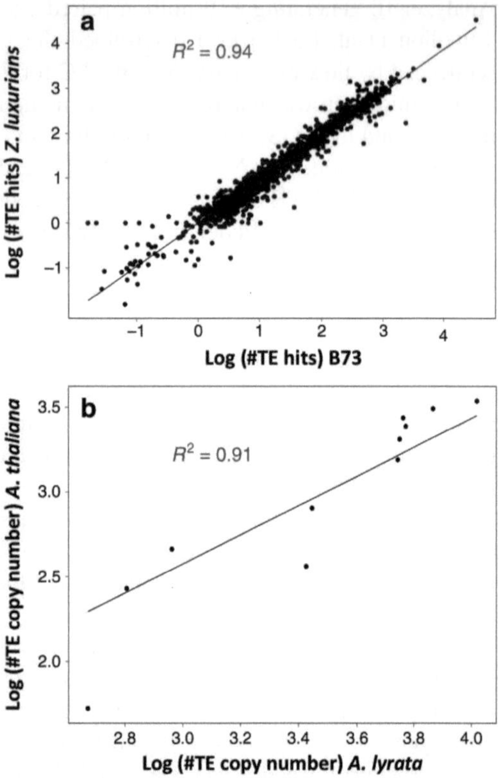

Fig. 3.4 Relative contribution of TE families to the genomes of two species pairs, the maize inbred line B73 and one accession of *Z. luxurians* (**a**), and the genomic sequences of *Arabidopsis thaliana* and *A. lyrata* (**b**). In (**a**), TE content was measured in 1,509 TE families as the number of Reads per Kilobase per Million mapped reads (RPKM) against the B73 Unique Transposable Element database (UTE). Values are shown on a log scale; the data are from Tenaillon et al. (2011). In (**b**), TE copy number was estimated from the annotation of the genomic sequence of *A. thaliana* and *A. lyrata* (Hollister et al. 2011)

et al. 2006) species divergence is much more ancient (on the order of a few million years) than in *Zea*, which may contribute to the difference between the observed patterns. For species with older divergence time, recurrent TE horizontal transfers between species are more likely to cause bursts of TE proliferation in the recipient species (Diao et al. 2006). This scenario seems less likely in *Zea*, not because there is no gene transfer among species but rather because there are likely no unique TEs among these recently diverged species that may easily escape the host suppression system.

3.3 Evolution of TE Profiles Through Evolutionary Times: A Comparison Between *Zea* and *Arabidopsis*

To date, the population dynamics of plant TEs have been studied primarily in the *Arabidopsis* species, *A. thaliana* and *A. lyrata*, which have relatively small genomes and for which reference genomes are available (Hu et al. 2011; the Arabidopsis Genome Initiative 2000). The two species diverged about 10 million years ago and exhibit several features that make their comparison especially interesting (Table 3.1). First, *A. lyrata* is a self-incompatible perennial while *A. thaliana* is a self-compatible annual species. Second, *A. lyrata* has $2n = 16$ chromosomes and its genome is larger than 200 Mb, whereas *A. thaliana* has $2n = 10$ chromosomes and one of the smallest angiosperm genomes at about 125 Mb. Third, Hu et al. (2011) have determined that more than 50 % of the *A. lyrata* genome appears to be missing from the *A. thaliana* reference genome but only about 25 % of the *A. thaliana* genome is absent from *A. lyrata*. Overall, *A. thaliana* exhibits a much higher ratio of genes to TEs than *A. lyrata*, and much of the genome size difference between these two species is likely caused by (1) reduced transposable element activity, (2) more efficient transposable element elimination in *A. thaliana*, especially near genes, or (3) systematic shortening of nontransposable element intergenic sequences and introns in *A. thaliana* (Fawcett et al. 2011; Hu et al. 2011).

Interestingly, Hollister et al. (2011) found a similar trend to the one observed in the *Zea* comparison (Tenaillon et al. 2011), which is that the relative contribution of TE families is well conserved between species (Fig. 3.4b, $r = 0.91$). Hence, in both interspecific comparisons, there are genome-wide differences in TE content rather than the proliferation of a small subset of TE families (as documented in *Gossypium* and *Oryza*). Two nonexclusive processes may help to explain this observation. First, there could be ongoing positive selection for genome shrinkage in both systems through the loss of TEs and genes. Supporting this idea, fewer insertions than deletions were found in a population of 95 individuals of *A. thaliana* among both segregating polymorphisms and fixed differences, with deletions longer on average than insertions (Hu et al. 2011). Moreover, a higher intron loss rate in *A. thaliana* than *A. lyrata* has been reported recently, reinforcing the hypothesis of selection for genome shrinkage (Fawcett et al. 2011). Additionally, simple calculations (Chevin and Hospital 2008) suggest that, in a species with a large effective population size similar to *Zea mays* (Fig. 3.2), even weakly beneficial mutations (TE deletions in this case) could increase to high frequency in timescales similar to the divergence between *luxurians* and *Zea mays* (Ross-Ibarra et al. 2009). If selection was driving this pattern, we would expect it to be more efficient in the species characterized by a greater effective population. While *A. thaliana* and *Z. mays* are thought to have higher effective population sizes than *A. lyrata* and *Z. luxurians* (Table 3.1), consistent with the observed differences in genome size, at least some estimates find weaker selection in *A. thaliana* than its congener (Wright et al. 2001; Lockton and Gaut 2010).

A second explanation is that closely related species may differ in aspects that control TE proliferation, such as the efficiency of epigenetic modification via pathways that include small interfering RNAs (siRNAs). Epigenetic mechanisms act by suppressing the expression of TEs (transcriptional silencing) or by cleaving TE mRNA (posttranscriptional silencing) (Slotkin et al. 2005; Matzke et al. 2009). Both pathways achieve site-specificity by homology between siRNA and their target sequences (Almeida and Allshire 2005). In plants, DICER-LIKE RNase enzymes produce 21–24-bp siRNA that guides ARGONAUTE and other downstream proteins to complementary DNA sequences, thereby promoting and maintaining DNA and histone methylation (Zhang 2008; Teixeira and Colot 2009). Hence, silenced TE sequences are generally characterized by identity with siRNAs and dense, even DNA methylation (Lippman et al. 2004; Zilberman and Henikoff 2007; Lister et al. 2008).

Differences in the efficiency of TE silencing by siRNAs has been investigated in *A. thaliana* and *A. lyrata* (Hollister et al. 2011). Sequences of siRNAs generated by NGS have been mapped to the two reference genomes and mapped siRNAs have been used as a proxy for TE methylation. Consistent with the hypothesis of differences in epigenetic control between the two species, the expression level of siRNAs was higher in *A. thaliana* by ~1.7-fold on average than in *A. lyrata*. The two species also exhibited a substantial difference in the ratio of uniquely- to multiply-mapping siRNAs. In fact a much higher proportion of TEs lacked uniquely mapping siRNA reads in *A. lyrata* (25 %) than in *A. thaliana* (10 %). Interestingly, Hollister et al. (2011) have shown that TEs targeted by uniquely mapping siRNAs are silenced more efficiently in both species. Altogether, lower TE expression levels, higher siRNA expression levels, and a higher ratio of unique/multiply-mapping siRNA signal more efficient silencing in *A. thaliana*, which correlates with its genomic characteristics: smaller genome and lower TE copy number. These phenomena should be evaluated in other pairs of closely related species with contrasting genome sizes, but reference genomes are still lacking in plant species to apply this approach.

Finally, it is also possible that genome size evolution is subject to a purely stochastic process in which the rate of genome size evolution (mean and variance) simply depends on current genome size, i.e., proportional evolution. Oliver et al. (2007) have supported this model by demonstrating the existence of a positive correlation between the rate of evolution and the average genome size in 20 eukaryotic taxonomic groups. The analysis of 68 eukaryotic sequenced genomes has revealed that the variation (as measured by standard deviation) of both the repetitive, i.e., masked, and unique, i.e., non-masked fraction, were proportional to the average repeat and unique fraction within a clade, suggesting that genome expansion is dominated by stochastic processes (Li et al. 2011). However, while genome size variation between closely related species such as described may be affected by drift, drift alone is difficult to reconcile with the observed ecological correlates of genome size.

3.4 Using NGS to Estimate TE Content and Diversity in Non-model Species

The examples presented above highlight how the availability of a reference genome and an exemplar TE database helps decipher the molecular origins of differences in TE content among species, by remapping short reads of DNA, RNA, or siRNAs.

But most species still lack a reference genome and are not closely related to a model species with a reference genome. When such a reference genome is not available, NGS can nonetheless serve to get a better understanding of TE content and diversity within a genome.

For species where BAC sequences are available, NGS can provide important help to refine TE annotation. Even though collections of TEs are now available for a vast number of species, these sequences may be too distant to the TEs of the species of interest. As a result, NGS reads from the focal species may match only to the most conserved regions of TEs from well-annotated species. For this reason, direct annotations of the focal species using computer tools such as Repeatmasker (Bedell et al. 2000) can lead to erroneous annotations, where TEs appear fragmented although they are not. The use of computer tools that look for specific structural features can provide de novo annotations in the focal genome. However, this approach is limited to TE families that harbor recognizable structural features (e.g., the terminal repeats of LTR retrotransposons) and to recent TE insertions that still harbor these features, leaving many TE copies unresolved.

This is where NGS may provide substantial help: TEs, which are repeated, are likely to show increased coverage as compared to unique sequences. Hence, mapping of NGS reads to a BAC sequence will delimit regions of high coverage (likely to be repeated) and regions with low coverage (likely low-copy). This, along with the annotation of conserved TE regions using TE databases from other species, may allow precise mapping of element boundaries. Of course, the detection of TE boundaries will be enhanced as sequencing coverage increases, but even low coverage may greatly facilitate annotation. NGS may thus be greatly valuable for TE annotation, which is the first step toward building a reference exemplar TE database for a given species. The quality and representation of the database will, however, depend on the number of BACs sequenced and whether they represent most or only a limited subset of TE families.

For species where no BAC sequences are available, NGS can still be used to generate consensus copies of the most abundant elements (exemplar TEs). For the same reasons presented above, highly repetitive elements will be represented by a large number of sequencing reads, which can then be used to reconstruct de novo consensus sequences of specific TE families. Such methodology has been implemented in the AAARF software (DeBarry et al. 2008) and has been successfully used on 454 reads. Adaptation of such tools to work on Illumina paired-end and mate-pair reads will likely provide improvements for TE detection. Note, however, that this approach will likely provide exemplar TE database of limited quality since the element builds may correspond to chimeric elements rather than a consensus sequence of several individual copies. For example, it may prove difficult to differentiate autonomous elements from their nonautonomous partners, because both may be merged in a single exemplar element. Nevertheless, such a database will be useful to determine a first approximation of TE content and diversity in the genomes of non-model species.

3.5 Conclusion

NGS technologies have enabled the generation of a vast amount of data. For complex genomes such as those of plants, their utilization has so far been limited to the analysis of the non-repetitive fraction of genomes, thus ignoring what is often the majority of the data. In this chapter, we illustrated how these data could be utilized to investigate the evolutionary processes driving variation in TE content, and hence genome size, among closely related species. The approach developed by Tenaillon et al. (2011) could, for species with a reference genome, be directly applied at the population level to assess the forces that determine TE content and the abundance of other heterochromatic repeats, as well as how repeat abundance relates to genome size variation. Coupled with NGS of siRNAs and mRNAs, such an approach may also provide substantial insights into the dynamics of TE methylation, its impact on gene expression (Hollister and Gaut 2009; Hollister et al. 2011), and more generally on the efficiency of the host response to TE invasion.

Application of this approach to species with no reference genome is more challenging. As a first step, we propose here to build exemplar TE databases using NGS to improve TE annotation from BAC sequences or for de novo TE assembly. Of course, these data will not provide a picture as complete as the one provided by a reference genome. In particular, it will not allow analysis of individual TE insertions, therefore hampering investigation of the distribution pattern of copies (e.g., between genic and nongenic regions) or the analysis of TE regulation by siRNAs. It nonetheless offers a first estimate of the most abundant elements and can be applied to many "orphan" species, thus providing a horizontal view of TE diversity among populations and species.

References

Almeida R, Allshire R (2005) RNA silencing and genome regulation. Trends Cell Biol 15:251–258

Ananiev EV, Phillips RL, Rines HW (1998) A knob-associated repeat in maize capable of forming fold-back DNA segments: Are chromosome knobs megatransposons? Proc Natl Acad Sci USA 95:10785–10790

Aury JM, Cruaud C, Barbe V, Rogier O, Mangenot S, Samson G, Poulain J, Anthouard V, Scarpelli C, Artiguenave F, Wincker P (2008) High quality draft sequences for prokaryotic genomes using a mix of new sequencing technologies. BMC Genomics 9:603

Baucom R, Estill J, Chaparro C, Upshaw N, Jogi A, Deragon J, Westerman R, Sanmiguel P, Bennetzen J (2009) Exceptional diversity, non-random distribution, and rapid evolution of retroelements in the B73 maize genome. PLoS Genet 5:e1000732

Bedell JA, Korf I, Gish W (2000) MaskerAid: a performance enhancement to RepeatMasker. Bioinformatics 16:1040–1041

Bennett MD, Leitch IJ, Hanson L (1998) DNA amounts in two samples of angiosperm weeds. Ann Bot 82:121–134

Biemont C (2008) Genome size evolution: within-species variation in genome size. Heredity 101:297–298

Brunner S, Fengler K, Morgante M, Tingey S, Rafalski A (2005) Evolution of DNA sequence nonhomologies among maize inbreds. Plant Cell 17:343–360

Cao J, Schneeberger K, Ossowski S, Gunther T, Bender S, Fitz J, Koenig D, Lanz C, Stegle O, Lippert C, Wang X, Ott F, Muller J, Alonso-Blanco C, Borgwardt K, Schmid KJ, Weigel D (2011) Whole-genome sequencing of multiple *Arabidopsis thaliana* populations. Nat Genet 43:956–960

Chevin LM, Hospital F (2008) Selective sweep at a quantitative trait locus in the presence of background genetic variation. Genetics 180:1645–1660

DeBarry JD, Liu R, Bennetzen JL (2008) Discovery and assembly of repeat family pseudomolecules from sparse genomic sequence data using the Assisted Automated Assembler of Repeat Families (AAARF) algorithm. BMC Bioinformatics 9:235

Devos KM, Brown JK, Bennetzen JL (2002) Genome size reduction through illegitimate recombination counteracts genome expansion in Arabidopsis. Genome Res 12:1075–1079

Diao XM, Freeling M, Lisch D (2006) Horizontal transfer of a plant transposon. PLoS Biol 4:119–128

Fawcett JA, Rouzé P, Van de Peer Y (2011) Higher intron loss rate in *Arabidopsis thaliana* than *A. lyrata* is consistent with stronger selection for a smaller genome. Mol Biol Evol 29:849–859

Galagan JE, Calvo SE, Cuomo C, Ma LJ, Wortman JR, Batzoglou S, Lee SI, Basturkmen M, Spevak CC, Clutterbuck J, Kapitonov V, Jurka J, Scazzocchio C, Farman M, Butler J, Purcell S, Harris S, Braus GH, Draht O, Busch S, D'Enfert C, Bouchier C, Goldman GH, Bell-Pedersen D, Griffiths-Jones S, Doonan JH, Yu J, Vienken K, Pain A, Freitag M, Selker EU, Archer DB, Penalva MA, Oakley BR, Momany M, Tanaka T, Kumagai T, Asai K, Machida M, Nierman WC, Denning DW, Caddick M, Hynes M, Paoletti M, Fischer R, Miller B, Dyer P, Sachs MS, Osmani SA, Birren BW (2005) Sequencing of *Aspergillus nidulans* and comparative analysis with *A. fumigatus* and *A. oryzae*. Nature 438:1105–1115

Gaut B, Ross-Ibarra J (2008) Selection on major components of angiosperm genomes. Science 320:484–486

Gaut B, Le Thierry d'Ennequin M, Peek A, Sawkins M (2000) Maize as a model for the evolution of plant nuclear genomes. Proc Natl Acad Sci USA 97:7008–7015

Gnerre S, MacCallum I, Przybylski D, Ribeiro FJ, Burton JN, Walker BJ, Sharpe T, Hall G, Shea TP, Sykes S, Berlin AM, Aird D, Costello M, Daza R, Williams L, Nicol R, Gnirke A, Nusbaum C, Lander ES, Jaffe DB (2011) High-quality draft assemblies of mammalian genomes from massively parallel sequence data. Proc Natl Acad Sci USA 108:1513–1518

Gonzalez GE, Poggio L (2011) Karyotype of *Zea luxurians* and *Z. mays* subsp *mays* using FISH/DAPI, and analysis of meiotic behavior of hybrids. Genome 54:26–32

Gossmann TI, Song BH, Windsor AJ, Mitchell-Olds T, Dixon CJ, Kapralov MV, Filatov DA, Eyre-Walker A (2010) Genome wide analyses reveal little evidence for adaptive evolution in many plant species. Mol Biol Evol 27:1822–1832

Grotkopp E, Rejmanek M, Sanderson MJ, Rost TL (2004) Evolution of genome size in pines (Pinus) and its life-history correlates: supertree analyses. Evolution 58:1705–1729

Guillin EA, Poggio L, Naranjo CA (1992) Genome size in annual species of Zea. Relation with cellular parameters and altitude. Maize Genet Coop Newslett 66:59–60

Hanson M, Gaut B, Stec A, Fuerstenberg S, Goodman M, Coe E, Doebley J (1996) Evolution of anthocyanin biosynthesis in maize kernels: the role of regulatory and enzymatic loci. Genetics 143:1395–1407

Hawkins J, Kim H, Nason J, Wing R, Wendel J (2006) Differential lineage-specific amplification of transposable elements is responsible for genome size variation in Gossypium. Genome Res 16:1252–1261

Hollister J, Gaut B (2009) Epigenetic silencing of transposable elements: a trade-off between reduced transposition and deleterious effects on neighboring gene expression. Genome Res 19:1419–1428

Hollister J, Smith L, Ott F, Guo Y-L, Weigel D, Gaut B (2011) Transposable elements and small RNAs contribute to gene expression divergence between *Arabidopsis thaliana* and *Arabidopsis lyrata*. Proc Natl Acad Sci USA 108:2322–2327

Hu TT, Pattyn P, Bakker EG, Cao J, Cheng JF, Clark RM, Fahlgren N, Fawcett JA, Grimwood J, Gundlach H, Haberer G, Hollister JD, Ossowski S, Ottilar RP, Salamov AA, Schneeberger K, Spannagl M, Wang X, Yang L, Nasrallah ME, Bergelson J, Carrington JC, Gaut BS, Schmutz J, Mayer KF, Van de Peer Y, Grigoriev IV, Nordborg M, Weigel D, Guo YL (2011) The *Arabidopsis lyrata* genome sequence and the basis of rapid genome size change. Nat Genet 43:476–481

Initiative TAG (2000) Analysis of the genome sequence of the flowering plant *Arabidopsis thaliana*. Nature 408:796–815

Knight CA, Molinari NA, Petrov DA (2005) The large genome constraint hypothesis: evolution, ecology and phenotype. Ann Bot 95:177–190

Laurie D, Bennett M (1985) Nuclear DNA content in the genera Zea and Sorghum—intergeneric, interspecific and intraspecific variation. Heredity 55:307–313

Lavergne S, Muenke NJ, Molofsky J (2010) Genome size reduction can trigger rapid phenotypic evolution in invasive plants. Ann Bot 105:109–116

Li R, Fan W, Tian G, Zhu H, He L, Cai J, Huang Q, Cai Q, Li B, Bai Y, Zhang Z, Zhang Y, Wang W, Li J, Wei F, Li H, Jian M, Li J, Zhang Z, Nielsen R, Li D, Gu W, Yang Z, Xuan Z, Ryder OA, Leung FC, Zhou Y, Cao J, Sun X, Fu Y, Fang X, Guo X, Wang B, Hou R, Shen F, Mu B, Ni P, Lin R, Qian W, Wang G, Yu C, Nie W, Wang J, Wu Z, Liang H, Min J, Wu Q, Cheng S, Ruan J, Wang M, Shi Z, Wen M, Liu B, Ren X, Zheng H, Dong D, Cook K, Shan G, Zhang H, Kosiol C, Xie X, Lu Z, Zheng H, Li Y, Steiner CC, Lam TT, Lin S, Zhang Q, Li G, Tian J, Gong T, Liu H, Zhang D, Fang L, Ye C, Zhang J, Hu W, Xu A, Ren Y, Zhang G, Bruford MW, Li Q, Ma L, Guo Y, An N, Hu Y, Zheng Y, Shi Y, Li Z, Liu Q, Chen Y, Zhao J, Qu N, Zhao S, Tian F, Wang X, Wang H, Xu L, Liu X, Vinar T, Wang Y, Lam TW, Yiu SM, Liu S, Zhang H, Li D, Huang Y, Wang X, Yang G, Jiang Z, Wang J, Qin N, Li L, Li J, Bolund L, Kristiansen K, Wong GK, Olson M, Zhang X, Li S, Yang H, Wang J, Wang J (2010a) The sequence and *de novo* assembly of the giant panda genome. Nature 463:311–317

Li Y, Vinckenbosch N, Tian G, Huerta-Sanchez E, Jiang T, Jiang H, Albrechtsen A, Andersen G, Cao H, Korneliussen T, Grarup N, Guo Y, Hellman I, Jin X, Li Q, Liu J, Liu X, Sparso T, Tang M, Wu H, Wu R, Yu C, Zheng H, Astrup A, Bolund L, Holmkvist J, Jorgensen T, Kristiansen K, Schmitz O, Schwartz TW, Zhang X, Li R, Yang H, Wang J, Hansen T, Pedersen O, Nielsen R, Wang J (2010b) Resequencing of 200 human exomes identifies an excess of low-frequency non-synonymous coding variants. Nat Genet 42:969–972

Li X, Zhu C, Lin Z, Wu Y, Zhang D, Bai G, Song W, Ma J, Muehlbauer GJ, Scanlon MJ, Zhang M, Yu J (2011) Chromosome size in diploid eukaryotic species centers on the average length with a conserved boundary. Mol Biol Evol 28:1901–1911

Lippman Z, Gendrel A, Black M, Vaughn M, Dedhia N, McCombie W, Lavine K, Mittal V, May B, Kasschau K, Carrington J, Doerge R, Colot V, Martienssen R (2004) Role of transposable elements in heterochromatin and epigenetic control. Nature 430:471–476

Lisch D (2009) Epigenetic regulation of transposable elements in plants. Annu Rev Plant Biol 60:43–66

Lister R, O'Malley R, Tonti-Filippini J, Gregory B, Berry C, Millar A, Ecker J (2008) Highly integrated single-base resolution maps of the epigenome in Arabidopsis. Cell 133:523–536

Lockton S, Gaut B (2010) The evolution of transposable elements in natural populations of self-fertilizing *Arabidopsis thaliana* and its outcrossing relative *Arabidopsis lyrata*. BMC Evol Biol 10:10

Ma J, Bennetzen J (2004) Rapid recent growth and divergence of rice nuclear genomes. Proc Natl Acad Sci USA 101:12404–12410

Matzke M, Kanno T, Daxinger L, Huettel B, Matzke A (2009) RNA-mediated chromatin-based silencing in plants. Curr Opin Cell Biol 21:367–376

Meagher TR, Vassiliadis C (2005) Phenotypic impacts of repetitive DNA in flowering plants. New Phytol 168:71–80

Meyers BC, Tingey SV, Morgante M (2001) Abundance, distribution, and transcriptional activity of repetitive elements in the maize genome. Genome Res 11:1660–1676

Ning ZM, Cox AJ, Mullikin JC (2001) SSAHA: a fast search method for large DNA databases. Genome Res 11:1725–1729

Oliver MJ, Petrov D, Ackerly D, Falkowski P, Schofield OM (2007) The mode and tempo of genome size evolution in eukaryotes. Genome Res 17:594–601

Peacock WJ, Dennis ES, Rhoades MM, Pryor AJ (1981) Highly repeated DNA-sequence limited to knob heterochromatin in maize. Proc Natl Acad Sci USA 78:4490–4494

Petrov DA, Sangster TA, Johnston JS, Hartl DL, Shaw KL (2000) Evidence for DNA loss as a determinant of genome size. Science 287:1060–1062

Piegu B, Guyot R, Picault N, Roulin A, Saniyal A, Kim H, Collura K, Brar D, Jackson S, Wing R, Panaud O (2006) Doubling genome size without polyploidization: dynamics of retrotransposition-driven genomic expansions in *Oryza australiensis*, a wild relative of rice. Genome Res 16:1262–1269

Piperno DR, Ranere AJ, Holst I, Iriarte J, Dickau R (2009) Starch grain and phytolith evidence for early ninth millennium BP maize from the Central Balsas River Valley, Mexico. Proc Natl Acad Sci USA 106:5019–5024

Poggio L, Rosato M, Chiavarino AM, Naranjo CA (1998) Genome size and environmental correlations in maize (*Zea mays* ssp. *mays*, Poaceae). Ann Bot 82:107–115

Ramakrishna W, Dubcovsky J, Park Y-J, Busso C, Emberton J, SanMiguel P, Bennetzen JL (2002) Different types and rates of genome evolution detected by comparative sequence analysis of orthologous segments from four cereal genomes. Genetics 162:1389–1400

Rayburn AL, Auger JA (1990) Nuclear-DNA content variation in the ancient indigenous races of mexican maize. Acta Bot Neerlandica 39(2):197–202

Rayburn A, Price H, Smith J, Gold J (1985) C-Band heterochromatin and DNA content in *Zea mays*. Am J Bot 72:1610–1617

Rayburn A, Biradar D, Bullock D, McMurphy L (1993) Nuclear DNA content in F1 hybrids of maize. Heredity 70:294–300

Ross-Ibarra J, Wright S, Foxe J, Kawabe A, DeRose-Wilson L, Gos G, Charlesworth D, Gaut B (2008) Patterns of polymorphism and demographic history in natural populations of *Arabidopsis lyrata*. PLoS One 3:e2411

Ross-Ibarra J, Tenaillon M, Gaut B (2009) Historical divergence and gene flow in the genus zea. Genetics 181:1399–1413

San Miguel P, Bennetzen JL (1998) Evidence that a recent increase in maize genome size was caused by the massive amplification of intergene retrotransposons. Nat Genet 82:37–44

Schmitz RJ, Schultz MD, Lewsey MG, O'Malley RC, Urich MA, Libiger O, Schork NJ, Ecker JR (2011) Transgenerational epigenetic instability is a source of novel methylation variants. Science 334:369–373

Schnable PS, Ware D, Fulton RS, Stein JC, Wei F, Pasternak S, Liang C, Zhang J, Fulton L, Graves TA, Minx P, Reily AD, Courtney L, Kruchowski SS, Tomlinson C, Strong C, Delehaunty K, Fronick C, Courtney B, Rock SM, Belter E, Du F, Kim K, Abbott RM, Cotton M, Levy A, Marchetto P, Ochoa K, Jackson SM, Gillam B, Chen W, Yan L, Higginbotham J, Cardenas M, Waligorski J, Applebaum E, Phelps L, Falcone J, Kanchi K, Thane T, Scimone A, Thane N, Henke J, Wang T, Ruppert J, Shah N, Rotter K, Hodges J, Ingenthron E, Cordes M, Kohlberg S, Sgro J, Delgado B, Mead K, Chinwalla A, Leonard S, Crouse K, Collura K, Kudrna D, Currie J, He R, Angelova A, Rajasekar S, Mueller T, Lomeli R, Scara G, Ko A, Delaney K, Wissotski M, Lopez G, Campos D, Braidotti M, Ashley E, Golser W, Kim H, Lee S, Lin J, Dujmic Z, Kim W, Talag J, Zuccolo A, Fan C, Sebastian A, Kramer M, Spiegel L, Nascimento L, Zutavern T, Miller B, Ambroise C, Muller S, Spooner W, Narechania A, Ren L, Wei S, Kumari S, Faga B, Levy MJ, McMahan L, Van Buren P, Vaughn MW, Ying K, Yeh CT, Emrich SJ, Jia Y, Kalyanaraman A, Hsia AP, Barbazuk WB, Baucom RS, Brutnell TP, Carpita NC, Chaparro C, Chia JM, Deragon JM, Estill JC, Fu Y, Jeddeloh JA, Han Y, Lee H, Li P, Lisch DR, Liu S, Liu Z, Nagel DH, McCann MC, SanMiguel P, Myers AM, Nettleton D, Nguyen J, Penning BW, Ponnala L, Schneider KL, Schwartz DC, Sharma A, Soderlund C, Springer NM, Sun Q, Wang H, Waterman M, Westerman R, Wolfgruber TK, Yang L, Yu Y, Zhang L, Zhou S, Zhu Q, Bennetzen JL, Dawe RK, Jiang J, Jiang N, Presting GG, Wessler SR, Aluru S, Martienssen RA, Clifton SW, McCombie WR, Wing RA, Wilson RK (2009) The B73 maize genome: complexity, diversity, and dynamics. Science 326:1112–1115

Schneeberger K, Ossowski S, Ott F, Klein JD, Wang X, Lanz C, Smith LM, Cao J, Fitz J, Warthmann N, Henz SR, Huson DH, Weigel D (2011) Reference-guided assembly of four diverse *Arabidopsis thaliana* genomes. Proc Natl Acad Sci USA 108:10249–10254

Slotkin R, Freeling M, Lisch D (2005) Heritable transposon silencing initiated by a naturally occurring transposon inverted duplication. Nat Genet 37:641–644

Teixeira FK, Colot V (2009) Gene body DNA methylation in plants: a means to an end or an end to a means? EMBO J 28:997–998

Tenaillon M, Hollister J, Gaut B (2010) A triptych of the evolution of plant transposable elements. Trends Plant Sci 15:471–478

Tenaillon MI, Hufford MB, Gaut BS, Ross-Ibarra J (2011) Genome size and transposable element content as determined by high-throughput sequencing in maize and *Zea luxurians*. Genome Biol Evol 3:219–229

Tenaillon O, Rodriguez-Verdugo A, Gaut RL, McDonald P, Bennett AF, Long AD, Gaut BS (2012) The molecular diversity of adaptive convergence. Science 335:457–461

Tito CM, Poggio L, Naranjo CA (1991) Cytogenetic studies in the genus Zea. 3. DNA content and heterochromatin in species and hybrids. Theor Appl Genet 83:58–64

Vitte C, Bennetzen J (2006) Analysis of retrotransposon structural diversity uncovers properties and propensities in angiosperm genome evolution. Proc Natl Acad Sci USA 103:17638–17643

Vitte C, Panaud O (2003) Formation of solo-LTRs through unequal homologous recombination counterbalances amplifications of LTR retrotransposons in rice *Oryza sativa* L. Mol Biol Evol 20:528–540

Wang Q, Dooner H (2006) Remarkable variation in maize genome structure inferred from haplotype diversity at the bz locus. Proc Natl Acad Sci USA 103:17644–17649

Whitney KD, Baack EJ, Hamrick JL, Godt MJW, Barringer BC, Bennett MD, Eckert CG, Goodwillie C, Kalisz S, Leitch IJ, Ross-Ibarra J (2010) A role for nonadaptive processes in plant genome size evolution? Evolution 64:2097–2109

Wright SI, Le QH, Schoen DJ, Bureau TE (2001) Population dynamics of an Ac-like transposable element in self- and cross-pollinating arabidopsis. Genetics 158:1279–1288

Xu JJ, Zhao QA, Du PN, Xu CW, Wang BH, Feng Q, Liu QQ, Tang SZ, Gu MH, Han B, Liang GH (2010) Developing high throughput genotyped chromosome segment substitution lines based on population whole-genome re-sequencing in rice (*Oryza sativa* L.). BMC Genomics 11:656

Zhang X (2008) The epigenetic landscape of plants. Science 320(5875):489–492

Zilberman D, Henikoff S (2007) Genome-wide analysis of DNA methylation patterns. Development 134:3959–3965

Chapter 4
Genome-Wide Analysis of Transposition Using Next Generation Sequencing Technologies

Moaine Elbaidouri and Olivier Panaud

Abstract Transposable elements (TEs) make a large part of most eukaryotic genomes and strongly impact their structure, function, and evolution. The identification of active TEs in a genome is, therefore, essential in order to fully understand its dynamics at both structural and functional levels. The recent advent of new sequencing technologies, often referred to as next generation sequencing (NGS) technologies, has opened new doors to study structural variations at full genome scale. Although restricted so far mostly to human studies, these new strategies have shown to be highly efficient and promising in few other model species, including the two plant species *Arabidopsis thaliana* and rice. This chapter describes the concepts and techniques of using NGS for the study of TE activity in eukaryotic genomes at large.

Keywords Next Generation sequencing • Paired-end mapping • Structural variation • Transposable elements • Genomics

4.1 Introduction

The last 15 years of genomic research have yielded considerable amounts of information regarding the structure, the function, and the evolution of many eukaryotic genomes (Messing and Bennetzen 2008). To date, the full genome sequences of 17 plant species are available (http://www.genomesonline.org). One of the common features of all these genomes is that (except for the unusually small genome of *Arabidopsis thaliana*) they are often largely composed of transposable elements (SanMiguel et al. 1996). TEs, defined as mobile genomic entities, were first described

M. Elbaidouri • O. Panaud (✉)
Laboratoire Génome et Développement des Plantes, UMR UPVD/CNRS 5096, Université de Perpignan Via Domitia, 52, Avenue Paul Alduy, 66860 Perpignan cedex, France
e-mail: panaud@univ-perp.fr

M.-A. Grandbastien and J.M. Casacuberta (eds.), *Plant Transposable Elements*, Topics in Current Genetics 24, DOI 10.1007/978-3-642-31842-9_4,

by B. McClintock in maize more than 50 years ago (McClintock 1953). At first, these elements were considered as mutagenic factors, i.e., with a propensity to inactivate genes upon insertion. Consequently, TEs were regarded mostly as deleterious agents, while more rarely as a source of genetic variation that eukaryotes could benefit from. Genomics has now completely changed this paradigm through the demonstration that their biological impact goes far beyond mutagenesis by contributing to a large extent to the structure, the evolution, and the function of the genomes of both plants and animals (Bennetzen 2005).

TEs can be classified into two main categories, with very distinct mechanisms of transposition (Wicker et al. 2007): Class I elements, or retrotransposons, have a copy and paste mode of transposition. Therefore, active class I elements can multiply their copy numbers in the genome without excision. On the other hand, class II, or transposons, have a cut and paste mode of transposition and are not expected to undergo genomic amplifications to the same extent as class I elements. Nevertheless, a particular type of class II elements, the miniature TEs (MITEs) is often found highly repeated in plant genomes, although the exact mechanisms of such amplification remains unclear. Class I-driven genomic expansions can reach such level in some lineages that it is now considered as the main factor of genome size variation in plants, besides polyploidy (Piegu et al. 2006). Several studies have shown that these expansions usually occur in a catastrophic manner, i.e. through strong transpositional activation of few families over short periods of time (Piegu et al. 2006), a process referred to as bursts of transposition. Moreover, such bursts are always found to have occurred in a recent past (within the last few million years) and often posterior to speciation, which leads to posit 1—that only recent, active families are responsible to the structural variations observed among genomes and 2—that more ancient bursts have been eliminated from the genome, which is indeed the case, due to a strong bias of mutations towards deletions in TE-related sequences (Vitte and Panaud 2005; Vitte et al. 2007). Following insertion, TEs can be involved in genomic rearrangements, such as translocations, inversions, and chromosome degeneration. Morgante et al. (2005), through a comparative genomic survey between inbred lines, showed that helitrons (a particular type of class II elements) can mediate gene movements in maize. Similarly, Jiang et al. (2004) showed that genes can be mobilized through a transposition-like mechanism, using the structure of Mutator-like elements. These TE/genes chimeric structures, referred to as pack-MULES, can be found in hundreds of copies in the genome of rice and may retain some functional activity. More recently, Wicker et al. (2010) showed that some gene movements observed when comparing three genomes of Poaceae were the result of double-strand break (DSB) repair through synthesis-dependent strand annealing that involve TE-related sequences. This suggests that some of the TE-associated structural variants (TEASVs) are not caused by transposition *per se*, but are the results of subsequent rearrangements where TEs play a role, although indirectly.

Active TEs can inactivate genes upon insertion into coding sequences. This has been evidenced in several instances following the original work of B. McClintock (Tsugane et al. 2006; Miclaus et al. 2011). In this regard, the mutagenic nature of

TEs can be considered as deleterious. This raises the question of their ubiquity in eukaryotic genomes, because one could expect that their propensity to cause loss of function at the genome scale would lead to their elimination from natural populations. This paradox has been solved with the recent progress in our understanding of the epigenetic pathways that control transposition at large (Slotkin and Martienssen 2007; Lisch 2009; Lisch and Bennetzen 2011). These concern both transcriptional gene silencing (TGS) and posttranscriptional gene silencing (PTGS). TGS pathways target TEs through methylation (of both DNA and histones), while PTGS target TEs that escaped TGS through the degradation of their mRNAs (Bourc'his and Voinnet 2010; Rigal and Mathieu 2011). In the TGS pathway small interfering RNAs (siRNAs) are generated from loci to be silenced, and they fuel the RNA-directed DNA methylation (RdDM) pathway that acts as a feedback loop to methylate the DNA at genomic regions harboring active copies, thus causing their transcriptional silencing (Law and Jacobsen 2010; Saze et al. 2012). As a result, the vast majority of the TEs that populate most plant genomes are under the strict control of various, complementary silencing pathways and therefore efficiently inactivated. Interestingly, several recent studies have unraveled a new functional impact of TEs which is associated with the process of silencing through methylation: The expression of a gene can be affected by the presence of a TE in its vicinity, because it induces changes of the epigenetic status of the chromatin in the region. In such cases, TEs act as epigenetic mediators, thus causing changes in gene expression. This was indirectly suggested for the fruit color in grape (Kobayashi et al. 2004) and more recently for the plant architecture in maize (Studer et al. 2011).

4.2 Genomic Approaches for the Study of Transposition

This brief overview shows that there is a need for a complete, genome-wide identification of active elements in a given species, in order to decipher their putative impact on several aspects of genome biology. Such information is not available, even in the case of model species, such as *Arabidopsis thaliana* or rice, the first two plant species for which a high quality genome sequence has been available (Arabidopsis Genome Initiative 2000; International Rice Genome Sequencing Project (IRGSP) 2005). Therefore, despite their propensity to invade and densely populate plant genomes, only few transpositionally active TEs have been identified so far in plants. As an example, until recently in rice, the only known active elements were the LTR retrotransposons *Tos17* (Hirochika et al. 1996) and *Lullaby* (Picault et al. 2009), the LINE *Karma* (Komatsu et al. 2003), and the transposons *dTok* (Moon et al. 2006), *nDart* (Tsugane et al. 2006), and *mPing*/*Pong* (Jiang et al. 2003). Altogether, these five TE families do not represent more than 100 kbp in Nipponbare genome, which contrasts with the fact that rice genome harbors several hundreds of TE families, totalling 250,000 copies that make up 130 Mbp (International Rice Genome Sequencing Project (IRGSP) 2005). One of the major difficulties of identifying active TEs is the use of a suitable screen for the detection of new insertions. *dTok* and *nDart* transposons in rice were discovered, because the mutation of the genes into which they inserted caused a change in the plant phenotype (Moon et al. 2006; Tsugane et al. 2006). Their identification was thus made possible by the positional cloning of these genes. Although there is no doubt that the genetic study of more mutant lines in plants will lead to the discovery of new active transposons, this approach remains tedious because of the cloning procedure. In the case of *mPing/Pong* elements, the authors have combined an *in silico* survey of recent MITE insertions in the rice genome sequence with a confirmation of the mobility of these TEs in cell culture using the transposon display procedure (Jiang et al. 2003). The efficiency of this double approach demonstrates that the availability of the full genomic sequence of rice considerably facilitates the discovery of new active TEs. However, not all recently inserted TEs are still transpositionally active. The discovery of both *Tos 17* and *Karma* elements was achieved through the cloning and sequencing of cDNAs that were amplified through PCR using primers designed in the conserved domains of the reverse transcriptase gene (Hirochika et al. 1996; Komatsu et al. 2003). Although this method is robust and straightforward, it may not be suitable for an exhaustive survey of all the transcriptionally active LTR retrotransposons because of the bias associated with the PCR amplification towards the most active elements. Recently, Picault et al. (2009) identified an active LTR retrotransposon (i.e., *Lullaby*) based on a genome-wide transcriptional survey of rice callus using a dedicated microarray harboring oligomers matching with all TE-related rice sequences. This postgenomic approach does not require to identify any mutant phenotype associated with the transposition of the element. However, most of the transcriptionally activated TE families did not actually transpose. This shows that only full genome sequencing could provide access to the exhaustive transpositional landscape of a mutant plant.

4.3 Use of Next Generation Sequencing Technologies to Study Transposition

The advent of new sequencing technologies has tremendously changed the field of genomics not only at technical level but also in terms of conceptual developments (Zhang et al. 2011). Most of the latest genome projects have been completed using one or a combination of Next Generation Sequencing (NGS), while conventional Sanger-based strategies have been abandoned for large genomes because of their cost (Argout et al. 2011). The latest technologies can generate Gigabases of sequences for much less than 100 US$ (Pareek et al. 2011). This dramatic reduction of sequencing cost opens new perspectives for genome-wide studies within species and for populations genomics in particular (Siol et al. 2010). While the first genome-wide studies of genetic diversity focussed on single nucleotide polymorphisms (SNPs), subsequent analyses showed that structural variations (SVs) may be identified at a much higher rate in natural populations, especially in species with a large genome like humans (Feuk et al. 2006). In addition, the importance of detecting structural variants in this species was emphasized with the discovery of their direct involvement of some disorders (Stankiewicz and Lupski 2010). The first genome-wide surveys of SVs in human relied upon the use of comparative genome hybridization (CGH) arrays (Sharp et al. 2005), which has the disadvantage of a low resolution (i.e., 50 kbp, below the resolution where TEASV can be detected). Korbel et al. (2007) conducted the first human genome-wide survey of structural variants using NGS. Their strategy, named paired-end mapping (PEM, Fig. 4.1), consisted in a deep sequencing of paired ends of 3 kb fragments from two individuals, followed by a mapping of the reads on the reference human genome. This analysis yielded 1,175 indels and 122 inversions. Indels were further characterized, which showed that 90% of the insertions were associated with the LINE L1 (the second most frequent TE in human genome). Therefore, this first pioneer study demonstrated that paired-end mapping could be successfully used for the identification of TEASVs. More recent studies followed the same strategy, but concerned larger samples of human populations (Stewart et al. 2011). NGS are now used routinely in this species and have significantly contributed to the building of comprehensive human SV databases (Iafrate et al. 2004). Although many studies had previously showed transpositional activity of L1 (in addition to that of the two other class I elements Alu and SVA), population studies such as that of Korbel et al. (2007) only provide indirect evidences of the activity of a TE family. Some of the TEASVs observed today may indeed originate from the activity of an element that has become inactivated long ago, because those can remain in natural populations for many generations through lineage sorting [for instance, the two individuals analyzed by Korbel et al. (2007) originated from Europe and Nigeria]. A more direct evidence of TE activity through genome-wide approaches was provided recently by Baillie et al. (2011). The authors used a combination of microarray hybridization and high throughput sequencing to unravel somatic transposition in the human brain. This strategy, named retrotransposon capture sequencing (RC-seq), allowed the authors to evidence 7,700, 13,700 and 1,350 new insertions of the retrotransposons *L1*, *Alu*, and *SVA* respectively, among the hypocampus samples of three distinct individuals. RC-seq is, strictly speaking, a knowledge-based approach, because it requires to properly identify putatively active elements prior to the experiments (in order to build the microarray). In this regard, it cannot be applied to the discovery of new active elements, although the authors clearly demonstrated the detection efficiency brought by NGS.

In plants, the first example of the use of paired-end mapping to detect TE movement was reported by Mirouze et al. (2009) in *Arabidopsis thaliana*. The authors first identified an active LTR retrotransposon *EVD* through the cloning of mutant alleles of several candidate genes in epigenetic-recombinant inbred lines (epi-RILs). Some of these lines, obtained through successive rounds of selfing of a cross between a wild-type and the *met1* homozygous mutant, exhibit aberrant phenotypes that the authors showed to be associated with the insertions of *EVD*.

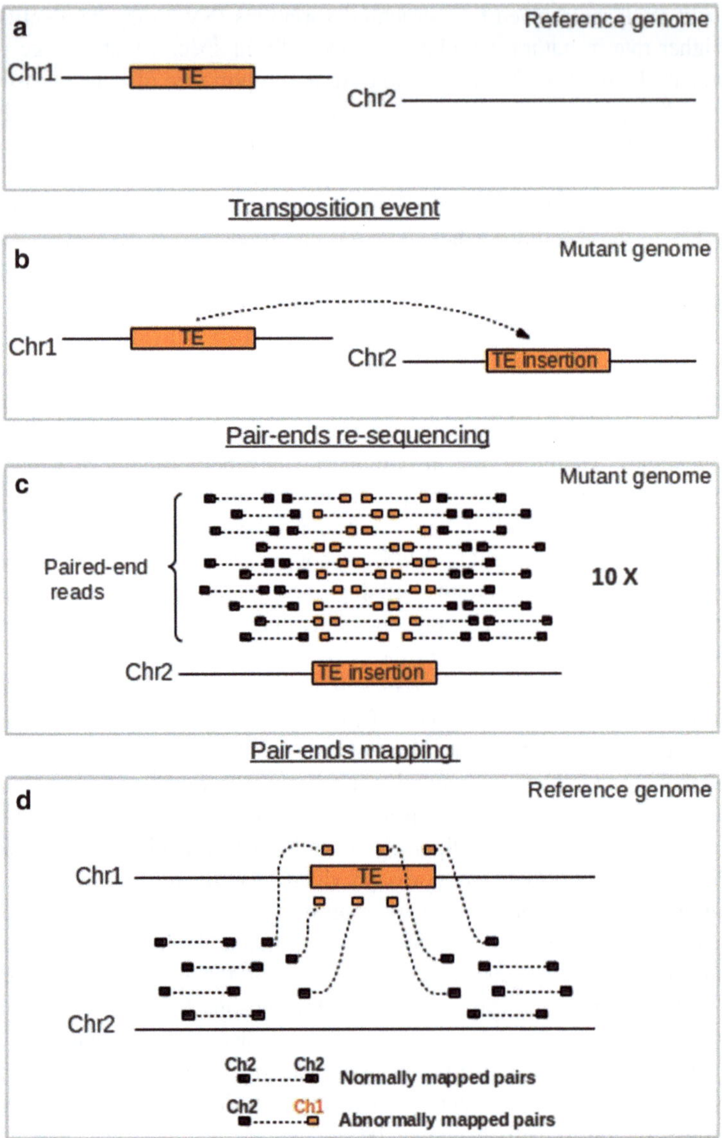

Fig. 4.1 Schematic view of paired-end mapping detection of TE-associated structural variants: (**a**) A close-up view of two genomic locations, one harboring a TE (chromosome 1) and the other harboring an empty site (chromosome 2). (**b**) The TE located on chromosome 1 transposes on chromosome 2. (**c**) sequencing of the mutant genome. The illumina amplicons are sequenced on both ends (*small boxes* represent paired-end reads). (**d**) paired-end mapping of the illumina reads. The mapping of the reads from amplicons spanning the insertion site will abnormally map on the reference sequence

In this report, paired-end mapping was used to search exhaustively the genome of several progenies of the double mutant (*met1*/*RdDM1*) in order to identify the TE families (other than *EVD*) that could be activated through the alteration of the corresponding silencing pathways. Even if, in this particular case, paired-end mapping did not allow to identify new active families, this report demonstrated the suitability of this approach to study genome dynamics in plants. More recently, Sabot et al. (2011) applied paired-end mapping to detect TE movements in a rice mutant line regenerated from callus culture. The LTR retrotransposon *Tos17* is known to be activated transpositionally in calli that have been cultivated *in vitro* for at least 12 weeks (Hirochika et al. 1996). This has been exploited to generate large collections of rice mutants (thus referred to as "Tos17 mutants") that constitute a valuable resource for rice genomics (Krishnan et al. 2009). Given that rice genome is composed of a large proportion of TEs, the authors anticipated that some families, in addition to *Tos17* elements, may be activated in these lines. To test their hypothesis, they sequenced the genome of a *Tos17* line harboring 11 new insertions of the element (estimation based on Southern hybridization experiment) and conducted a genome-wide search of TE movements using paired-end mapping. They first validated their approach by mapping the 11 new insertions of *Tos17*. Moreover, they found a total of 23 new TE insertions not caused by the activity of *Tos17*. Eleven were from LTR retrotransposons that belong to 7 distinct families and 12 from miniature inverted transposable elements (MITEs) that belong to 5 distinct families. Interestingly, among these 12 TE families, only one (the MITE *mPing*) had been previously reported as being active (Jiang et al. 2003). This study, therefore, suggests that genome-wide surveys of transposition using NGS can speed up the discovery of active elements in sequenced genome. *Tos17* is the best known TE in rice. Several studies have clearly shown that its activation in cultured calli is caused by demethylation of the active copy. This element was also shown to be activated in rice mutants that are deficient in histone methylation (Qin et al. 2010). The results obtained by Sabot, Picault et al. confirm that *Tos 17* is the most active element in calli. However, this first rice genome-wide survey of transposition also raises several questions regarding the control of transposition, either for the other families found to be transpositionally active (the 7 LTR retrotransposons and the 5 MITE families) or for the families previously reported to be transcriptionally active in *calli* (e.g., the LTR retrotransposon *Lullaby*) but that did not transpose in the sequenced line. As mentioned above, transposition is controlled by several distinct and complementary pathways. Upon impediment of one of these pathways, either genetically or physiologically, several TE families can be reactivated, as shown by many studies (Mirouze et al. 2009; Qin et al. 2010). However, such reactivation remains a stochastic process, and one should expect that not all the TE families known to be activated under certain conditions would actually transpose in a single generation. Therefore, in order to provide a comprehensive and relevant list of the active TE families of a given species, one should survey its transpositional landscape through the analysis of several individuals for each of the genetic background or physiological condition tested.

More recently, Fiston-Lavier et al. (2011) and Kofler et al. (2012) conducted some surveys of TEASVs in *Drosophila* genome using paired-end mapping of NGS data. This shows that this concept can be applied to any plant or animal species, given that a good reference genome sequence is available and that TEs have been correctly annotated (see below).

4.4 Technical Discussion

The four requirements for the genome-wide extensive identification of TEASV in any organism are (1) a good reference genome sequence (i.e., a good assembly of high-quality sequence reads), (2) a comprehensive annotation of the TEs from the genomics sequence, (3) a suitable dataset of NGS, preferably as paired-end (PE) sequences, and (4) a suitable software for TEASV detection.

In all published work cited above, the first requirement was obviously met, because all concerned model species for which high-quality genome sequence has been available for several years (i.e., human, Arabidopsis, rice, and Drosophila). However, most draft genome sequences published over the past few years for many plant and animals were obtained through the assembly of whole genome shotgun (WGS) using NGS. Although considerably more cost efficient than Sanger-based and physical map-based genome sequences, these strategies often lead to poor assemblies and thus incomplete reference sequences. This does not impede the detection of TEASVs, but rather restrains it to nonrepetitive, gene-rich regions. In fact, even in the case of high-quality sequences, PEM detection is often only reliable when one PE is anchored on a nonrepetitive region of the genome. Most programs systematically eliminate sequence reads that map in multiple loci in the genome (Medvedev et al. 2009).

The second requirement and probably the most challenging among the four is a correct annotation of TEs from the reference genome sequence use for paired-end mapping. Eukaryotic TEs exhibit a tremendous diversity of forms that makes difficult their automated annotation from sequenced genomes (Wicker et al. 2007). Moreover, the advent of NGS and the decrease in sequencing costs lead to the availability to ever increasing amount of genomic data, from which new TEs are continuously found. These include new TE families from known class I and class II types, but also new types, the mode of transposition of which remain unknown. Even in the case of well-known TE types, such as LTR retrotransposons, for which bioinformatic tools are available, several conceptual aspects of their annotation are still debated (Flutre et al. 2011). The basic PEM strategies to detect TE movements is based on the mapping of one PE on a unique sequence and of the other on a known TE located in a different locus on the reference sequence (Medvedev et al. 2009). The efficiency of such strategy will thus depend on the availability of a TE database of the species and moreover will be proportional to the quality of such database.

The third requirement is a sequence dataset which is suitable for paired-end mapping, i.e., with a good coverage of the genome at a sufficient depth (minimum

Table 4.1 Current tools available for TEASV detection

Name of the program	Reference	Methodology
PEMer	Korbel et al. (2009)	PEM
Variation hunter	Hormozdiari et al. (2009)	PEM
Breakdancer	Chen et al. (2009)	PEM
SVdetect	Zeitouni et al. (2010)	PEM
T-lex	Fiston-Lavier et al. (2010)	SRM
inGAPsv	Qi and Zhao (2011)	PEM, SRM, DOC

PEM paired-end mapping, *SRM* split read mapping, *DOC* depth of coverage

of 5×). The Illumina platform offers today the best cost efficiency and generates Gigabases of PE reads of 100 nucleotides, from insert sizes ranging from few hundred to thousands base pairs. This is the method of choice for SV detection (Mirouze et al. 2009; Sabot et al. 2011). However, one should keep in mind that technologies in the field of DNA sequencing are moving fast. Single molecule sequencing technologies (Pacific Biosciences, Nanopore Oxford) are now considered as being the future of genome sequencing. These produce longer reads (although with higher error rate at the moment). They will probably offer new perspectives in genome dynamics studies and in particular for the detection of TEASVs. Longer reads will enable direct identification of homology breakpoints on a single read, a method referred to as split read mapping (SRM, Medvedev et al. 2009). Although not often used because of the short size of the illumina reads, SR mapping may prove to be superior than paired-end mapping in detecting TEASVs for longer reads.

The fourth and last requirement for TEASV detection is the availability of dedicated softwares. Since the first pioneer publication of Korbel et al. (2007), many bioinformatic teams have developed tools to detect structural variants from NGS data (Medvedev et al. 2009) but not all of them are suitable for TEASV screening. Table 4.1 presents the programs freely available today for the detection of TEASVs. Most of these tools are based on paired-end mapping, but a program integrating PEM, SRM, and depth of coverage (DOC) methods was recently proposed (Qi and Zhao 2011). DOC is based on the variation of the number of reads covering a given sequence. If such region is either deleted, or if its copy number decrease (also through deletion), then the DOC of the corresponding sequence will vary accordingly. Although this method does not allow to precisely map the structural variation, it increases the efficiency of their detection when combined with paired-end mapping.

4.5 Conclusion

The genome-wide characterization of active TEs in both plant and animals is now accessible to many species thanks to the newest developments of NGS. This will be of particular interest in Biology at large, not only because it will lead to a better understanding of genome dynamics in terms of structural variations either in mutants or in natural populations but also because it will provide new ways to establish links between these structural variations and the functional diversity in eukaryotes. Moreover, when similar, high-throughput technologies will be available to study genome-wide epigenetic diversity, then the knowledge of TEASVs in individuals and in populations will become relevant as one of the key process for the generation of epigenetic diversity.

References

Arabidopsis Genome Initiative (2000) Analysis of the genome sequence of the flowering plant *Arabidopsis thaliana*. Nature 408:796–815

Argout X, Salse J, Aury JM, Guiltinan MJ, Droc G, Gouzy J, Allegre M, Chaparro C, Legavre T, Maximova SN, Abrouk M, Murat F, Fouet O, Poulain J, Ruiz M, Roguet Y, Rodier-Goud M, Barbosa-Neto JF, Sabot F, Kudrna D, Ammiraju JS, Schuster SC, Carlson JE, Sallet E, Schiex T, Dievart A, Kramer M, Gelley L, Shi Z, Bérard A, Viot C, Boccara M, Risterucci AM, Guignon V, Sabau X, Axtell MJ, Ma Z, Zhang Y, Brown S, Bourge M, Golser W, Song X, Clement D, Rivallan R, Tahi M, Akaza JM, Pitollat B, Gramacho K, D'Hont A, Brunel D, Infante D, Kebe I, Costet P, Wing R, McCombie WR, Guiderdoni E, Quetier F, Panaud O, Wincker P, Bocs S, Lanaud C (2011) The genome of *Theobroma cacao*. Nat Genet 43:101–108

Baillie JK, Barnett MW, Upton KR, Gerhardt DJ, Richmond TA, De Sapio F, Brennan PM, Rizzu P, Smith S, Fell M, Talbot RT, Gustincich S, Freeman TC, Mattick JS, Hume DA, Heutink P, Carninci P, Jeddeloh JA, Faulkner GJ (2011) Somatic retrotransposition alters the genetic landscape of the human brain. Nature 479:534–537

Bennetzen JL (2005) Transposable elements, gene creation and genome rearrangement in flowering plants. Curr Opin Genet Dev 15:621–627

Bourc'his D, Voinnet O (2010) A small-RNA perspective on gametogenesis, fertilization, and early zygotic development. Science 330:617–622

Chen K, Wallis JW, McLellan MD, Larson DE, Kalicki JM, Pohl CS, McGrath SD, Wendl MC, Zhang Q, Locke DP et al (2009) BreakDancer: an algorithm for high-resolution mapping of genomic structural variation. Nat Methods 6:677–681

Feuk L, Carson AR, Scherer SW (2006) Structural variation in the human genome. Nat Rev Genet 7:85–97

Fiston-Lavier AS, Carrigan M, Petrov DA, González J (2011) T-lex: a program for fast and accurate assessment of transposable element presence using next-generation sequencing data. Nucleic Acids Res 39:e36

Fiston-Lavier AS, Carrigan M, Petrov DA, Gonzalez J (2010) T-lex: a program for fast and accurate assessment of transposable element presence using next-generation sequencing data. Nucleic Acids Res 39:e36

Flutre T, Duprat E, Feuillet C, Quesneville H (2011) Considering transposable element diversification in de novo annotation approaches. PLoS One 6:e16586

Hormozdiari F, Alkan C, Eichler EE, Sahinalp SC (2009) Combinatorial algorithms for structural variation detection in high-throughput sequenced genomes. Genome Res 19:1270–1278

Hirochika H, Sugimoto K, Otsuki Y, Tsugawa H, Kanda M (1996) Retrotransposons of rice involved in mutations induced by tissue culture. Proc Natl Acad Sci USA 93:7783–7788

Iafrate AJ, Feuk L, Rivera MN, Listewnik ML, Donahoe PK, Qi Y, Scherer SW, Lee C (2004) Detection of large-scale variation in the human genome. Nat Genet 36:949–951

International Rice Genome Sequencing Project (IRGSP) (2005) The map-based sequence of the rice genome. Nature 436:793–800

Jiang N, Bao Z, Zhang X, Hirochika H, Eddy SR, McCouch SR, Wessler SR (2003) An active DNA transposon family in rice. Nature 421:163–167
Jiang N, Bao Z, Zhang X, Eddy SR, Wessler SR (2004) Pack-MULE transposable elements mediate gene evolution in plants. Nature 431(7008):569–573
Kobayashi S, Goto-Yamamoto N, Hirochika H (2004) Retrotransposon-induced mutations in grape skin color. Science 304:982
Kofler R, Betancourt AJ, Schlötterer C (2012) Sequencing of pooled DNA samples (Pool-Seq) uncovers complex dynamics of transposable element insertions in *Drosophila melanogaster*. PLoS Genet 8(1):e1002487
Komatsu M, Shimamoto K, Kyozuka J (2003) Two-step regulation and continuous retrotransposition of the rice LINE-type retrotransposon Karma. Plant Cell 15:1934–1944
Korbel JO, Urban AE, Affourtit JP, Godwin B, Grubert F, Simons JF, Kim PM, Palejev D, Carriero NJ, Du L, Taillon BE, Chen Z, Tanzer A, Saunders AC, Chi J, Yang F, Carter NP, Hurles ME, Weissman SM, Harkins TT, Gerstein MB, Egholm M, Snyder M (2007) Paired-end mapping reveals extensive structural variation in the human genome. Science 318:420–426
Korbel JO, Abyzov A, Mu XJ, Carriero N, Cayting P, Zhang Z, Snyder M, Gerstein MB (2009) PEMer: a computational framework with simulation-based error models for inferring genomic structural variants from massive paired-end sequencing data. Genome Biol 10:R23
Krishnan A, Guiderdoni E, An G, Hsing YI, Han CD, Lee MC, Yu SM, Upadhyaya N, Ramachandran S, Zhang Q, Sundaresan V, Hirochika H, Leung H, Pereira A (2009) Mutant resources in rice for functional genomics of the grasses. Plant Physiol 149:165–170
Law JA, Jacobsen SE (2010) Establishing, maintaining and modifying DNA methylation patterns in plants and animals. Nat Rev Genet 11:204–220
Lisch D (2009) Epigenetic regulation of transposable elements in plants. Annu Rev Plant Biol 60:43–66
Lisch D, Bennetzen JL (2011) Transposable element origins of epigenetic gene regulation. Curr Opin Plant Biol 14:156–161
McClintock B (1953) Induction of instability at selected loci in maize. Genetics 38:579–599
Medvedev P, Stanciu M, Brudno M (2009) Computational methods for discovering structural variation with next-generation sequencing. Nat Methods 6:S13–S20
Messing J, Bennetzen JL (2008) Grass genome structure and evolution. Genome Dyn 4:41–56
Miclaus M, Wu Y, Xu JH, Dooner HK, Messing J (2011) The maize high-lysine mutant opaque7 is defective in an acyl-CoA synthetase-like protein. Genetics 189:1271–1280
Mirouze M, Reinders J, Bucher E, Nishimura T, Schneeberger K, Ossowski S, Cao J, Weigel D, Paszkowski J, Mathieu O (2009) Selective epigenetic control of retrotransposition in Arabidopsis. Nature 461:427–430
Moon S, Jung KH, Lee DE, Jiang WZ, Koh HJ, Heu MH, Lee DS, Suh HS, An G (2006) Identification of active transposon dTok, a member of the hAT family, in rice. Plant Cell Physiol 47:1473–1483
Morgante M, Brunner S, Pea G, Fengler K, Zuccolo A, Rafalski A (2005) Gene duplication and exon shuffling by helitron-like transposons generate intraspecies diversity in maize. Nat Genet 37:997–1002
Pareek CS, Smoczynski R, Tretyn A (2011) Sequencing technologies and genome sequencing. J Appl Genet 52:413–435
Picault N, Chaparro C, Piegu B, Stenger W, Formey D, Llauro C, Descombin J, Sabot F, Lasserre E, Meynard D, Guiderdoni E, Panaud O (2009) Identification of an active LTR retrotransposon in rice. Plant J 58:754–765
Piegu B, Guyot R, Picault N, Roulin A, Saniyal A, Kim H, Collura K, Brar DS, Jackson S, Wing RA, Panaud O (2006) Doubling genome size without polyploidization: dynamics of retrotransposition-driven genomic expansions in *Oryza australiensis*, a wild relative of rice. Genome Res 16:1262–1269

Qin FJ, Sun QW, Huang LM, Chen XS, Zhou DX (2010) Rice SUVH histone methyltransferase genes display specific functions in chromatin modification and retrotransposon repression. Mol Plant 3:773–782

Qi J, Zhao F (2011) inGAP-sv: a novel scheme to identify and visualize structural variation from paired and mapping data. Nucleic Acids Res 39:W567–W575

Rigal M, Mathieu O (2011) A "mille-feuille" of silencing: epigenetic control of transposable elements. Biochim Biophys Acta 1809:452–458

Sabot F, Picault N, Elbaidouri M, Llauro C, Chaparro C, Piegu B, Roulin A, Guiderdoni E, Delabastide M, Mccombie R, Panaud O (2011) Transpositional landscape of rice genome revealed by Paired-End Mapping of high-throughput resequencing data. Plant J 66:241–246

SanMiguel P, Tikhonov A, Jin YK, Motchoulskaia N, Zakharov D, Melake-Berhan A, Springer PS, Edwards KJ, Lee M, Avramova Z, Bennetzen JL (1996) Nested retrotransposons in the intergenic regions of the maize genome. Science 274:765–768

Saze H, Tsugane K, Kanno T, Nishimura T (2012) DNA methylation in plants: relationship with small RNAs and histone modifications, and functions in transposon inactivation. Plant Cell Physiol 53:766–784

Sharp AJ, Locke DP, McGrath SD, Cheng Z, Bailey JA, Vallente RU, Pertz LM, Clark RA, Schwartz S, Segraves R, Oseroff VV, Albertson DG, Pinkel D, Eichler EE (2005) Segmental duplications and copy-number variation in the human genome. Am J Hum Genet 77:78–88

Siol M, Wright SI, Barett SC (2010) The population genomics of plant adaptation. New Phytol 188:313–332

Slotkin RK, Martienssen RA (2007) Transposable elements and the epigenetic regulation of the genome. Nat Rev Genet 8:272–285

Stankiewicz P, Lupski JR (2010) Structural variation in the human genome and its role in disease. Annu Rev Med 61:437–455

Stewart C, Kural D, Strömberg MP, Walker JA, Konkel MK, Stütz AM, Urban AE, Grubert F, Lam HY, Lee WP, Busby M, Indap AR, Garrison E, Huff C, Xing J, Snyder MP, Jorde LB, Batzer MA, Korbel JO, Marth GT, 1000 Genomes Project (2011) A comprehensive map of mobile element insertion polymorphisms in humans. PLoS Genet 7:e1002236

Studer A, Zhao Q, Ross-Ibarra J, Doebley J (2011) Identification of a functional transposon insertion in the maize domestication gene tb1. Nat Genet 43:1160–1163

Tsugane K, Maekawa M, Takagi K, Takahara H, Qian Q, Eun CH, Iida S (2006) An active DNA transposon nDart causing leaf variegation and mutable dwarfism and its related elements in rice. Plant J 45:46–57

Vitte C, Panaud O (2005) LTR retrotransposons and flowering plant genome size: emergence of the increase/decrease model. Cytogenet Genome Res 110:91–107

Vitte C, Panaud O, Quesneville H (2007) LTR retrotransposons in rice (*Oryza sativa*, L.): recent burst amplifications followed by rapid DNA loss. BMC Genomics 8:218

Wicker T, Sabot F, Hua-Van A, Bennetzen JL, Capy P, Chalhoub B, Flavell A, Leroy P, Morgante M, Panaud O, Paux E, SanMiguel P, Schulman AH (2007) A unified classification system for eukaryotic transposable elements. Nat Rev Genet 8:973–982

Wicker T, Buchmann JP, Keller B (2010) Patching gaps in plant genomes results in gene movement and erosion of colinearity. Genome Res 20:1229–1237

Zeitouni B, Boeva V, Janoueix-Lerosey I, Loeillet S, Legoixne P, Nicolas A, Delattre O, Barillot E (2010) SVDetect: a tool to identify genomic structural variations from paired-end and mate-pair sequencing data. Bioinformatics 26:1895–1896

Zhang J, Chiodini R, Badr A, Zhang G (2011) The impact of next-generation sequencing on genomics. J Genet Genomics 38:95–109

Chapter 5
Hitching a Ride: Nonautonomous Retrotransposons and Parasitism as a Lifestyle

Alan H. Schulman

Abstract Large genomes in plants are composed primarily of long terminal repeat (LTR) retrotransposons, which replicate and propagate by a "copy-and-paste" mechanism dependent on enzymes encoded by the retrotransposons themselves. The enzymes direct a life cycle involving transcription, translation, packaging, reverse transcription, and integration. Loss of any coding capacity will render a retrotransposon incapable of completing its life cycle autonomously. Nevertheless, retrotransposons lacking complete open reading frames for one or more of their proteins are abundant in the genome. These nonautonomous retrotransposons can, however, be complemented in *trans* by proteins expressed by another retrotransposon, restoring mobility. It is sufficient for a nonautonomous LTR retrotransposon to retain the signals needed for recognition by the transcription machinery and the proteins of autonomous elements. The degree to which nonautonomous retrotransposons interfere with the propagation of autonomous elements has major evolutionary consequences for the genome, affecting the relative rate of gain versus loss of retrotransposons and thereby genome size.

Keywords Retrotransposon • Replication • Integration • Reverse transcription • Genome dynamics

A.H. Schulman (✉)
Institute of Biotechnology, University of Helsinki, P.O. Box 65, Viikinkaari 1, FIN-00014 Helsinki, Finland

Biotechnology and Food Research, MTT Agrifood Research, Jokioinen, Finland
e-mail: alan.schulman@helsinki.fi

M.-A. Grandbastien and J.M. Casacuberta (eds.), *Plant Transposable Elements*, Topics in Current Genetics 24, DOI 10.1007/978-3-642-31842-9_5,

5.1 Retrotransposons

5.1.1 Retrotransposons, Drivers of Genome Evolution

As described in elsewhere in this volume (Chap. 1), transposable elements (TEs) can be grouped into 2 major Classes, 9 Orders and 29 Superfamilies (Wicker et al. 2007). Class I, the retrotransposons, is composed of TEs that replicate via an RNA intermediate by a "copy-and-paste" mechanism. Class II elements move generally by "cut-and-paste" as DNA segments. However, Subclass 2 of Class II includes as well the *Helitron* (Kapitonov and Jurka 2007) and *Maverick/Polinton* elements that propagate by what could be called "cut and copy" (Fischer and Suttle 2011). This chapter will be focused on retrotransposons.

The most abundant TEs in plant genomes are the long terminal repeat (LTR) retrotransposons, the structures of which are described below. Most plant genomes contain hundreds of LTR retrotransposon families, each in low or moderate copy numbers. However, the large plant genomes contain a few very abundant and replicatively successful retrotransposon families. In the Triticeae (barley, wheat, and relatives), the *BARE1*, *WIS*, and *Angela* elements account for more than 10 % of the genome (Vicient et al. 1999a; Kalendar et al. 2000; Soleimani et al. 2006; Wicker et al. 2009). A whole-genome survey of barley showed that 50 % of the genome is comprised of only 14 TE families, 12 being LTR retrotransposons (Wicker et al. 2009). Why certain LTR retrotransposon families have been able to expand to large numbers while others have not is unknown, though of great interest. Some abundant LTR retrotransposon families are activated by stresses such as drought (Kalendar et al. 2000) or UV light (Ramallo et al. 2008), but so are other retrotransposons that are nevertheless rare in the genome (Grandbastien et al. 2005). Moreover, it is also a reasonable conjecture that selective forces act to drive copy numbers down for some families because of their propensity, for example, to insert into genes.

As a consequence of their overall abundance, LTR retrotransposons are responsible for major variations in genome size other than those explained by genome duplication and polyploidization. For example, *Arabidopsis thaliana* and sorghum, respectively, having 120 Mbp and 700 Mbp genomes, contain a similar amount of Class II transposons, with the difference in their genome size explained mainly by the differential abundance of LTR retrotransposons (Arabidopsis Genome Initiative 2000; Paterson et al. 2009). In barley, a whole-genome survey showed that less than a dozen LTR retrotransposon families account for almost half of the genome, while Class II elements contribute about 5 % (Wicker et al. 2009). Earlier, we showed that the difference in genome size between two particular *Hordeum* species can be explained primarily by the difference in *BARE1* abundance (Vicient et al. 1999b).

5.1.2 Replication of Autonomous Retrotransposons

The Class I transposable elements all employ a replication cycle in which transcribed RNA is copied into dsDNA by reverse transcriptase. The two largest orders of Class I

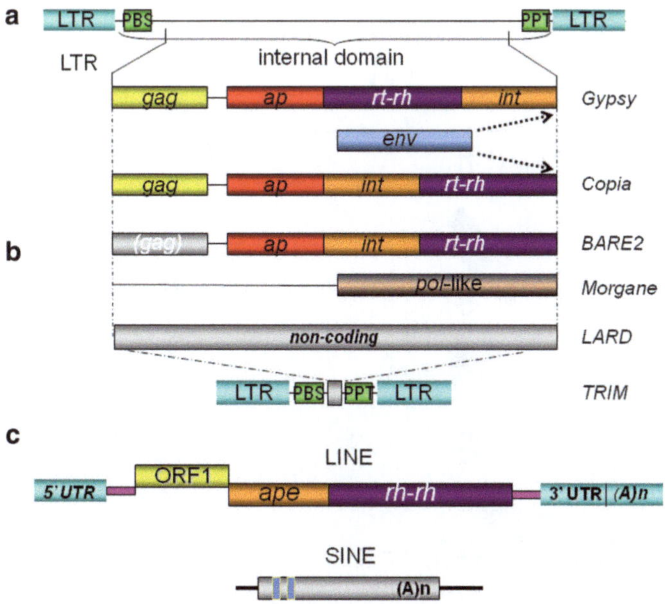

Fig. 5.1 Main groups of autonomous and nonautonomous retrotransposons. (**a**) Autonomous LTR retrotransposons. Above, the basic structure of an LTR retrotransposon, comprising: the long terminal repeats (LTRs); the primer binding site (PBS), which is the (−)-strand priming site for reverse transcription; the polypurine tract (PPT), which is the (+)-strand priming site for reverse transcription; the PBS and PPT are part of the internal domain, which in autonomous elements includes the protein-coding open reading frame(s). Below, the major superfamilies of LTR retrotransposons, *Gypsy* and *Copia*. The open reading frame(s) of the internal domain are *gag*, encoding the capsid protein Gag; *ap*, aspartic proteinase; *rt–rh*, reverse transcriptase–RNase H; *int*, integrase. The position of the *env* domain encoding the envelope protein in those *Gypsy* and *Copia* clades that contain it is shown. (**b**) Nonautonomous retrotransposons. *BARE2* is an example of a major conserved group having a specific deletion that generates a nonautonomous subfamily. Elements like *Morgane* have a degenerate or truncated, but still recognizable open reading frame. *LARD* elements have a long internal domain with conserved structure but lacking coding capacity. *TRIM* elements have virtually no internal domain except for the PBS and PPT signals. (**c**) Autonomous and nonautonomous non-LTR retrotransposons. Shown are the autonomous order LINE of the L1 superfamily (*ape* = apurinic endonuclease) and the nonautonomous order SINE. A *gray bar* indicates a noncoding domain

TEs are named by the presence or absence of an LTR at either end of the retrotransposon (Fig. 5.1). The LINEs (Long Interspersed Nuclear Elements; Goodier and Kazazian 2008) are generally seen as the canonical non-LTR retrotransposons, though the DIRS (Dictyostelium Intermediate Repeat Sequence), PLE (Penelope-like element), and SINE (Short Interspersed Nuclear Elements) retrotransposons also lack LTRs (Wicker et al. 2007). The non-LTR retrotransposons are found throughout the clades of eukaryotes. While they predominate in the genomes of vertebrates and some fungi (Spanu et al. 2010), they are generally much less abundant in plants.

The LINEs are considered to be the primordial Class I elements due to their simple structure, specifying only reverse transcriptase and endonuclease activities in the basic forms. Not only lacking LTRs, the non-LTR retrotransposons also function

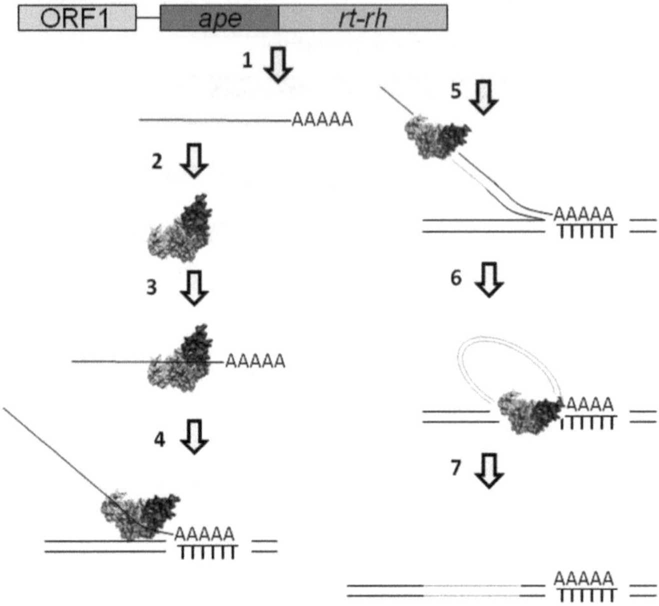

Fig. 5.2 Replication mechanism of a non-LTR retrotransposon. Replication of a LINE of superfamily L1 is shown. The element contains ORF1, specifying an RNA-binding protein, and an open reading frame encoding an apurinic endonuclease (*ape*) and reverse transcriptase–RNase H (*rt–rh*). During replication, the LINE is transcribed (Step 1), the open reading frames translated (Step 2; for simplicity only the RT is shown), assembled into a ribonucleoprotein particle (Step 3), and transported into the nucleus (step not shown). The APE nicks the target site, at which point the RNA anneals (Step 4). The free 3′ hydroxyl group of the nicked target is used to prime reverse transcription by a process called target-primed reverse transcription (Step 5). The other strand of the target DNA is also nicked, and the second strand of the LINE is synthesized by the RT (Step 6). The process is completed and the new copy is now inserted at the target site (Step 7). The process is reviewed by Han and Boeke (2005)

without an integrase gene (Figs. 5.1 and 5.2). Instead, the reverse transcriptase primes DNA synthesis from the poly-A tail of the element's transcript (Fig. 5.2), later ligating the end of the newly synthesized DNA into the insertion point.

The first step of replication of an LTR retrotransposon (Fig. 5.3) is transcription of an integrated element. The LTRs both drive transcription, by providing a promoter at the 5′ end of the retrotransposon, and specify RNA termination and polyadenylation, using signals in the LTR that are operational at the 3′ end of the inserted element. Transcription by pol II thus begins within the 5′ LTR and terminates within the 3′ LTR before its 3′ end. The RNA transcripts meet two fates: they are translated to form the protein products needed for the retrotransposon life cycle; they are packaged into virus-like particles (VLPs) and later reverse transcribed into cDNA. If the same RNA serves in both pathways, translation must precede reverse transcription for two reasons. First, packaging removes the RNA from access to the translation machinery. Second, during reverse transcription the RNA is hydrolyzed by the action of RNaseH.

Packaging into VLPs is mediated by two signals present in the untranslated leader (UTL) between the PBS and the beginning of *gag*. These are the PSI

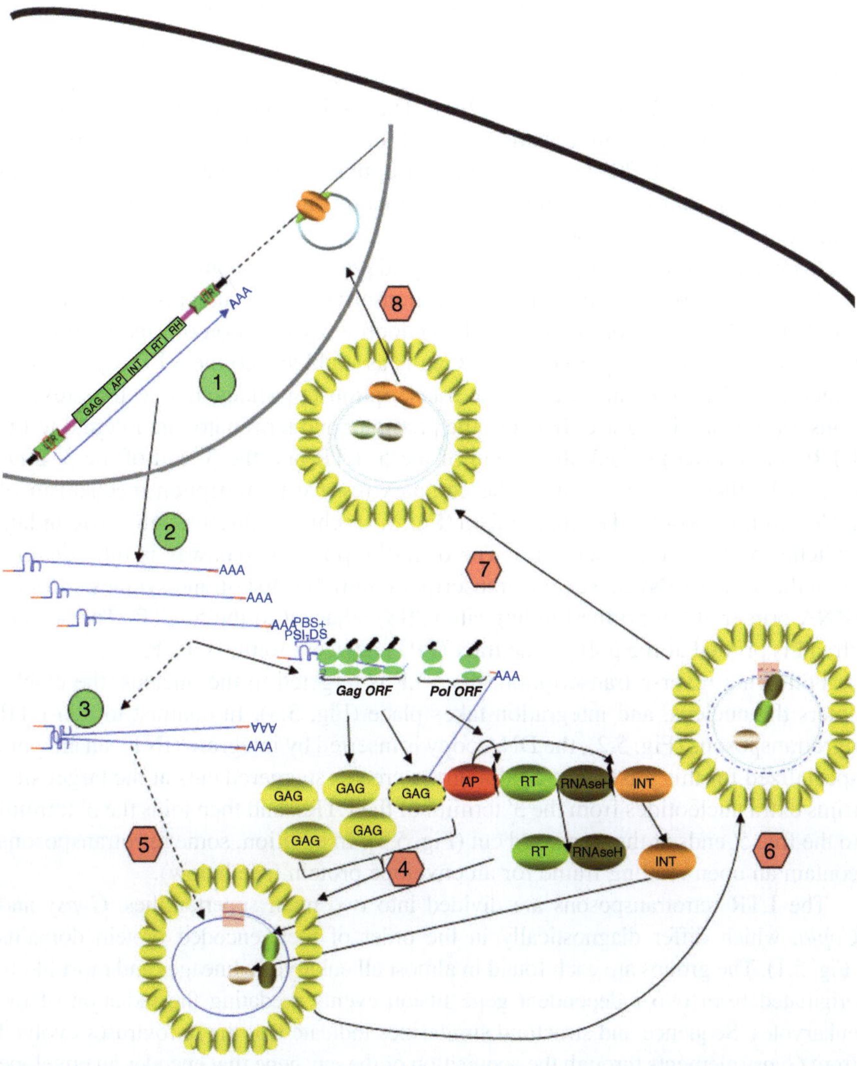

Fig. 5.3 Lifecycle of LTR retrotransposons. An element of superfamily *Copia* with a single open reading frame (ORF) is depicted diagrammatically, integrated into the genome, within the nucleus (*gray curve*). The plasma membrane is represented as a *black curve*. The major steps of the life cycle are shown in *green circles*. If the step depends on the proteins encoded by the retrotransposon and is therefore potentially blocked in a nonautonomous retrotransposon in the absence of complementation, it is shown in a *red hexagon*. The steps are (1) transcription of a copy integrated into the genome, from the promoter in the long terminal repeat (LTR); (2) nuclear export; (3) alternative translation or buckling of two transcripts destined for packaging and reverse transcription; (4) translation either of separate *gag* and *pol* ORFs or of one common ORF to produce the capsid protein Gag and a polyprotein containing aspartic proteinase (AP), reverse transcriptase (RT), RNAseH, and integrase (INT), the order of the protein units being shown being as for elements of superfamily *Gypsy*; (5) assembly of a virus-like particle (VLP) from Gag containing RNA transcripts, integrase, reverse transcriptase–RnaseH; (6) reverse transcription by RT; (7) localization of the VLP to the nucleus; (8) passage of the cDNA–integrase complex into the nucleus and integration of the cDNA into the genome

(Packaging SIgnal) and DIS (DImerization Signal) motifs, which form conserved secondary structures in the RNA as stem–loops. In retroviruses, and by extension in retrotransposons, PSI mediates packaging of the transcript into its specific particle (Lu et al. 2011; Miyazaki et al. 2011). The DIS directs so-called kissing-loop interactions leading to dimerization of the transcripts during, or just before, packaging (Paillart et al. 2004). Such signals are highly important for propagation of retroviruses, because any change in their structures may severely weaken both the replication and the infection processes.

Translation of the RNA produces the capsid protein Gag, sometimes in a separate reading frame from the enzymes reverse transcriptase and integrase. The proteins are derived from the polyprotein by the endoproteolytic action of aspartic proteinase, also part of the polyprotein. The Gag is assembled into the VLP capsids, into which the RNA template for reverse transcription is packaged as well as reverse transcriptase and integrase. Because the promoter and terminator are internal to the LTRs, the transcripts lack the 5′ end of the 5′ LTR and the 3′ end of the 3′ LTR (Fig. 5.4); these are restored by the complex reverse transcription mechanism of LTR retrotransposons. The mechanism (Fig. 5.4) achieves this through two template switches by reverse transcriptase. The overall replication pathway is fully distinct from that of the LINES. Reverse transcriptase initiates first-strand synthesis from a tRNA primer at the primer binding site (PBS) adjacent to the 5′ LTR. The second strand is primed at the polypurine tract (PPT) adjacent to the 3′ LTR.

Following reverse transcription, the VLP is targeted to the nucleus, the cDNA enters the nucleus, and integration takes place (Fig. 5.3). In contrast to non-LTR retrotransposons (Fig. 5.2), the DNA copy is inserted by integrase (INT), an enzyme specialized for this job (Fig. 5.5). Integrase creates staggered cuts at the target site, trims extra nucleotides from the 3′ termini of the LTRs, and then joins the 3′ termini to the free 5′ ends at the staggered cut (Fig. 5.5). In addition, some retrotransposons contain an open reading frame for an envelope protein (see below).

The LTR retrotransposons are divided into two main superfamilies, *Gypsy* and *Copia*, which differ diagnostically in the order of their encoded protein domains (Fig. 5.1). The groups are each found in almost all eukaryotic lineages and most likely originated from two independent gene fusion events predating the radiation of the eukaryotes. Sequence and structural similarities indicate that the retroviruses evolved from *Gypsy* elements through the acquisition of the *env* gene that encodes an envelope protein with transmembrane domains. The protein mediates the formation of an envelope, derived from the plasma membrane, around retroviruses, which consequently can bud from the plasma membrane, leave the host cell, and go on to infect other cells. The *gypsy* family of *Drosophila*, the type element of the superfamily, has retroviral-like properties because it can be infectious under laboratory conditions (Kim et al. 1994).

In fact, the *env* domain is not restricted to animal retroviruses; an *env*-bearing clade of *Gypsy* elements is widespread in plants (Vicient et al. 2001). Moreover, *env* domains can be found in a clade of *Copia* retrotransposons (Laten et al. 2005; see also a review on this topic, Chap. 6). The likely early division of the *Copia* and *Gypsy* lineages and the distinct position of *env* in the clades of the two superfamilies argues for independent gain of function in both cases and begs a function in the organisms where an extracellular segment of the life cycle has not been demonstrated.

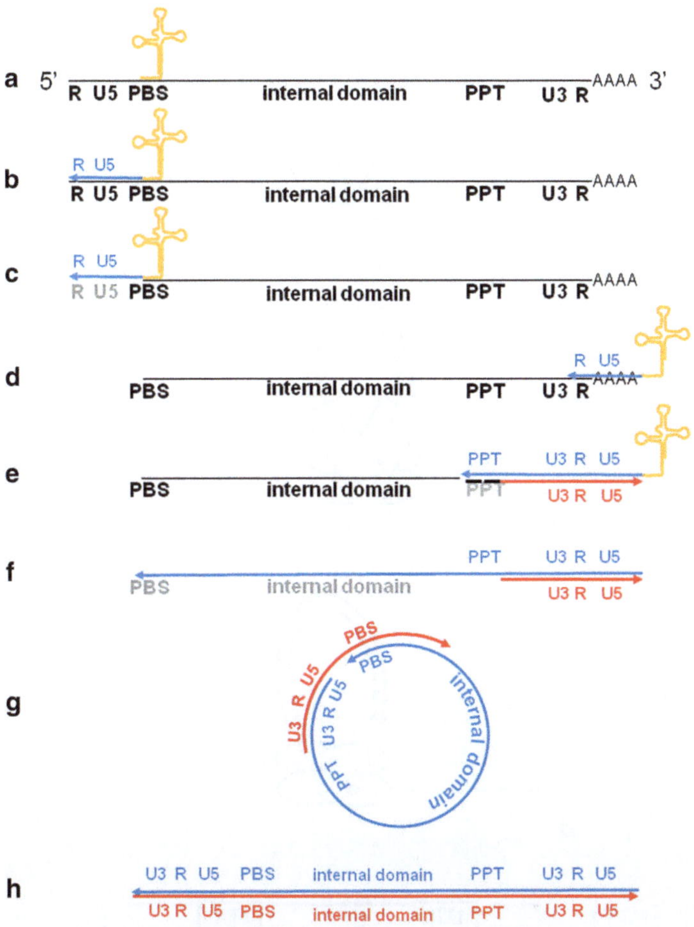

Fig. 5.4 Reverse transcription of LTR retrotransposons. Diagrammatically represented are the major steps. (**a**) Attachment of a tRNA primer at the primer-binding site (PBS) of the retrotransposon transcript (*black line*), adjacent to the 3′ end of the 5′ LTR regions R and U5, to initiate reverse transcription. (**b**) Extension of the minus-strand cDNA (shown as a *gray line*) to the end of the transcript to form minus-strand strong-stop DNA (−sssDNA); (**c**) Degradation of the RNA from the RNA/DNA hybrid by RNaseH, exposing the repeat (R) domain that is present at both ends of the transcript. (**d**) Transfer of the exposed −sssDNA to the 3′ end of the transcript by hybridization of the R domain. (**e**) Extension of the minus-strand and concomitant degradation of the hybridized regions of the transcript by RNase H until the polypurine tract (PPT) of the cDNA is exposed, whereupon plus-strand cDNA (*dotted line*) synthesis is initiated from RNA fragments (*short black lines*) as primers. The plus strand is extended to the 5′ end of the minus-strand cDNA, and generating a complementary copy of the PBS, and forms plus-strand strong-stop DNA (+sssDNA). (**f**) The RNA primers are removed by RNAseH, exposing the PBS on the +sssDNA. (**g**) Transfer of the +sssDNA, mediated by hybridization of the PBS domain, and continuation of cDNA synthesis requiring strand displacement, each strand serving as a template for the other. (**h**) Completion of cDNA synthesis to generate a double-stranded linear molecular with intact LTRs at either end. The details and representation are essentially as presented earlier (Telesnitsky and Goff 1997)

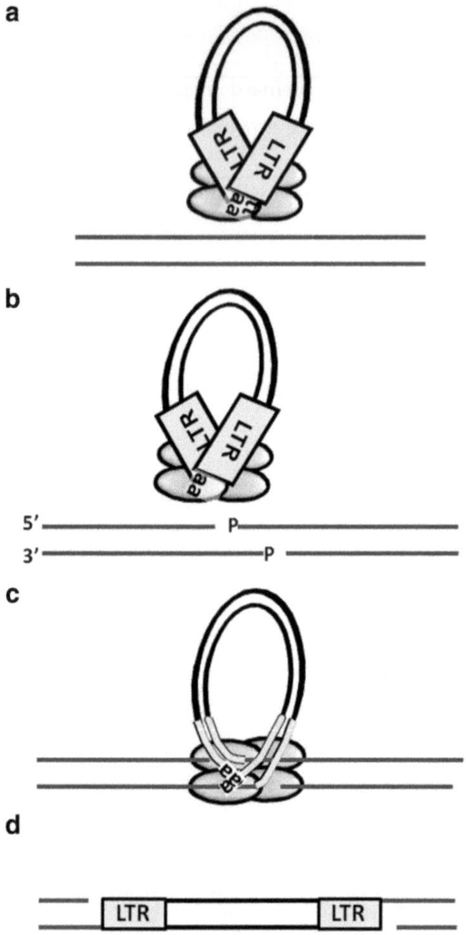

Fig. 5.5 Integration mechanism of an LTR retrotransposon. The retrotransposon is represented as a loop bounded by two LTRs. Each LTR is flanked by an extra dinucleotide basepair (in this case AA/TT, as found in retrotransposon *BARE* of barley), which is copied by RT from the dinucleotide found between the PBS and the 3′ end of the 5′ LTR during reverse transcription. The integrase is represented, bound to the LTRs, as a tetramer (Dolan et al. 2009; Cherepanov et al. 2011), forming a pre-integration complex together with the retrotransposon. The genomic DNA target is shown as a *pair of gray lines* beneath the preintegration complex. (**a**) The pre-integration complex and target site. (**b**) The integrase makes a 4- to 6-bp staggered cut in the genomic DNA and trims the dinucleotide from the 3′ end of each LTR, generating 5′ overhangs on both the retrotransposon and at the target site (shown as "P" for 5′ phosphate). (**c**) Integration of the LTR retrotransposon. The 3′ ends of the LTR are joined to the 5′ overhangs of the target. The *trans*-esterification reaction, in which the target is cleaved and retrotransposon joined, proceeds as a single-step. (**d**) Following the integration reaction and removal of the remaining dinucleotide from the 5′ end of each LTR, the gaps generated by the staggered cut remain. The repair of these gaps generates the target-site duplication (TSD) flanking the retrotransposon

5.2 Nonautonomous Transposable Elements

Retrotransposons play a major role in genome size variation over evolutionary time (discussed above) and are dynamic in their induction both by biotic and abiotic stresses (Wessler 1996; Kalendar et al. 2000; Grandbastien et al. 2005; Ramallo et al. 2008) as well as by "genome stress" (McClintock 1984; Kashkush et al. 2003; Belyayev et al. 2010). Nevertheless, most copies of retrotransposons encountered in a random segment of the genome contain deletions or mutations affecting their open reading frames (ORFs), if they have them at all. These elements, which appear at first glance to be incapable of replicating, can form the majority of the retrotransposon population. This observation may lead the casual onlooker to conclude that Ohno was correct when he referred to the nongenic component of the genome as "junk" (Ohno 1972). However, while the genome may contain "fossils," or no-longer active transposable elements, these are no more junk, an anthropomorphic term, than are pseudogenes or dinosaur bones.

Many of the apparently fossilized TEs, in fact, can be brought back to life when mobilized by another element; it takes more than a few mutations to kill a TE. This was recognized early on when McClintock observed both autonomous and nonautonomous controlling elements, respectively, *Ac* and *Ds* (McClintock 1948; Jones 2005). The canonical autonomous elements contain intact open reading frames and promoters, as well as the structural motifs that are recognized by the TE enzymes and processing signals recognized by the enzymes of general cellular DNA and RNA metabolism.

The nonautonomous but active mobile elements can still be transcribed and mobilized in *trans* by proteins from autonomous elements; others may have lost the motifs required for *trans* activation and are both nonautonomous and nonmobile. Among the Class II transposons such as those studied by McClintock, the term "nonautonomous element" has referred to those that cannot express transposase and catalyze their own transposition. They form binary systems with the autonomous elements able to drive their transposition. The classical examples of these include the *Ac–Ds* (McClintock 1948; Fedoroff et al. 1983; Jones 2005) and *Suppressor–Mutator* (*Spm*; Fedoroff 1999) systems, although similar ones are widespread (Hartl et al. 1992).

5.2.1 Nonautonomous Retrotransposons

For Class I elements, the phenomenon of non-autonomy has several additional facets because of the complexity of their replicative life cycle (Figs. 5.2, 5.3, 5.4, and 5.5; Sabot and Schulman 2006). In Class II transposons, a nonautonomous element can be mobilized as long as its termini are recognized by transposase. The LTR retrotransposons must be transcribed and translated, then transcripts packaged, together with integrase and reverse transcriptase, into VLPs formed from self-encoded Gag (Fig. 5.3). Reverse transcription, targeting and entering of the nucleus, and finally integration must occur. While any of these steps may be blocked by lack of a self-encoded protein (Fig. 5.3), all potentially can be complemented in *trans* if a translationally or enzymatically defective LTR retrotransposon nevertheless possesses the correct recognition signals for proteins encoded by an autonomous and competent element.

5.2.2 Types of Nonautonomous Retrotransposons

The many nonautonomous TEs fall into several categories. The first group, referred to here as Type 1, is comprised of previously autonomous elements that have been variously mutated or deleted so that one or more of their motifs or encoded proteins are no longer functional. In many cases, parts of their protein coding domains may still be recognizable even if they are rendered nonfunctional by substitutions, stop codons, or both. Because retrotransposons encode a polyprotein, any upstream mutation generating a frameshift or stop codon will have polar effects, knocking out expression of the downstream proteins until an efficient start codon is reached. Therefore, nonautonomous elements encompass not only those where some or all of their coding capacity has been deleted but also otherwise autonomous elements with a point mutation leading to polar truncation of translation. The diverse Type 1 are, therefore, expected to be very widespread among the retrotransposons and could still be activated in *trans* by autonomous elements. A particular nonautonomous copy may have been integrated as a fully functional, autonomous copy and accumulated mutations thereafter, or may have been propagated from a genomic copy that was already nonautonomous.

A second category, Type 2, more interesting than the first because it sheds light on what is minimally required for transposition, consists of groups of nonautonomous mobile elements that have conserved structures or deletions in which one, several, or all protein-coding domains are missing. Type 2 elements have made a successful "lifestyle" of being nonautonomous. Members of this category likely arose from among the variety of mutated forms in the first category. Effective, repeated replication and propagation of particular individual elements gave rise to families or subfamilies of elements with conserved deletions. Further, stepwise deletions and cycles of replication and propagation may lead to conserved groups of elements lacking all protein-coding domains.

Type 3, like Type 2, contains nonautonomous elements of conserved structure, but these are not derived from transposable elements. Instead, they coincidentally possess the signals required for replication due to their role in other or earlier cellular functions. Classic examples of this category are the SINE elements, which will be discussed in more detail below.

Type 4 contains many elements that can no longer be mobilized in *trans* without restoring mutations. These are both nonautonomous and inactive and may be derived from members of either of the first two categories. These are the true fossils of the genome. Further insertions, deletions, and point mutations may render them unrecognizable as derivatives of transposable elements.

5.2.3 Examples of Type 2 Nonautonomous Retrotransposons

A good example of a Type 2 nonautonomous element is the *BARE2* retrotransposon of barley (Tanskanen et al. 2007), a member of the *Copia* superfamily. *BARE2* is a conserved, abundant, and insertionally polymorphic subfamily of the *BARE* family of retrotransposons and has most of its protein-coding domains intact. However, it has a small, conserved deletion that removes the *gag* start codon, so that it cannot produce this protein. Instead, the capsid protein is supplied to it by *BARE1* for packaging (Tanskanen et al. 2007). Further along the pathway of ORF loss are the *Morgane* elements of wheat and its relatives (Sabot et al. 2006). *Morgane* lacks the Gag entirely; the degenerate polyprotein is, however, still recognizable as belonging to the *Gypsy* superfamily, though it is riddled with stop codons. Nevertheless, *Morgane* possesses the PBS and PPT motifs needed for reverse transcription.

An endpoint of ORF degeneration, on a continuum from *BARE2* through *Morgane* and onward to complete loss of coding capacity, is represented by the Large Retrotransposon Derivative (*LARD*) elements. LARDs code for no protein, but possess a long internal domain with a predicted well-conserved RNA structure (Kalendar et al. 2004). The LARDs were found to be abundant (estimated 1.3×10^3 full-length copies and 1.16×10^4 solo LTRs in barley), polymorphic in their insertion sites, and widespread within the grass tribe Triticeae, possessing 4.4-kb LTRs and ~3.5-kb internal domains flanked by the PBS and PPT priming sites for reverse transcriptase. The conserved RNA structure and priming sites suggests that LARDs have evolved to be reverse transcribed and packaged by the proteins of another retrotransposon, apparently of the *Gypsy* superfamily.

If a retrotransposon can replicate without encoding proteins, the internal domain may be dispensed with as well, providing that the RNA template for cDNA still can be packaged. This requires retention of the PSI and DIS motifs, described above. Such reduced elements, where the signals for replication have been retained but the rest of the internal domain virtually completely deleted, are exemplified by the Terminal Repeat retrotransposon In Miniature (TRIM; Witte et al. 2001; Kalendar et al. 2008). These lack protein-coding capacity and have only very short internal domains, but nevertheless are abundant and conserved in plants.

Among the TRIM retrotransposons, *Cassandra* is a particularly interesting family (Kalendar et al. 2008). These elements are 565–860 bp overall, comprising 240–350 bp LTRs flanking a PBS, PPT, and as little as 34 bp in between these signals. Their LTRs all contain conserved 5S RNA sequences and associated RNA polymerase (pol) III promoters and terminators. These resemble the 5S RNA components of ribosomes. The predicted *Cassandra* RNA 5S secondary structures resemble those of cellular 5S rRNA, with high information content specifically in the pol III promoter region. *Cassandra* thus appears both to have adapted a ubiquitous cellular gene for ribosomal RNA for use as a promoter and to co-opt an as-yet-unidentified group of retrotransposons for the proteins needed in its lifecycle. The occurrence of *Cassandra* in the ferns, tree ferns, and in all the angiosperms that have been investigated to date places their origin at least in the Permian, 250 MYA, and suggests that their means of replication as nonautonomous elements has been highly successful for a very long time.

5.2.4 Examples of Type 3 Nonautonomous Retrotransposons

Similar to the TRIMs in their degree of reduction are the short interspersed elements (SINEs), nonautonomous Class I elements that are mobilized by non-LTR retrotransposons. Rather than being derived from LINEs by reduction or mutation, SINEs comprise a diverse group of sequences, sharing the ability to be recognized by the enzymatic machinery of the LINEs (Goodier and Kazazian 2008). They are highly abundant in mammalian genomes, with numbers ranging from 10^4 to 10^6 (Kramerov and Vassetzky 2005), but are also found in plants and elsewhere (Deragon and Zhang 2006). Although sharing a mechanism of propagation and a classification as a Order of Class I elements (Wicker et al. 2007), SINEs are polyphyletic in origin and are derived variously from tRNA, rRNA, and other pol III transcripts (Kramerov and Vassetzky 2005). They are generally 150–200 bp; those originating from tRNA possess the tRNA sequence at their 5′ ends and homology at their 3′ ends to a LINE from the same genome, which is thought to provide binding sites for LINE-encoded proteins. The 3′ tails are generally AT rich, betraying origins as reverse-transcribed gene transcripts. Although the enzymology of SINE retroposition is not fully understood, at least for the *Alu* SINE element of humans, one of the LINE L1 proteins, ORF2p, is needed while the other, ORF1p, may aid the movement (Kroutter et al. 2009).

5.2.5 Classification of Nonautonomous Retrotransposons

Classification of nonautonomous retrotransposons, and nonautonomous transposable elements in general, can be problematic. The current consensus classification (Wicker et al. 2007; see also a review on this topic, Chap. 1) hierarchically divides TEs, respectively, by the presence of an RNA transposition intermediate (Class), mobility during reverse transcription and the number of DNA strands cut at the TE donor site (Subclass), major differences in insertion mechanism (Order), large-scale features such as the structure of protein or noncoding domains (Superfamily), and DNA sequence conservation (Families and Subfamilies). Type 1 nonautonomous elements are relatively easy to fully classify down to the family level. Type 2 elements such as *BARE2*, if their internal domains retain coding capacity, can generally be placed as subfamilies within TE families. Highly reduced elements, such as the TRIMs discussed below, may be impossible to define below the level of subclass on the basis of sequence analysis and may require experimental data such as evidence for packaging or interactions with the gene products of autonomous elements for more precise phylogenetic placement.

Type 3 elements present a special problem for classification because they can be polyphyletic in origin. Moreover, while some SINEs, for example, may rely on a particular partner for mobilization, others are relatively nonspecific (Kajikawa and Okada 2002). The same may be the case for highly reduced nonautonomous LTR retrotransposons such as TRIMs. For such elements, association to the level of order based on mechanistic considerations may be the limit to what is possible. Depending on their origin or degree of degeneracy, Type 4 nonautonomous elements may or may not be possible to classify. The scheme of Wicker et al. (2007) allows for an "X" to denote ambiguity in the classification of a TE by the three-letter code defining its phylogenetic position.

5.3 Population Structure of Nonautonomous Elements

A thought-provoking feature of the highly reduced, nonautonomous TEs, such as SINEs and TRIMs among the retrotransposons and MITEs among the DNA transposons, is their exceptional abundance. One can view the great abundance of small nonautonomous elements and the comparative rarity of large autonomous elements metaphorically, as abundant but small parasites carried by individual large organisms. While the relative numbers of organismal hosts and parasites reflect an ecosystem's carrying capacity as related to size and niche, the meaning of this model for replicating entities within a genome is far from clear. The mechanisms behind the differences in abundance between autonomous TEs and their small, nonautonomous derivatives or partners are likewise opaque. However, the high probability of formation and the low cost or the selective advantage of the symbiotic lifestyle of nonautonomous elements may be the factors affecting their prevalence.

5.4 Evolution of Autonomous and Nonautonomous Retrotransposons

The minimalist SINEs and TRIMs illustrate the principal that so long as processing and recognition signals such as, for TRIMs, the PBS, PPT, PSI, and DIS remain present in *cis*, all of the proteins needed for propagation can be supplied in *trans*. Hence, the nonautonomous TEs provide a model for the *de novo* evolution of mobile elements. Today, the proteins for replication and packaging are supplied in *trans* to nonautonomous elements. In the deep past, the proteins ancestral to those of modern TEs could have acted in *trans* to mobilize nascent Class I or Class II elements. The various coding domains and replication signals need not have been assembled simultaneously but could have been captured or added sequentially. The respective likelihoods of TEs arising *de novo* and nonautonomous derivatives appearing are not equal, however. The abundance of nonautonomous elements in the genome demonstrates that the loss of coding capacity occurs often. Independent evolution of new types of TEs, based on the presence of relatively few (two classes, nine orders; Wicker et al. 2007) types of transposable elements in the eukaryotes, appears to happen rarely.

One can nevertheless begin to model the evolution of TEs based on the nonautonomous elements as the minimal functional unit needing to be assembled in *cis*. Focusing on the retrotransposons, mobility requires propagation of a copy, which requires an integrase enzymatic function to break the genomic DNA and integrate a mobile DNA segment into the chromosome. The LTR retrotransposon integrases are part of a large range of DNA-active enzymes that share the DDD or DDE motif at the active site, including the V(D)J recombinases and the bacterial transposases (Keith et al. 2008). This implies a common origin; recent structural studies of the enzymes strongly support this view (Hickman et al. 2010; Montaño and Rice 2011). Early on, it was noticed that retrotransposons, retroviruses, and bacteriophage Mu all share the terminal TG. . .CA ends that are found within LTRs (Temin 1980). The formation of terminal inverted repeats (TIRs) flanking a promoter within the ancestral retrotransposon provided recognition and binding sites for the primordial integrase, allowing its propagation. Research to identify the amino acid residues of integrase that interact with the LTR (Dolan et al. 2009) should eventually allow a clear picture to emerge of the coevolution of integrases and their recognition sites.

An LTR is, in essence, a pair of TIRs flanking a promoter, terminator, and polyadenylation signal, the whole of which is then repeated twice. The short TIRs recognized by the integrase almost universally share the 5′ TG. . .CA 3′ termini that form the outer nucleotides of the TIRs. Promoters are plentiful in the genome, and terminators, polyadenylation signals, and 5- or 6-bp repeats are short enough to occur with high frequency. In between the two LTRs, one needs the PBS and PPT signals as a minimum for reverse transcription. Although it seems at first glance to be unlikely that two LTR repeat units would occur close to one another in the genome by sheer chance, the process of replication by reverse transcriptase, involving two strand jumps, homogenizes the two ends of the final double-stranded cDNA, creating the LTRs. It is not so implausible to imagine that the acquisition of a tRNA gene near a promoter and of a purine-rich tract near a terminator, together with the presence of a stretch of a few 10s of bases of similar nucleotides at either end, would have permitted reverse transcription to create two LTRs, each possessing the promoter and terminator flanking the genes.

The reverse transcriptase itself appears to be derived from an ancient family of enzymes involved in nucleic acid metabolism, in this case polymerization. This view is supported by the presence in plants, animals, fungi, protists, and bacteria of a conserved family of genes, *rvt*, which encode polymerases able to incorporate both ribonucleotides and deoxyribonucleotides (Gladyshev and Arkhipova 2011). All retrotransposon reverse transcriptases have in their catalytic center a highly conserved motif, generally YVDD, which is surrounded by several small hydrophobic amino acids, together referred to as the reverse transcriptase signature.

The eukaryotic telomerase enzyme, which adds telomeres to the ends of chromosomes through reverse transcription of an RNA template, contains a similar motif in its catalytic center (Autexier and Lue 2006; Lue et al. 2005; Lingner et al. 1997). Structure-based alignments indicate that the *rvt* enzymes most closely resemble modern LINE reverse transcriptases and belong with them in a larger family including the reverse transcriptases of LTR retrotransposons, retroviruses, pararetroviruses, telomerases, and the *PLE* order of Class I elements. Thus, Class I reverse transcriptase and telomerase are descendants of a common ancestral enzyme. The earliest retrotransposon reverse transcriptase probably then fused with an RNaseH gene. Subsequent acquisition of regulatory sequences gave rise to the structurally simplest known Class I elements, the non-LTR retrotransposons. Once a template is primed, reverse transcriptases are generally nonspecific. Hence, reverse transcription of a primordial retrotransposon could well have been carried out in *trans* by an enzyme not encoded by the TE itself.

Autonomous LTR retrotransposons appear to have arisen as a fusion of a reverse transcriptase and an integrase. Such a fusion event appears to have occurred at least twice, each leading to the formation of the two main LTR retrotransposon superfamilies, *Gypsy* and *Copia* (Fig. 5.1). The LTRs of *Gypsy* and *Copia* elements are very similar in their overall structure and function and in the presence of TG....CA ends. The similarities are unsurprising, considering both the similarity in the integrases that recognize the LTRs and the reliance of all LTRs on conserved transcriptional machinery. As argued above, LTRs may arise relatively easily over evolutionary time. Hence, if the primeval *Gypsy* and *Copia* elements evolved independently, they could have acquired LTRs independently. Alternatively, both have evolved from an ancestral LTR-containing intermediate.

5.5 Conclusions

The life cycle of retrotransposons involves stages of transcription, translation, packaging, reverse transcription, and integration. Loss of any of the functions will render a retrotransposon incapable of completing its life cycle autonomously. However, complementation in *trans* by proteins expressed by another retrotransposon can restore the ability of nonautonomous elements to transpose. Nonautonomous elements may be unable to express one or more proteins, or they may lack coding capacity entirely. It appears that all that needs to be retained are the signals required in *cis*, respectively, within the element residing in the genome, for transcription, termination, and polyadenylation, within the transcript for dimerization, packaging, and reverse transcription, and within the cDNA copy for integration. The signals are enough for transcripts of nonautonomous elements to hitch a ride in the VLPs of an autonomous retrotransposon and be carried as cDNA to elsewhere in the genome.

Because of the many ways in which full function can be lost from an autonomous retrotransposon, the nonautonomous elements probably form the majority of all TEs. Moreover, major groups of nonautonomous elements have highly conserved, but deleted internal domains where the open reading frame normally resides; these tend to be abundant. These groups have become specialized as effectively propagating nonautonomous elements. Besides clarifying how much of the genomic DNA that does not code for long ORFs may nevertheless be mobile, the *trans*-complementation model helps explain how autonomous retrotransposons may have evolved through sequential gain of function.

An important question which remains unanswered is the effect of nonautonomous retrotransposons on their autonomous partners: are they propagating at the expense of the partners providing proteins in *trans*? For example, if nonautonomous elements are freer to optimize very efficient packaging structures in the absence of constraints to maintain open reading frames, will they block replication of the autonomous partners, leading to their ultimate demise? While scenarios can be modeled, the question will need to be addressed by finding and studying the partnerships experimentally. The answer has major evolutionary consequences for the genome, affecting the relative rates of gain versus loss of retrotransposons and thereby genome size (Hawkins et al. 2009).

A related question is to what extent a nonautonomous retrotransposon group is dependent on a particular autonomous family for replication, and to what extent the nonautonomous elements are generalists and can be complemented by many or all autonomous elements. A specialist group will disappear if its autonomous partners in the genome should all become nonautonomous or inactive. A third alternative over evolutionary time is, like a surfing sailboat moving from wave to wave, to develop specificity for a new, active group as the older one declines. This is conceivable given the high mutation rates of retrotransposon replication. Despite the importance of retrotransposons to genome dynamics and gene activity (e.g., through epigenetic effects), our understanding of their biology is still in a primitive state.

Acknowledgment Research on which this review is based was carried out under a grant from the Academy of Finland, Decision 123074.

References

Arabidopsis Genome Initiative (2000) Analysis of the genome sequence of the flowering plant *Arabidopsis thaliana*. Nature 408:796–815

Autexier C, Lue NF (2006) The structure and function of telomerase reverse transcriptase. Annu Rev Biochem 75:493–517

Belyayev A, Kalendar R, Brodsky L, Nevo E, Schulman AH, Raskina O (2010) Transposable elements in a marginal plant population: temporal fluctuations provide new insights into genome evolution of wild diploid wheat. Mob DNA 1:6

Cherepanov P, Maertens GN, Hare S (2011) Structural insights into the retroviral DNA integration apparatus. Curr Opin Struct Biol 2:249–256

Deragon J, Zhang X (2006) Short interspersed elements (SINEs) in plants: origin classification and use as phylogenetic markers. Syst Biol 55:949–956

Dolan J, Chen A, Weber IT, Harrison RW, Leis J (2009) Defining the DNA substrate binding sites on HIV-1 integrase. J Mol Biol 385:568–579

Fedoroff NV (1999) The *supressor-mutator* element and the evolutionary riddle of transposons. Genes Cells 4:11–19

Fedoroff N, Wessler S, Shure M (1983) Isolation of the transposable maize controlling elements Ac and Ds. Cell 35:235–242

Fischer MG, Suttle CA (2011) A virophage at the origin of large DNA transposons. Science 332:231–234

Gladyshev EA, Arkhipova IR (2011) A widespread class of reverse transcriptase-related cellular genes. Proc Natl Acad Sci USA 108:20311–20316

Goodier JL, Kazazian HHJ (2008) Retrotransposons revisited: the restraint and rehabilitation of parasites. Cell 135:23–35

Grandbastien MA, Audeon C, Bonnivard E, Casacuberta JM, Chalhoub B, Costa AP, Le QH, Melayah D, Petit M, Poncet C, Tam SM, Van Sluys MA, Mhiri C (2005) Stress activation and genomic impact of Tnt1 retrotransposons in Solanaceae. Cytogenet Genome Res 110:229–241

Han JS, Boeke JD (2005) LINE-1 retrotransposons: modulators of quantity and quality of mammalian gene expression? Bioessays 27:775–784

Hartl DL, Lozovskaya ER, Lawrence JG (1992) Nonautonomous transposable elements in prokaryotes and eukaryotes. Genetica 86:47–53

Hawkins JS, Proulx SR, Rapp RA, Wendel JF (2009) Rapid DNA loss as a counterbalance to genome expansion through retrotransposon proliferation in plants. Proc Natl Acad Sci USA 106:17811–17816

Hickman AB, Chandler M, Dyda F (2010) Integrating prokaryotes and eukaryotes: DNA transposases in light of structure. Crit Rev Biochem Mol Biol 45:50–69

Jones RN (2005) McClintock's controlling elements: the full story. Cytogenet Genome Res 109:90–103

Kajikawa M, Okada N (2002) LINEs mobilize SINEs in the eel through a shared 3′ sequence. Cell 111:433–444

Kalendar R, Tanskanen J, Immonen S, Nevo E, Schulman AH (2000) Genome evolution of wild barley (*Hordeum spontaneum*) by *BARE-1* retrotransposon dynamics in response to sharp microclimatic divergence. Proc Natl Acad Sci USA 97:6603–6607

Kalendar R, Vicient CM, Peleg O, Anamthawat-Jonsson K, Bolshoy A, Schulman AH (2004) LARD retroelements: novel non-autonomous components of barley and related genomes. Genetics 166:1437–1450

Kalendar R, Tanskanen JA, Chang W, Antonius K, Sela H, Peleg P, Schulman AH (2008) *Cassandra* retrotransposons carry independently transcribed 5S RNA. Proc Natl Acad Sci USA 105:5833–5838

Kapitonov VV, Jurka J (2007) Helitrons on a roll: eukaryotic rolling-circle transposons. Trends Genet 23:521–529

Kashkush K, Feldman M, Levy AA (2003) Transcriptional activation of retrotransposons alters the expression of adjacent genes in wheat. Nat Genet 32:102–106

Keith JH, Schaeper CA, Fraser TS, Fraser MJ Jr (2008) Mutational analysis of highly conserved aspartate residues essential to the catalytic core of the piggyBac transposase. BMC Mol Biol 9:73

Kim A, Terzian C, Santamaria P, Pélisson A, Prud'homme N, Bucheton A (1994) Retroviruses in invertebrates: the *gypsy* retrotransposon is apparently an infectious retrovirus of *Drosophila melanogaster*. Proc Natl Acad Sci USA 91:1285–1289

Kramerov D, Vassetzky N (2005) Short retroposons in eukaryotic genomes. Int Rev Cytol 247:165–221

Kroutter EN, Belancio VP, Wagstaff BJ, Roy-Engel AM (2009) The RNA polymerase dictates ORF1 requirement and timing of LINE and SINE retrotransposition. PLoS Genet 5:e1000458

Laten HM, Havecker ER, Farmer LM, Voytas DF (2005) SIRE1, an endogenous retrovirus family from *Glycine max*, is highly homogeneous and evolutionarily young. Mol Biol Evol 20:1222–1230

Lingner J, Hughes TR, Shevchenko A, Mann M, Lundblad V, Cech TR (1997) Reverse transcriptase motifs in the catalytic subunit of telomerase. Science 276:561–567

Lu K, Heng X, Summers MF (2011) Structural determinants and mechanism of HIV-1 genome packaging. J Mol Biol 410:609–633

Lue NF, Bosoy D, Moriarty TJ, Autexier C, Altman B, Leng S (2005) Telomerase can act as a template- and RNA-independent terminal transferase. Proc Natl Acad Sci USA 102:9778–9783

McClintock B (1948) Mutable loci in maize. Year B Carnegie Inst Wash 47:155–169

McClintock B (1984) The significance of responses of the genome to challenge. Science 226:792–801

Miyazaki Y, Miyake A, Nomaguchi M, Adachi A (2011) Structural dynamics of retroviral genome and the packaging. Front Microbiol 2:264

Montaño SP, Rice PA (2011) Moving DNA around: DNA transposition and retroviral integration. Curr Opin Struct Biol 21:370–378

Ohno S (1972) So much 'junk' in our genome. Brookhaven Symp Biol 23:366–370

Paillart JC, Shehu-Xhilaga M, Marquet R, Mak J (2004) Dimerization of retroviral RNA genomes: an inseparable pair. Nat Rev Microbiol 2:461–472

Paterson AH, Bowers JE, Bruggmann R, Dubchak I, Grimwood J, Gundlach H, Haberer G, Hellsten U, Mitros T, Poliakov A, Schmutz J, Spannagl M, Tang H, Wang X, Wicker T, Bharti AK, Chapman J, Feltus FA, Gowik U, Grigoriev IV, Lyons E, Maher CA, Martis M, Narechania A, Otillar RP, Penning BW, Salamov AA, Wang Y, Zhang L, Carpita NC, Freeling M, Gingle AR, Hash CT, Keller B, Klein P, Kresovich S, McCann MC, Ming R, Peterson DG, Mehboob ur R, Ware D, Westhoff P, Mayer KFX, Messing J, Rokhsar DS (2009) The *Sorghum bicolor* genome and the diversification of grasses. Nature 457:551–556

Ramallo E, Kalendar R, Schulman AH, Martínez-Izquierdo JA (2008) *Reme1*: a *Copia* retrotransposon in melon is transcriptionally induced by UV light. Plant Mol Biol 66:137–150

Sabot F, Schulman AH (2006) Parasitism and the retrotransposon life cycle in plants: a hitchhiker's guide to the genome. Heredity 97:381–388

Sabot F, Sourdille P, Chantret N, Bernard M (2006) *Morgane*, a new LTR retrotransposon group, and its subfamilies in wheats. Genetica 128:439–447

Soleimani VD, Baum BR, Johnson DA (2006) Quantification of the retrotransposon *BARE-1* reveals the dynamic nature of the barley genome. Genome 49:389–396

Spanu PD, Abbott JC, Amselem J, Burgis TA, Soanes DM, Stüber K, Loren V, van Themaat E, Brown JK, Butcher SA, Gurr SJ, Lebrun MH, Ridout CJ, Schulze-Lefert P, Talbot NJ, Ahmadinejad N, Ametz C, Barton GR, Benjdia M, Bidzinski P, Bindschedler LV, Both M, Brewer MT, Cadle-Davidson L, Cadle-Davidson MM, Collemare J, Cramer R, Frenkel O, Godfrey D, Harriman J, Hoede C, King BC, Klages S, Kleemann J, Knoll D, Koti PS, Kreplak J, López-Ruiz FJ, Lu X, Maekawa T, Mahanil S, Micali C, Milgroom MG, Montana G, Noir S, O'Connell RJ, Oberhaensli S, Parlange F, Pedersen C, Quesneville H, Reinhardt R, Rott M, Sacristán S, Schmidt SM, Schön M, Skamnioti P, Sommer H, Stephens A, Takahara H, Thordal-Christensen H, Vigouroux M, Wessling R, Wicker T, Panstruga R (2010) Genome expansion and gene loss in powdery mildew fungi reveal tradeoffs in extreme parasitism. Science 330:1543–1546

Tanskanen JA, Sabot F, Vicient C, Schulman AH (2007) Life without GAG: The BARE-2 retrotransposon as a parasite's parasite. Gene 390:166–174

Telesnitsky A, Goff SP (1997) Reverse transcriptase and the generation of retroviral DNA in retroviruses. In: Coffin JM, Hughes SH, Varmus HE (eds) Retroviruses. Cold Spring Harbor Laboratory Press, Cold Spring Harbor, NY, pp 121–160

Temin HM (1980) Origin of retroviruses from cellular moveable genetic elements. Cell 21:599–600

Vicient CM, Kalendar R, Anamthawat-Jonsson K, Schulman AH (1999a) Structure functionality and evolution of the *BARE-1* retrotransposon of barley. Genetica 107:53–63

Vicient CM, Suoniemi A, Anamthawat-Jónsson K, Tanskanen J, Beharav A, Nevo E, Schulman AH (1999b) Retrotransposon *BARE*-1 and its role in genome evolution in the genus *Hordeum*. Plant Cell 11:1769–1784

Vicient CM, Kalendar R, Schulman AH (2001) Envelope-containing retrovirus-like elements are widespread transcribed and spliced and insertionally polymorphic in plants. Genome Res 11:2041–2049

Wessler SR (1996) Turned on by stress: plant retrotransposons. Curr Biol 6:959–961

Wicker T, Sabot F, Hua-Van A, Bennetzen JL, Capy P, Chalhoub B, Flavell A, Leroy P, Morgante M, Panaud O, Paux E, SanMiguel P, Schulman AH (2007) A unified classification system for eukaryotic transposable elements. Nat Rev Genet 8:973–982

Wicker T, Taudien S, Houben A, Keller B, Graner A, Platzer M, Stein N (2009) A whole-genome snapshot of 454 sequences exposes the composition of the barley genome and provides evidence for parallel evolution of genome size in wheat and barley. Plant J 59:712–722

Witte CP, Le QH, Bureau T, Kumar A (2001) Terminal-repeat retrotransposons in miniature (TRIM) are involved in restructuring plant genomes. Proc Natl Acad Sci USA 98:13778–13783

Chapter 6
Plant Endogenous Retroviruses? A Case of Mysterious ORFs

Howard M. Laten and Garen D. Gaston

Abstract Endogenous retroviruses have traditionally been defined as descendants of extinct retroviruses that infected and integrated into the chromosomes of host germ-line cells and were thereafter transmitted vertically as part of host genomes. Most retain at least the vestiges of genes once required for infectious horizontal transfer, namely envelope genes. In contrast, the long evolutionary histories of retrotransposons are presumed not to have included infectious ancestors. With the characterization of the Gypsy retrotransposon in *Drosophila melanogaster* as an infectious, endogenous retrovirus, these distinctions have blurred. A number of plant LTR retroelements possess coding regions whose conceptual translations produce hypothetical proteins with predicted structural elements found in viral envelope proteins, and the term endogenous retrovirus began to be applied to these elements. The question of whether any of the many plant retroelement genes now annotated as "*env*-like" generate proteins that have or had envelope functions remains unanswered. This review reevaluates the available data.

Keywords LTR retrotransposon • Endogenous retrovirus • Envelope protein • Transmembrane • Coiled coil • Sirevirus • Env-like

6.1 Beyond *gag* and *pol*: Plant Retroelements with Extra ORFs

While plant LTR retrotransposons are generally easily identified by conserved domains in the POL polyprotein [retropepsin (PROT), integrase (INT), reverse transcriptase (RT), and RNase H (RH)], and to a lesser extent by zinc knuckle RNA-binding motifs in GAG, there are a significant number of families among both

H.M. Laten (✉) • G.D. Gaston
Department of Biology and Program in Bioinformatics, Loyola University Chicago, 1032 W. Sheridan Road, Chicago, IL 60660, USA
e-mail: hlaten@luc.edu

M.-A. Grandbastien and J.M. Casacuberta (eds.), *Plant Transposable Elements*, Topics in Current Genetics 24, DOI 10.1007/978-3-642-31842-9_6,

Fig. 6.1 Structure of LTR retroelements with extra ORFs. LTRs, *black triangles*; extra 5′ ORF, *blue arrow*; *gag*, *brown arrow*; pol, *green arrow*; 3′ extra ORF, *red arrow*; *black dots*, noncoding regions. *gag* and *pol* may be fused and translated in a single reading frame, separated by a stop codon or in different reading frames. Distances between elements are variable. Not to scale

Ty3/Gypsy and Ty1/Copia superfamilies that possess additional or extra open reading frames (eORFs) (Fig. 6.1). The conceptual translations of most of these eORFs produce novel proteins with no definitive homology to proteins with known functions (Peterson-Burch et al. 2000; Wicker and Keller 2007; Grandbastien 2008; Steinbauerová et al. 2012), nor have protein products from these eORFs been isolated, let alone functionally assayed. In most cases, these regions are found between *pol* and the 3′ LTR (3′ eORFs) (Fig. 6.1), but there are several exceptions (Steinbauerová et al. 2012). Members of the Ogre lineage, best characterized in legumes, possess conserved, intact 5′ eORFs between the 5′ LTR and *gag* (Fig. 6.1) (Neumann et al. 2003; Macas and Neumann 2007; Steinbauerová et al. 2012).

There are a few instances of small numbers of elements containing fragments of recognizable host genes, the probable result of transcriptional readthrough or recombinational capture (Jin and Bennetzen 1994; Du et al. 2006; SanMiguel and Vitte 2009; Steinbauerová et al. 2012). It is doubtful these host genes played any functional role, and these elements will not be addressed here. Interestingly, Steinbauerová et al. (2012) reported partial sequence similarities in eORFs to the plant mobile domain, a member of a group of conserved zinc finger motifs found in a large superfamily of eukaryotic transcription factors and shown to be associated with MULE transposases (Babu et al. 2006). These similarities were found within 5′ eORFs or 3′ eORFs in a single clade of Ty3/Gypsy elements that included the Ogre family. Finally, the DIRS-1 retrotransposon family is characterized by a domain encoding a tyrosine recombinase at the 3′ end of *pol* (Poulter and Goodwin 2005; Wicker and Keller 2007), but no representatives have been found in plants (Piedöel et al. 2011).

The partial conservation of the conceptual translations of some 3′ eORFs in several retroelement families in species as distantly related as *Arabidopsis*, tomato, soybean, maize, and barley strongly suggests that these proteins play or have played an important role in the proliferation of these elements. What that role or roles may be is open to speculation, but for reasons that will be discussed below, many of these eORFs were described as "envelope-like" based on varying degrees of predicted secondary structure similarity of their conceptual translation products to viral envelope proteins (Laten et al. 1998; Peterson-Burch et al. 2000; Vicient et al. 2001; Wright and Voytas 2002; Boeke et al. 2005b; Holligan et al. 2006; Hafez et al. 2009; Laten and Bousios 2012). By extension, it has been suggested that these retrotransposon families are analogous to animal endogenous retroviruses (Kumar 1998; Laten et al. 1998; Peterson-Burch et al. 2000; Wright and Voytas 2002), the integrated vestiges of ancient infectious retroviruses.

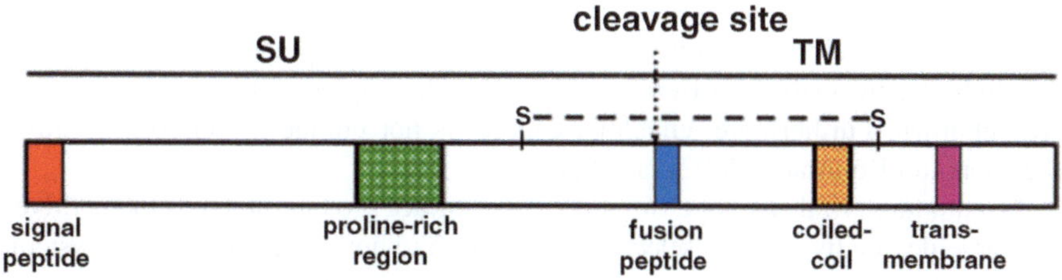

Fig. 6.2 General structural elements of viral envelope proteins. *SU* surface protein, *TM* transmembrane protein, *S–S* disulfide bridge

6.2 Viral Envelope Proteins

Viral envelope proteins are a diverse family of glycoproteins that sponsor attachment, entry into, and exit from infected cells by "enveloped" viruses like Influenza A, Hepatitis C, SARS Coronavirus, and HIV (Harrison 2008; Cosset and Lavillette 2011). These processes include peptide cleavage, receptor binding, intracellular targeting and transport, disulfide bond formation, glycosylation, membrane fusion, and oligomerization (Cosset and Lavillette 2011). In the case of many, including those of retroviruses, structural features of the envelope protein may include a signal peptide, a proline-rich region, transmembrane domains, a coiled coil, a fusion peptide, and a conserved cleavage site (Wu et al. 1998; Harrison 2008; Cosset and Lavillette 2011) (Fig. 6.2). Many viral envelope proteins, including those of retroviruses, are translated as precursors that are cleaved into a surface glycoprotein and a transmembrane glycoprotein (Hunter 1997) (see Fig. 6.2).

6.2.1 Envelope Protein Variation

While structural and functional elements are shared by diverse groups of viral envelope proteins, amino acid sequence variation is high, and it remains unclear if the three major classes of envelope proteins—based on their fusion peptides—are related by descent from a single ancestral gene (Kadlec et al. 2008; Cosset and Lavillette 2011). For mammalian retroviruses and endogenous retroviruses, even in cases where clear evolutionary relationships are inferred from phylogenetic trees based on RT alignments, the corresponding envelope sequences may have diverged to the extent that homology cannot be deduced from global sequence-based analyses (Benit et al. 2001). However, by restricting multi-sequence alignments to transmembrane subunits, homology has been inferred across retroviral genomes and those of other enveloped viruses, such as Ebola and Marburg, with Class I fusion proteins (Benit et al. 2001). Although not addressed by Benit et al. (2001), the observed localized sequence similarities could have been the result of convergent evolution or localized domain capture.

Kim et al. (2004) suggested that the homology between distantly related retroviruses is the result of envelope capture, and this hypothesis is supported by the phylogenetic analysis of Benit et al. (2001). The origins of these envelope genes are unknown. Furthermore, envelope capture is not unique to vertebrate viruses (Pearson and Rohrmann 2002) (see below).

In mammals, viral envelope variation in the surface protein subunit is likely driven to a large degree by positive selection in response to host adaptive immune systems (Caffrey 2011). While innate immune responses in vertebrates, invertebrates, and plants have been shown to contribute to the evolution of virulence/effector proteins in pathogens that attenuate these responses (Finlay and McFadden 2006; Nishimura and Dangl 2010), there is no evidence that antigenic variation is employed as a mechanism to escape innate immunity (Finlay and McFadden 2006). Nor is there any evidence that envelope variants are responsible for suppression or evasion of silencing of viral gene expression by host siRNAs in plants or animals (Li and Ding 2006; Obbard et al. 2009).

6.3 Endogenous Retroviruses

Endogenous retroviruses (ERV) are the integrated remains of extinct retroviruses that infected and reinfected host germ-line cells, inserting into germ-line chromosomes and consequently vertically inherited by generations of host descendants (Bannert and Kurth 2006; Jern and Coffin 2008; Ribet et al. 2008; Feschotte and Gilbert 2012).

6.3.1 Human and Other Vertebrate Endogenous Retroviruses

With the possible exception of the highest copy-number families (Belshaw et al. 2005), few endogenous human retroviruses appear to be capable of autonomous retrotransposition in germ-line cells (Belshaw et al. 2004), most likely because of debilitating mutations and/or epigenetic silencing (Belshaw et al. 2005; Maksakova et al. 2008). However, some murine ERVs are far more active (Maksakova et al. 2006). In most ERV families, the envelope gene sequences are riddled with nonsense mutations and deletions. It has been suggested that most, but not all, vertebrate multi-copy ERV families arose by short bursts of multiple germ-line infections, not by retrotransposition (Belshaw et al. 2004; Bannert and Kurth 2006; Jern and Coffin 2008). While there is no evidence for recent retrotransposition of human ERVs, mobilization of ERVs in other mammals has been reported (Maksakova et al. 2006, 2008; Ribet et al. 2007; Stocking and Kozak 2008; Zhang et al. 2008; Wang et al. 2010), and the expression of ERV mRNA and production of proteins in somatic tissue has been associated with some cancers (Moyes et al. 2007; Howard et al. 2008; Maksakova et al. 2008).

6.3.2 Invertebrate Endogenous Retroviruses

Env-like genes downstream of *pol* have been reported for several invertebrate LTR-retroelements. Most notably, Gypsy from *D. melanogaster* has long been recognized as an endogenous retrovirus (Kim et al. 1994; Song et al. 1994) with strong evidence that it retains infectivity (Kim et al. 1994; Song et al. 1994; Teysset et al. 1998; Pelisson et al. 2002; Misseri et al. 2004). While transfer of Gypsy elements from somatic to germ-line tissue does not require a functional *env* gene (Chalvet et al. 1999), the Gypsy envelope glycoprotein has been shown to sponsor cell–cell fusion in cell culture assays (Misseri et al. 2004). Other invertebrate retroelements that contain envelope-like coding regions include several additional Drosophilid elements (Mejlumian et al. 2002; Llorens et al. 2008, 2011), TED, a lepidopteran element from *Trichoplusia ni* (Friesen and Nissen 1990; Ozers and Friesen 1996), yoyo from the Mediterranean fruit fly, *Ceratitis capata* (Zhou and Haymer 1998), Tas from the parasitic nematode *Ascaris lumbricoides* (Felder et al. 1994), Cer7 (Bowen and McDonald 1999) from *C. elegans*, and two elements, Juno and Vesta, from bdelloid rotifers (Gladyshev et al. 2007).

The *env*-like regions of the insect elements have been shown to be homologous (Terzian et al. 2001). Many of the hypothetical ENV-like proteins contain multiple structural features common to viral envelope proteins. Based on sequence similarities, Eickbush and Malik (2002) suggested that the *env*-like genes in Tas and Cer7 were derived from a Phlebovirus and a Herpesvirus, respectively. With the exception of Gypsy, invertebrate retroelements have not been demonstrated to be infectious. Gypsy and related arthropod elements have been designated as Errantiviruses (Boeke et al. 2005a).

Several phylogenetic and functional analyses strongly suggest that the genes encoding the Errantivirus envelope-like proteins are derived from Baculoviral *env* genes (Malik et al. 2000; Rohrmann and Karplus 2001; Pearson and Rohrmann 2002, 2004, 2006; Misseri et al. 2003; Kim et al. 2004). However, any homology to vertebrate retroviral envelope proteins is only weakly supported at best (Lerat and Capy 1999; Malik et al. 2000), and the very small number of short blocks of amino acid similarity between conserved Errantivirus envelope proteins and those of vertebrate retroviruses could be fortuitous, or the result of convergent evolution or recombinational domain capture.

6.3.3 Are There Plant Endogenous Retroviruses?

Animal endogenous retroviruses have been defined as vertically transmitted, retroviral-related DNAs distinguished from LTR retrotransposons by the presence of at least vestiges of an envelope-coding region downstream of *pol* and/or a close phylogenetic relationship to extant retroviruses (Boeke and Stoye 1997; Bannert and Kurth 2006; Jern and Coffin 2008; Feschotte and Gilbert 2012). In the case of plants, infectious retroviruses have not been reported. However, integrated, vertically transmitted copies of plant pararetroviral genomes are widespread in both dicots and monocots (Staginnus and Richert-Poggeler 2006; Hohn et al. 2008). Plant pararetroviruses, like the Caulimoviruses, are DNA viruses characterized by genomes encoding GAG, PROT, RT, and RH, as well as additional essential proteins (Lazarowitz 2007). Unlike retroviruses, pararetroviruses are not enveloped, and their infectious cycles do not normally include integration into the host genome (Lazarowitz 2007). Integration appears to be extremely rare, and integrated viral sequences are generally incomplete, rearranged and mutated, and not known to be infectious or capable of autonomous retrotransposition (Staginnus and Richert-Poggeler 2006; Hohn et al. 2008).

The first suggestions that plant genomes might contain endogenous retroviruses were made based on the presence of predicted ENV-like structural features in the conceptual translations of LTR elements with 3′ eORFs of several hundreds to over 2,000 bp (Laten et al. 1998; Wright and Voytas 1998). Four families of Athila elements, members of the Ty3/Gypsy superfamily from *A. thaliana*, were initially shown to contain extended ORFs downstream of *int* with conceptual translation products containing one or more predicted transmembrane regions (Wright and Voytas 1998). These sequences were not considered to be homologous to retroviral *env* genes, but the suggestion was made that the encoded proteins might once have promoted membrane fusion (Wright and Voytas 1998).

Predicted structural similarities between viral envelope proteins and the conceptual translation of a 3′ eORF of an unrelated element, SIRE1 from *Glycine max*, were far more extensive (Laten et al. 1998). The suggestion that SIRE1, a member of the Ty1/Copia superfamily, encoded an envelope-like protein was derived from several features of the conceptual translation of the long, uninterrupted 3′ eORF in the same reading frame as *pol* but separated from *pol* by a single stop codon. The conceptual translation of this ORF produced a 70 kDa, 650-amino acid polypeptide (Laten et al. 1998). This hypothetical protein was predicted to contain transmembrane domains at positions corresponding to the signal and fusion domains of viral envelope proteins and a strongly predicted coiled coil in a region corresponding to those containing coiled coils in several viral envelope proteins, including that of HIV (Laten et al. 1998) (Fig. 6.2). While the conceptual translation contained only two N-glycosylation motifs, there were several serines and threonines in contexts known to promote O-glycosylation, a characteristic of many viral envelope proteins (Pinter and Honnen 1988; Wilson et al. 1991). In addition, there was an extended proline-rich region from amino acid 60 to 128. The overall amino acid composition of this region was remarkably similar to those found in the neutralization domains of some mammalian retroviruses (Laten et al. 1998).

Retroviral envelope proteins are known to be expressed from spliced transcripts (Rabson and Graves 1997). However, there are no recognizable splice acceptor sites in SIRE1 or in related elements that would fuse this ORF with an upstream start codon (Peterson-Burch and Voytas 2002). Nor are there AUG codons downstream of the *pol* stop codon that might support translational initiation at an internal ribosomal entry site (Peterson-Burch and Voytas 2002). However, Havecker and Voytas (2003) showed that the SIRE1 *pol* stop codon was embedded in a hexanucleotide motif that had previously been shown to sponsor developmentally regulated stop codon suppression in tobacco mosaic virus and in yeast. They demonstrated that the SIRE1 sequence supported low levels of stop codon suppression (5%) in in vivo readthrough assays and that suppression was lost with single base-pair changes in the sequence (Havecker and Voytas 2003).

Once the potential characteristics of these unusual elements were recognized, analyses of previously reported plant retrotransposons with long uncharacterized regions between *pol* and the 3′ LTR revealed that conceptual translation of these interrupted 3′ eORFs could generate hypothetical proteins with highly significant sequence similarity to those described above (Laten 1999; Peterson-Burch et al. 2000) (see Table 6.1). Three of these hypothetical proteins were aligned to highlight their similarities (Fig. 6.3). The extent and degree of sequence identity was variable but in the case of SIRE1 and Endovirl encompassed most of the sequence. The densities of sequence matches were far greater in the second half of the alignment. The distances between the *pol* stop codon and the beginning of the *env*-like coding region were also highly variable, ranging from 0 to over 1,000 bp (Peterson-Burch and Voytas 2002; Laten et al. 2003; Havecker et al. 2005; Weber et al. 2010).

The phylogenetic relationships among groups of retroelements with and without eORFs are illustrated in Fig. 6.4. A fusion of the network analyses of Llorens et al. (2009) and the more classical approach illustrated in Eickbush and Jamburuthugoda (2008), this consensus tree illustrates the widespread acquisition of primarily 3′ eORFs with both known, as in the case of vertebrate retroviruses and Gypsy, and unknown function.

6.3.3.1 Ty1/Copia Sireviruses

The SIRE1 element family in soybean, with as many as 1,350 copies per genome (Laten and Morris 1993; Du et al. 2010b; Bousios et al. 2012b), is highly conserved and recently amplified (Laten et al. 2003; Du et al. 2010b; Bousios et al. 2012b). Nearly all copies have inserted into their present genomic positions in the last 750,000 years, with as many as 10% having done so in the last 30,000 (Du et al. 2010b; Bousios et al. 2012b). SIRE1 has been designated as the Type Species for the Genus Sirevirus (Boeke et al. 2005b), and based on reverse transcriptase sequences constitutes a monophyletic group within the Ty1/Copia superfamily (Boeke et al. 2005b; Du et al. 2010b; Bousios et al. 2012a). This group has been alternatively designated as the Maximus lineage (Du et al. 2010b) or the Sirevirus lineage (Bousios et al. 2012a). Not all members of the lineage contain 3′ eORFs that encode hypothetical proteins with ENV-like features (Havecker et al. 2005; Pearce 2007; Bousios et al. 2010, 2012a, b), but those that do have been found in the genomes of most eudicots and monocots for which extensive sequence data are available (see Table 6.1). Many of the hypothetical proteins are truncated or heavily mutated and have not been annotated. The initial recognition and discovery of some of these 3′ eORFs required tBLASTn searches of nucleotide databases using previously reported ENV-like proteins as queries (Laten 1999; Havecker et al. 2005; Wicker and Keller 2007; Du et al. 2010b; Laten and Bousios 2012).

Table 6.1 *Env*-containing plant retroelements. Only elements with full-length or disrupted ORFs with extended 3′ eORFs that give statistically significant hits to other ENV-like sequences are listed

Family	Species	References
Ty1/Copia		
SIRE1	*Glycine max*	Laten et al. (1998, 2003)
Endovir1	*Arabidopsis thaliana*	Kapitonov and Jurka (1999), Laten (1999), Peterson-Burch et al. (2000)
ToRTL	*Solanum lycopersicum*	Daraselia et al. (1996), Laten (1999)
Hopie	*Zea mays*	Nagaki et al. (2003), Havecker et al. (2005)
Ji9009/Jienv	*Zea mays*	SanMiguel et al. (1996), Baucom et al. (2009), Bousios et al. (2012a)
Giepum	*Zea mays*	Bousios et al. (2012a)
Tnd-1	*Nicotiana debneyi*	Kenward et al. (1999), Havecker et al. (2005)
Osr9, Osr10	*Oryza sativa*	McCarthy et al. (2002), Havecker et al. (2005)
SIRE-like	*Medicago truncatula*	Vitte and Bennetzen (2006), Laten and Bousios (2012)
Lotus1,2, Lj1-3	*Lotus japonicus*	Havecker et al. (2005), Holligan et al. (2006), Du et al. (2010b)
Maximus	*Triticum aestivum*	Wicker and Keller (2007)
Inga	*Triticum aestivum*	Wicker and Keller (2007)
Usier	*Triticum aestivum*	Wicker and Keller (2007)
Barbara_B	*Triticum aestivum*	Wicker and Keller (2007)
SIRE-like	*Vitis vinifera*	Wicker and Keller (2007), Bousios et al. (2010, 2012b)
SIRE-like	*Musa acuminata*	Hribova et al. (2010)
MguSIRV	*Mimulus guttatus*	Laten and Bousios (2012)
Cotzilla1	*Beta vulgaris*	Weber et al. (2010)
BraSIRV	*Brassica rapa*	Laten and Bousios (2012), Wang et al. (2011)
SIRE-like	*Brassica oleracea*	Laten, unpublished
PsaSIRV	*Pisum sativum*	Macas et al. (2007), Laten and Bousios (2012)
AF464952[a]	*Vicia faba*	Chen, Chen, Wang, and Wang, unpublished
SIRE-like	*Brachypodium distachyon*	Bousios et al. (2012b)
SIRE-like	*Theobroma cocoa*	Bousios et al. (2012b)
SIRE-like	*Trifolium repens*	Laten, unpublished
SIRE-like	*Trifolium pratense*	Laten, unpublished
SIRE-like	*Antirrhinum hispanicum*	Laten, unpublished
Pyrubu	*Sorghum bicolor*	Ramakrishna et al. (2002), Havecker et al. (2005)
SIRE-like	*Cucumis melo*	Gonzalez et al. (2010)
Pt copia-like B	*Poncirus trifoliata*	Yang et al. (2003), Havecker et al. (2005)
Ty3/Gypsy		
Athila1-6,9	*Arabidopsis thaliana*	Wright and Voytas (1998), Wright and Voytas (2002)
Calypso	*Glycine max*	Wright and Voytas (2002)
Bagy-2	*Hordeum vulgare*	Vicient et al. (2001)
PIGY	*Pisum sativum*	Neumann et al. (2005)
MEGY, Mtr60,64	*Medicago truncatula*	Neumann et al. (2005), Du et al. (2010b)

(continued)

Table 6.1 (continued)

Family	Species	References
Lj18	*Lotus japonicus*	Du et al. (2010b)
Rigy-2	*Oryza sativa*	Vicient et al. (2001)
Cyclops-2	*Pisum sativum*	Chavanne et al. (1998)
GmOgre/ SNARE	*Glycine max*	Laten et al. (2009), Du et al. (2010a)
Unnamed	*Gossypium* sp.	Hafez et al. (2009)
FIDEL	*Arachis* sp.	Nielen et al. (2010)

[a]Based on a 177 nt *env*-like cDNA that is 74 % identical at the DNA level and 71 % identical at the amino acid level to genomic SIRE1

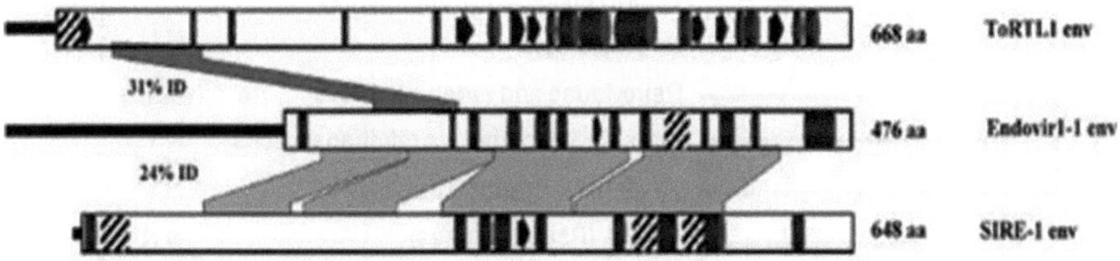

Fig. 6.3 Alignment of ENV-like regions from ToRTL1 from *S. lycopersicum*, Endovir1-1 from *A. thaliana*, and SIRE1 from *G. max*. The *env*-like ORFs are represented by *white bars* and are drawn to scale. *Black lines* depict noncoding sequences between *pol* and the start of the *env*-like ORF. Regions of amino acid similarity between elements are connected by *shading*. Percentages on the *left* represent the total amino acid similarity over the shaded regions. The numbers of amino acids in the *env*-like ORFs are given for each element. Predicted features are denoted as follows: α-helices, *dark gray boxes*; β-sheets, *arrows*; transmembrane domains, *slanted line boxes*. Adapted from Peterson-Burch and Voytas (2002) with permission

Recognizable conservation of the ENV-like peptide sequences extends to a broad range of eudicot taxa and includes members in the order Fabales, Vitales, Brassicales, Solanales, Lamiales, and Caryophyllales. Most of the extended sequence identities and similarities shared by these hypothetical proteins would correspond to the carboxyl half of a retroviral protein encompassing the transmembrane protein and part of the surface protein (see Fig. 6.2). However, not all of these hypothetical proteins contain predicted transmembrane domains (Havecker et al. 2005) (Fig. 6.5), and, not unexpectedly, multi-sequence alignments generated few positions with consensus residues (Havecker et al. 2005). Weaker sequence similarity corresponding to the first 300 amino acids of the SIRE1 ENV-like hypothetical protein has only been detected in short regions of the related elements in *L. japonicus* (Laten, unpublished). Additional members of the same lineage, based on their RT sequences, possess several hundred bp between the *pol* stop codon and the 3′LTR, including PREM-2, Opie-2, and most members of the Ji lineage from maize, and Osr7 and Osr8 from rice. These elements have no discernible 3′ eORFs, although the maize Jienv clade does (Bousios et al. 2012a).

Even among the elements for which *env*-like ORFs have been deduced, few Sireviruses with intact *env*-like regions with greater than 500 contiguous codons

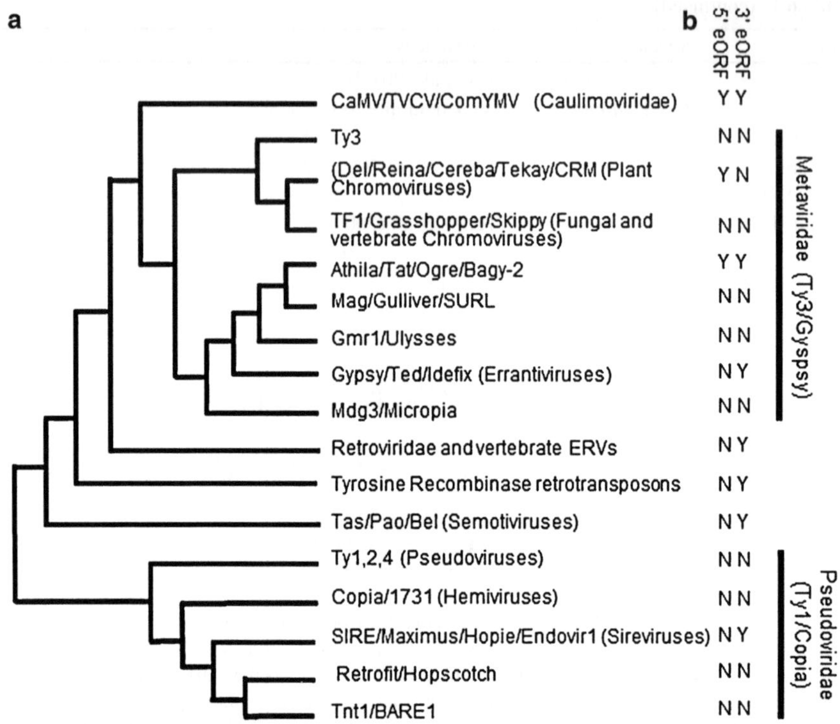

Fig. 6.4 (**a**) Simplified, unrooted phylogeny of LTR-related retroelements. Modeled with modification after Eickbush and Jamburuthugoda (2008) and Llorens et al. (2009). Branch lengths do not represent distances. (**b**) Presence of eORFs in one or more members within terminal clades representing groups of related subfamilies indicated with Y. Absence of eORF in all subfamilies within a terminal clade indicated with N. Metaviridae family defined by Boeke et al. (2005a). Pseudoviridae family defined by Boeke et al. (2005b). Data sources for B: Llorens et al. 2011; Steinbauerová et al. 2012; King et al. 2012 (http://ictvonline.org/index.asp)

have been found. The recognition of others are often derived from consensus sequences generated from multi-sequence alignments (Wicker and Keller 2007; Laten et al. 2009). Among those that possess long intact 3′ eORFs, the *G. max*, *L. japonicus*, *B. vulgaris*, and *M. guttatus* Sireviruses encode hypothetical ENV proteins of 648–680, 630–949, 606, and 780 amino acids, respectively, for SIRE1 (Laten et al. 2003), Lotus2 (Holligan et al. 2006), Cotzilla1 (Weber et al. 2010), and MguSIRV (Laten and Bousios 2012).

Neighbor joining trees of Sirevirus RT domains showed that those elements containing intact or vestiges of "ENV-like" domains appear to be monophyletic (Bousios et al. 2010, 2012a; Du et al. 2010b). Members of the Maximus lineage (Wicker and Keller 2007) all fall within the Sirevirus clade based on their RT domains (Fig. 6.6) (Bousios et al. 2010; Du et al. 2010b) and most are characterized by extended GAG regions with multiple RNA binding motifs and predicted coiled

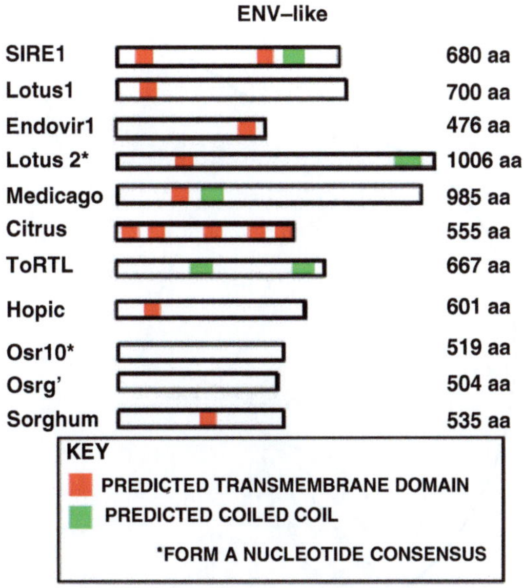

Fig. 6.5 Predicted structural elements found in translated 3′ ORFs of selected members of the Sirevirus family. Adapted from Havecker et al. (2005) with permission

coils (Peterson-Burch and Voytas 2002; Havecker et al. 2005). Bousios et al. (2010) have also described a number of highly conserved features in Sirevirus noncoding regions in the LTR and immediately upstream of the 3′ LTR.

The Sirevirus group in *L. japonicus* is the predominant Ty1/Copia lineage in *L. japonicus*, constituting 40% of these retroelements (Holligan et al. 2006). This group is also among the most recently amplified in the *L. japonicus* genome, with many members possessing identical LTR sequences (Holligan et al. 2006). As in the case of SIRE1, most of the full-length elements in this lineage contain intact 3′ eORFs ranging in length from 630 to 949 codons. The conceptual translation products in two of three sub-lineages contained predicted transmembrane domains and the product of one sub-lineage also contained a predicted coiled coil (Holligan et al. 2006). However, Holligan et al. (2006) reported that significant similarities among the ENV-like sequences were restricted to the individual sub-lineages.

SIRE is also the predominant retroelement in the Ty1/Copia lineage in *G. max* (Du et al. 2010b), and the Maximus lineage is the predominant retroelement group in banana, constituting 13% of that genome (Hribova et al. 2010). The Osr8 lineage in the Sirevirus clade (Fig. 6.6) is also the most abundant Ty1/Copia lineage in the rice genome (McCarthy et al. 2002).

In the maize genome, retroelement families identified as members of the Sirevirus lineage with ENV-like domains, Hopie, Giepum, and Jienv, and those without, Opie and Ji, are represented by >10,600 intact and approximately 28,000 degenerate copies (Bousios et al. 2012a). This constitutes as much as 90% of the total population of Ty1/Copia elements in maize. Many of these insertions occurred within the last 600,000 years (Bousios et al. 2012a).

Cotzilla1 from *B. vulgaris* is another recently reported member of the Sirevirus genus (Weber et al. 2010). Conceptual translation of its *env*-like gene generates a

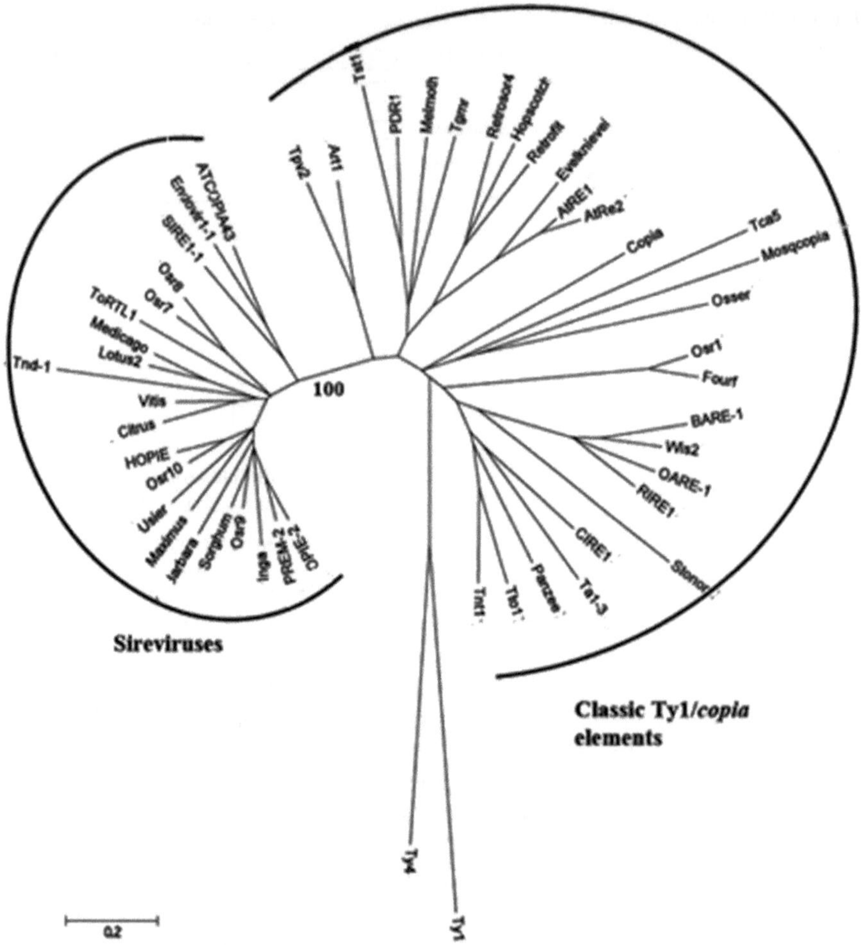

Fig. 6.6 Neighbor joining phylogenetic tree based on shared RT/RH domains highlighting the Sirevirus clade. From Bousios et al. (2010)

proline-rich region and a predicted coiled coil near the carboxyl terminal but no predicted transmembrane domains (Weber et al. 2010). The 606-codon *env*-like ORF begins 561 bp downstream from the end of *pol*. With an estimated copy number of 2,100 and members as young or younger than 290,000 years, Cotzilla may be the youngest and most abundant retroelement family in the sugar beet genome (Weber et al. 2010).

The lineages containing *G. max* and *L. japonicus* are estimated to have separated from each other over 50 million years ago (Lavin et al. 2005). In addition to the genus *Lotus*, the latter lineage contains the genera *Medicago*, *Pisum*, and *Trifolium*. While the species in these genera contain Sirevirus-like sequences with at least fragments of homologous *env*-like ORFs, fully intact *env*-like ORFs have not been reported. The relative youth of the apparently functional copies of the Sireviruses in *G. max* and *L. japonicus* suggests that significant amplification of one or a few ancestral copies with preexisting intact *env*-like ORFs occurred over the last few hundreds of thousands of years, with integration of some copies of diverged sub-lineages within the last tens of thousands years (Laten et al. 2003; Holligan et al. 2006; Du et al. 2010b). The presence of intact or nearly intact retroelement 3′ eORFs that have retained and/or acquired shared predicted structural elements over such a broad range of taxa argues strongly for function. However, expression of these elements has not been unequivocally demonstrated.

In the case of SIRE1, transcripts were not detected in northern blots, but *gag*, *rt* and *env* transcripts were detected by RT-PCR of leaf and/or root tissue (Lin 2001). However, amplification of RNAs derived from high copy-number elements does not signify functional expression because of the strong possibility of cryptic transcriptional initiation or readthrough sponsored by adjacent promoters. The 30 EST sequences containing SIRE1 fragments in the Genbank database as of May 2011 are equally distributed among sense and antisense transcripts (Gaston 2011).

The SIRE1 *env*-like ORF has been expressed from fusion constructs in *S. cerevisiae* (Gouvas and Laten, unpublished) and in *E. coli* (Gaston 2011). In the case of the former, yeast two-hybrid screens suggested that the protein self-associates and forms protein–protein interactions with at least two other soybean proteins with transmembrane domains (Gouvas and Laten, unpublished). In preliminary experiments, polyclonal antibodies raised against a sub-region expressed in *E. coli* bound to a 65-kDa protein isolated from soybean callus tissue (Gaston 2011). The protein has not been identified, but is only slightly smaller than the 70 kDa predicted for the SIRE1-4 ENV.

6.3.3.2 Plant Ty3/Gypsy "Endogenous Retroviruses"

The number of plant Ty3/Gypsy elements characterized as encoding ENV-like proteins is presently fewer than that in the Sirevirus lineage but just as widely distributed among taxa (Grandbastien 2008). As in the case of the Sireviruses, there is considerable variation in the amino acid sequences of the conceptually translated ORFs and in the possession of ENV-like secondary structures in elements from *Arabidopsis* (Wright and Voytas 1998, 2002), soybean (Wright and Voytas 2002; Du et al. 2010b), pea (Neumann et al. 2005), and barley (Vicient et al. 2001). These include transmembrane domains, coiled coils, cleavage sites, and N-glycosylation motifs. Many other elements within the same lineages possess vestiges of these regions that can be shown to be related through tBLASTn searches (e.g., Neumann et al. 2005). With the exception of one family (see below), all fall within the Athila clade based on their RT sequences.

The Athila family itself was the first among plant elements in the Ty3/Gypsy superfamily to be labeled as possible endogenous retroviruses based on the presence of 3′ eORFs whose conceptual translation produced hypothetical proteins with strongly predicted, transmembrane domains (Wright and Voytas 1998, 2002).

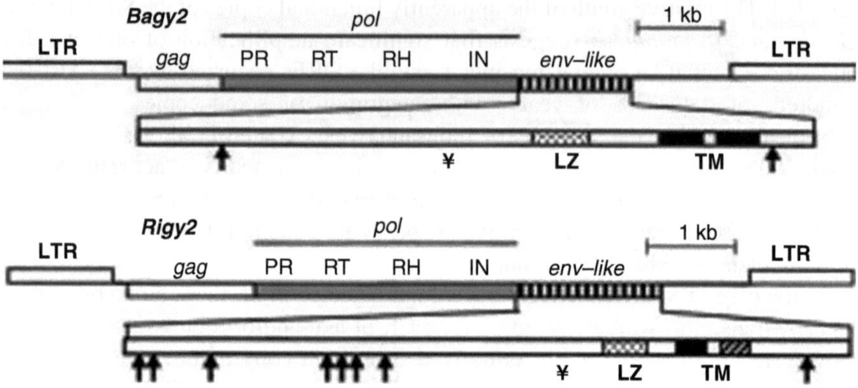

Fig. 6.7 Features of the *Bagy-2* and *Rigy-2* retrovirus-like retrotransposons and their predicted ENV-like attributes. Putative N-glycosylation sites (↑), proteinase cleavage site (¥), leucine zipper (LZ), and transmembrane domains (TM). From Vicient et al. (2001) with permission

Although highly degenerate, consensus elements were constructed for seven subfamilies, and all contained ENV-like hypothetical proteins with at least one predicted transmembrane domain (Wright and Voytas 2002). In addition, splice acceptor sites were predicted near the beginning of the 3′ eORF (Wright and Voytas 1998, 2002). The Athila4 consensus generated a 619-amino acid ENV-like hypothetical protein (Wright and Voytas 2002).

With the recognition that 3′ eORFs in Ty3/Gypsy elements might encode ENV-like proteins based on shared predicted secondary structural elements, related elements were sought and found in a broad range of taxa beginning with two related element families: Cyclops-2 in *P. sativum* (Chavanne et al. 1998; Peterson-Burch et al. 2000) and the Calypso family in *G. max* (Peterson-Burch et al. 2000; Wright and Voytas 2002) (see Table 6.1). The *env*-like ORF in the former was 423 codons and 420 in the latter. As in the case of Athila, Calypso had a strongly predicted splice acceptor site near the 5′ end of the *env*-like ORF (Wright and Voytas 2002). Analyses of the transmembrane domains suggested targeting to the plasma membrane in the case of Calypso2 and the endoplasmic reticulum in the case of Athila4 (Wright and Voytas 2002).

While individual members of the Athila and Calypso families are degenerate and appear to be nonfunctional, a related family in barley, Bagy-2, contains copies with intact ORFs for *gag* and *pol*, and an intact *env*-like ORF whose conceptual translation produces a 47-kDa protein (Fig. 6.7) (Vicient et al. 2001). Furthermore, RT-PCR amplification from several tissues with 3′ eORF-specific primers suggested that Bagy-2 is transcribed and that transcripts are spliced (Vicient et al. 2001). In addition, insertional polymorphisms among a number of related barley cultivars suggested that Bagy-2 copies have recently transposed (Vicient et al. 2001). A consensus sequence for a closely related element with an ENV-like hypothetical protein in rice, Rigy-2, was generated from an alignment of four copies interrupted by other nested elements (Fig. 6.6). The 3′ eORFs in the Rigy-2 consensus sequence contained both nonsense and frameshift mutations (Vicient et al. 2001). Related elements have also been reported in cultivated allotetraploid cotton and their diploid progenitors, and the hypothetical ENV-like proteins are strongly predicted to possess transmembrane domains (Hafez et al. 2009).

TBLASTn searches using the Bagy-2 ENV hypothetical protein retrieved statistically significant hits ($e < 10^{-8}$) to sequences in several legume species (*M. truncatula, L. japonicus, G. max, V. radiata and V. unguiculata, T. pratense, A. duranensis, C. cajan, P. vulgaris,* and *T. labialis*), and in carrot (*D. carota*), monkey flower (*M. guttatus* and *M. lewisii*), tobacco (*N. tabacum*), and ginseng (*P. ginseng*) (Laten, unpublished).

The PIGY family from *P. sativum* also contains members with 3′ eORFs whose conceptual translations produce hypothetical proteins with predicted transmembrane domains. These showed significant amino acid similarity to the Athila ENV-like hypothetical proteins (Neumann et al. 2005). A related but highly disrupted family, MEGY, was also found in *M. truncatula* (Neumann et al. 2005).

Another related element family, FIDEL, has recently been characterized from peanut (Nielen et al. 2010). The 3′ LTR of FIDEL is separated from the end of *pol* by 2.1 kb, but no members of this family contained an extended ORF in this region (Nielen et al. 2010). However, conceptual translations of this region in a FIDEL consensus sequence generated multiple, strongly predicted transmembrane domains (Laten, unpublished).

As in the case of the Sireviruses, most of the 3′ eORFs from these elements—all members of the Athila clade (Llorens et al. 2011)—are interrupted by multiple stop codons and/or frameshifts, and recognition of amino acid sequence conservation across families is often difficult. Nonetheless, these regions appear to have been under some degree of negative selection during their evolutionary history (Vicient et al. 2001; Wright and Voytas 2002; Neumann et al. 2005).

Families in the Tat clade, which include Grande1, Tat4, RIRE2, Ogre, RetroSort, and Cinful-1 (Llorens et al. 2011), also contained regions between the end of *pol* and the 3′-LTR but none with detectable vestiges of ORFs. However, there is a family of soybean elements within the Ogre lineage that, despite its close evolutionary relationship to other legume Ogre families that have no detectable *env*-like coding regions (Neumann et al. 2003; Macas and Neumann 2007), possesses an *env*-like 3′ eORF. GmOgre/SNARE is a family from *G. max* that shares the unusual features of Ogre lineage members—a conserved, intact 5′ eORF upstream of *gag*, a conserved intron in *pol*, and a minisatellite repeat region adjacent to the 3′-LTR (Laten et al. 2009; Du et al. 2010a). It is the most abundant transposon family in the soybean genome (Du et al. 2010b). But unlike all other members of the Ogre lineage, a GmOgre/SNARE consensus sequence from the end of *pol* to the minisatellite repeats contains an intact, 425-codon ORF whose conceptual translation generates a hypothetical protein with patches of significant similarity to the ENV-like hypothetical proteins from Cyclops-2 and Endovir1 (Laten et al. 2009). tBLASTn searches identified homologous coding regions in *M. truncatula* and *L. japonicus* in disrupted ORFs (Laten et al. 2009). What makes the GmOgre/SNARE ENV protein especially intriguing is the fact that Cyclops-2 is a member of the Ty3/Gypsy superfamily and Endovir1 is a member of the Ty1/Copia superfamily. This suggests that the ENV-like protein in GmOgre/SNARE may be a chimera. Du et al. (2010a) suggested that the GmOgre/SNARE *env*-like region represents a relatively recent capture event, but it also may reflect the maintenance of selective pressure in the *G. max* lineage and the relaxation of this pressure in the other lineages.

6.4 Origin of Plant *env*-Like Genes

Because of highly disrupted ORFs and the great diversity of conceptually translated *env*-like sequences, even from intact ORFs, homology that extends beyond closely related families, let alone to functionally characterized envelope proteins, is difficult to infer. Nor have these sequences been shown unequivocally to be homologous to any other characterized genes in plant or viral genomes. Nonetheless, it has been proposed and widely presumed that *env*-like coding regions were independently acquired or captured (from an unknown source or sources) by ancestral Ty1/Copia and/or Ty3/Gypsy retrotransposons (Peterson-Burch et al. 2000; Du et al. 2010b). The putative chimeric *env*-like region or GmOgre/SNARE might represent a more recent fusion event (Laten et al. 2009; Du et al. 2010b). A less likely but not inconceivable scenario is the possibility that some and perhaps many retrotransposons are actually the descendants of ancient enveloped retroviruses (Eickbush and Jamburuthugoda 2008) and that genomes, including those of plants (Yano et al. 2005), have recorded the history of the demise of *env* genes.

Based on a multi-sequence alignment of an unprecedentedly broad range of ENV sequences, Du et al. (2010b) created a neighbor joining tree linking sequences from plant Ty1/Copia and Ty3/Gypsy retroelements rooted to the *Drosophila* 17.6 ENV protein (Fig. 6.8). Conservation of ENV sequences between the superfamilies in the alignment is limited to a small number of identical residues and a larger number that are similar. But these similarities could also reflect convergent evolution and not evolutionary homology. Nonetheless, assuming homology, the Ty1/Copia sequences appeared to be monophyletic but the Ty3/Gypsy sequences were not. Instead, one clade of ENV sequences from Ty3/Gypsy elements in soybean, Lotus, and Medicago was the sister group to a subset of ENV sequences associated with elements belonging to the Ty1/Copia superfamily. The neighbor joining trees of the corresponding RT sequences did not generate this tree topology and conformed to the expected segregation of all members of the two superfamilies into two sister clades (Du et al. 2010b). The authors inferred that the Ty1/Copia *env*-like gene was acquired from an ancestral member of its sister Ty3/Gypsy clade, long after the capture of the *env*-like sequence by an ancestral Ty3/Gypsy retrotransposon near the crown of the tree (Fig. 6.8). However, this conclusion was based, in part, on the questionable rooting of the tree to the ENV sequence of a Drosophila element. Removal of the root generates an unrooted tree whose topology leaves open the question of the origin of the ENV sequences.

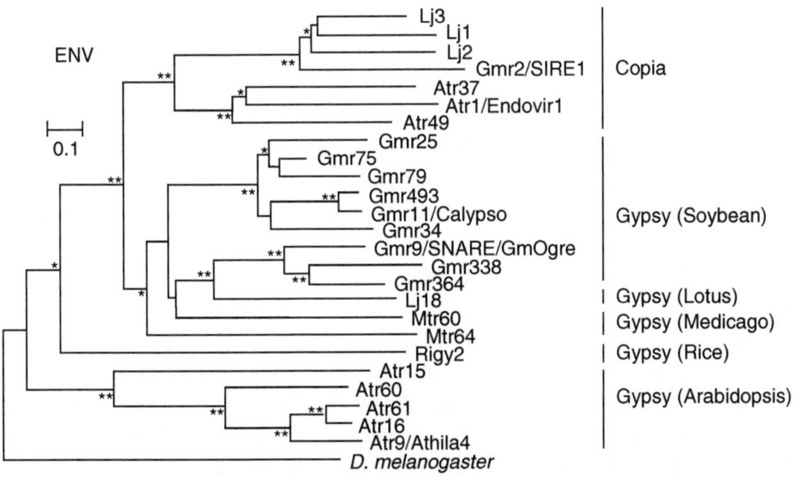

Fig. 6.8 Neighbor joining tree generated from plant retroelement ENV-like sequences. *Double asterisk* represent nodes with 86–100 % bootstrap support; *asterisk* represent nodes with 64–75 % bootstrap support. Rooted to the putative Env protein from the Gypsy-like element, 17.6, in *D. melanogaster*. From Du et al. (2010b)

6.5 Function of Plant ENV Hypothetical Proteins

There can be little dispute that large numbers of plant retroelement families have possessed genes encoding transmembrane proteins sometime during their evolutionary history, and that in a few cases what have been called *env*-like genes still encode what appear to be potentially functional proteins. However, it seems unlikely that the expression of an *env*-like ORF was essential to the proliferation of most families in the Athila and Tat clades, although traces of their widespread distribution suggests an important function, even if that function was transient. The presence of highly conserved, intact *env*-like ORFs in the hundreds of copies of Sireviruses in *G. max* and *L. japonicus* could be due to strong selection or to their recent explosive amplification. One can only speculate whether those *env*-like genes that appear to have retained function are the products of continuing, lineage-specific, purifying selection, or resurrected Phoenixes that have emerged from the ashes of degenerate copies by a variety of mutational processes.

The possible function of plant retroelement ENV-like proteins has been the subject of much speculation in the nearly total absence of experimental data (Kumar 1998; Laten et al. 1998; Wright and Voytas 1998, 2002; Peterson-Burch et al. 2000; Vicient et al. 2001; Grandbastien 2008). Based on predicted secondary structural elements, and the suggested parallels to endogenous retroviruses in mammals and invertebrates, membrane fusion has been the most promoted candidate.

Membrane fusion might be an unlikely choice, however, since cell walls would preclude this mechanism as an efficient mode of transmission and systemic infection in plants. Most plant viruses are transmitted by insect vectors in which the viruses do not propagate in their insect hosts (Lazarowitz 2007). But in the case of a few, the viruses also infect the cells of their hosts (propagative viruses) and could just as well be considered animal viruses (Lazarowitz 2007). This latter group includes members of two families of enveloped viruses: *Rhabdoviridae* and *Bunyaviridae*. The former includes Sonchus Yellow Net Virus (SYNV) that generates a virion composed of a lipid envelope embedded with virally encoded glycoproteins, while the latter includes tospoviruses like Tomato Spotted Wilt Virus (TSWV) with a genome that encodes two envelope glycoproteins (Lazarowitz 2007; Whitfield et al. 2005). In their plant hosts, intracellular SYNV and TSWV particles appear to associate with the nuclear and ER membranes, respectively (Lazarowitz 2007). In the case of TSWV single-enveloped particles are formed and transferred to feeding thrips (Kikkert et al. 1999). In thrip hosts, TSWV virions are associated with the plasma membrane and are released from infected cells by fusion with the cell membrane (Whitfield et al. 2005). However, there are no reports of detected homology between any plant retroelement ENV-like hypothetical protein and those of plant enveloped viruses.

The maintenance of envelope-encoding sequences in these viruses appears to be directly related to infectivity in their animal hosts, not in their plant hosts. When maintained solely by serial mechanical inoculations from one infected plant to another, non-enveloped mutant isolates accumulate (Goldbach and Peters 1996). These isolates are fully capable of mounting a systemic infection in plants after mechanical transfer (Goldbach and Peters 1996). However, non-enveloped isolates with mutations in the glycoprotein genes have been shown to be incapable of reinfecting the thrip host (Nagata et al. 2000). These observations provide an attractive, albeit highly speculative, model for the existence of endogenous retrovirus lineages in plants with nonfunctional and functional *env*-like genes. Confirming this model would require at a minimum the discovery of related elements in invertebrate vectors and demonstrating that virions from plants could fuse with the plasma membranes of the invertebrate host. Attempts to detect SIRE1 using PCR amplification in several known vectors including several species of thrips and aphids were unsuccessful (Laten, unpublished). Nor have tBLASTn or BLASTn searches of the Genbank database retrieved any animal DNA or mRNA with significant similarity to plant *env*-like genes. (Laten, unpublished).

6.6 Concluding Remarks

While much is now known about the structure and evolutionary relationships of the large collection of plant retroelements in both the Ty1/Copia and Ty3/Gypsy superfamilies that possess a "mysterious" 3′ eORF downstream of *pol*, hard evidence for the function(s) of the encoded protein(s) remains elusive. Regardless of whether or not transcripts, spliced or otherwise, represent functional expression, no reports of protein products have been published, let alone the results of functional assays. Potentially functional ENV-like proteins need to be isolated, either from plant tissue or from cloned constructs. Assays need to be developed and optimized for the evaluation of not only putative functions, e.g., membrane fusion, but also for alternative functions. Viral envelope proteins are just one of the many classes of proteins characterized by transmembrane and/or coiled coil domains, although the model set by the structure and evolution of animal endogenous retroviruses has greatly influenced the annotations of these elements. Continuing to annotate as "*env*-like" 3′ eORFs whose conceptual translations produce hypothetical proteins with transmembrane domains seems ill-advised at the present time, and the question of the existence of plant retroviruses, endogenous or infectious, remains unanswered. Function notwithstanding, the *env*-like genes in plant genomes are arguably the most abundant protein coding regions in the genomes of higher plants for which no function has been determined.

References

Babu MM, Iyer LM, Balaji S, Aravind L (2006) The natural history of the WRKY-GCM1 zinc fingers and the relationship between transcription factors and transposons. Nucleic Acids Res 34:6505–6520

Bannert N, Kurth R (2006) The evolutionary dynamics of human endogenous retroviral families. Annu Rev Genom Hum Genet 7:149–173

Baucom RS, Estill JC, Chaparro C, Upshaw N, Jogi A, Deragon JM, Westerman RP, SanMiguel PJ, Bennetzen JL (2009) Exceptional diversity, non-random distribution, and rapid evolution of retroelements in the B73 maize genome. PLoS Genet 5:e1000732

Belshaw R, Pereira V, Katzourakis A, Talbot G, Paces J, Burt A, Tristem M (2004) Long-term reinfection of the human genome by endogenous retroviruses. Proc Natl Acad Sci USA 101:4894–4899

Belshaw R, Katzourakis A, Paces J, Burt A, Tristem M (2005) High copy number in human endogenous retrovirus families is associated with copying mechanisms in addition to reinfection. Mol Biol Evol 22:814–817

Benit L, Dessen P, Heidmann T (2001) Identification, phylogeny, and evolution of retroviral elements based on their envelope genes. J Virol 75:11709–11719

Boeke JD, Stoye JP (1997) Retrotransposons, endogenous retroviruses, and the evolution of retroelements. In: Coffin JM, Hughes SH, Varmus HE (eds) Retroviruses. Cold Spring Harbor Laboratory Press, Plainview, NY, pp 343–435

Boeke JD, Eickbush TH, Sandmeyer SB, Voytas DF (2005a) Family Metaviridae. In: Fauquet CM, Mayo MA, Maniloff J, Desselberger U, Ball LA (eds) Virus taxonomy: VIIIth report of the International Committee on Taxonomy of Viruses. Elsevier Academic, San Diego, CA, pp 409–420

Boeke JD, Eickbush TH, Sandmeyer SB, Voytas DF (2005b) Family Pseudoviridae. In: Fauquet CM, Mayo MA, Maniloff J, Desselberger U, Ball LA (eds) Virus taxonomy: VIIIth report of the International Committee on Taxonomy of Viruses. Elsevier Academic, San Diego, CA, pp 397–407

Bousios A, Darzentas N, Tsaftaris A, Pearce SR (2010) Highly conserved motifs in non-coding regions of Sirevirus retrotransposons: the key for their pattern of distribution within and across plants? BMC Genomics 11:89

Bousios A, Kourmpetis YAI, Pavlidis P, Minga E, Tsaftaris A, Darzentas N (2012a) The turbulent life of Sirevirus retrotransposons and the evolution of the maize genome: more than ten thousand elements tell the story. Plant J 69:475–488

Bousios A, Minga E, Kalitsou N, Pantermali M, Tsaballa A, Darzentas N (2012b) MASiVEdb: the Sirevirus plant retrotransposon database. BMC Genomics 13:158

Bowen NJ, McDonald JF (1999) Genomic analysis of *Caenorhabditis elegans* reveals ancient families of retroviral-like elements. Genome Res 9:924–935

Caffrey M (2011) HIV envelope: challenges and opportunities for development of entry inhibitors. Trends Microbiol 19:191–197

Chalvet F, Teysset L, Terzian C, Prud'homme N, Santamaria P, Bucheton A, Pelisson A (1999) Proviral amplification of the Gypsy endogenous retrovirus of *Drosophila melanogaster* involves env-independent invasion of the female germline. EMBO J 18:2659–2669

Chavanne F, Zhang DX, Liaud MF, Cerff R (1998) Structure and evolution of Cyclops: a novel giant retrotransposon of the Ty3/Gypsy family highly amplified in pea and other legume species. Plant Mol Biol 37:363–375

Cosset FL, Lavillette D (2011) Cell entry of enveloped viruses. Adv Genet 73:121–183

Daraselia ND, Tarchevskaya S, Narita JO (1996) The promoter for tomato 3-hydroxy-3-methylglutaryl coenzyme A reductase gene 2 has unusual regulatory elements that direct high-level expression. Plant Physiol 112:727–733

Du C, Swigonova Z, Messing J (2006) Retrotranspositions in orthologous regions of closely related grass species. BMC Evol Biol 6:62

Du J, Tian Z, Bowen NJ, Schmutz J, Shoemaker RC, Ma J (2010a) Bifurcation and enhancement of autonomous-nonautonomous retrotransposon partnership through LTR swapping in soybean. Plant Cell 22:48–61

Du J, Tian Z, Hans CS, Laten HM, Cannon SB, Jackson SA, Shoemaker RC, Ma J (2010b) Evolutionary conservation, diversity and specificity of LTR-retrotransposons in flowering plants: insights from genome-wide analysis and multi-specific comparison. Plant J 63:584–598

Eickbush TH, Jamburuthugoda VK (2008) The diversity of retrotransposons and the properties of their reverse transcriptases. Virus Res 134:221–234

Eickbush TH, Malik HS (2002) Origins and evolution of retrotransposons. In: Craig NL, Craigie R, Gellert M, Lambowitz AM (eds) Mobile DNA II. ASM Press, Washington, pp 1111–1144

Felder H, Herzceg A, de Chastonay Y, Aeby P, Tobler H, Muller F (1994) Tas, a retrotransposon from the parasitic nematode *Ascaris lumbricoides*. Gene 149:219–225

Feschotte C, Gilbert C (2012) Endogenous viruses: insights into viral evolution and impact on host biology. Nat Rev Genet 13:283–296

Finlay BB, McFadden G (2006) Anti-immunology: evasion of the host immune system by bacterial and viral pathogens. Cell 124:767–782

Friesen PD, Nissen MS (1990) Gene organization and transcription of TED, a lepidopteran retrotransposon integrated within the Baculovirus genome. Mol Cell Biol 10:3067–3077

Gaston GD (2011) Detection of SIRE1 ENV, a potential retroviral like protein in soybean. M.S. Thesis, Loyola University, Chicago

Gladyshev EA, Meselson M, Arkhipova IR (2007) A deep-branching clade of retrovirus-like retrotransposons in bdelloid rotifers. Gene 390:136–145

Goldbach R, Peters D (1996) Molecular and biological aspects of tospoviruses. In: Elliot RM (ed) The Bunyaviridae. Plenum Press, New York, pp 129–157

Gonzalez VM, Benjak A, Henaff EM, Mir G, Casacuberta JM, Garcia-Mas J, Puigdomenech P (2010) Sequencing of 6.7 Mb of the melon genome using a BAC pooling strategy. BMC Plant Biol 10:246

Grandbastien MA (2008) Retrotransposons in plants. In: Mahy BWJ, Van Regenmortel MHV (eds) Encyclopedia of plants. Elsevier, Oxford, pp 428–436

Hafez EE, Abdel Ghany AA, Paterson AH, Zaki EA (2009) Sequence heterogeneity of the envelope-like domain in cultivated allotetraploid Gossypium species and their diploid progenitors. J Appl Genet 50:17–23

Harrison SC (2008) Viral membrane fusion. Nat Struct Mol Biol 15:690–698

Havecker ER, Voytas DF (2003) The soybean retroelement SIRE1 uses stop codon suppression to express its envelope-like protein. EMBO Rep 4:274–277

Havecker ER, Gao X, Voytas DF (2005) The Sireviruses, a plant-specific lineage of the Ty1/copia retrotransposons, interact with a family of proteins related to dynein light chain 8. Plant Physiol 139:857–868

Hohn T, Richert-Poggeler KR, Staginnus C, Harper G, Schwarzacher T, Teo CH, Teycheney P-Y, Iskra-Caruana M-L, Hull R (2008) Evolution of integrated plant viruses. In: Roossinck MJ (ed) Plant virus evolution. Springer, Berlin, pp 53–81

Holligan D, Zhang X, Jiang N, Pritham EJ, Wessler SR (2006) The transposable element landscape of the model legume *Lotus japonicus*. Genetics 174:2215–2228

Howard G, Eiges R, Gaudet F, Jaenisch R, Eden A (2008) Activation and transposition of endogenous retroviral elements in hypomethylation induced tumors in mice. Oncogene 27:404–408

Hribova E, Neumann P, Matsumoto T, Roux N, Macas J, Dolezel J (2010) Repetitive part of the banana (*Musa acuminata*) genome investigated by low-depth 454 sequencing. BMC Plant Biol 10:204

Hunter E (1997) Viral entry and receptors. In: Coffin JM, Hughes SH, Varmus HE (eds) Retroviruses. Cold Spring Harbor Laboratory Press, Plainview, NY, pp 71–119

Jern P, Coffin JM (2008) Effects of retroviruses on host genome function. Annu Rev Genet 42:709–732

Jin YK, Bennetzen JL (1994) Integration and nonrandom mutation of a plasma membrane proton ATPase gene fragment within the Bs1 retroelement of maize. Plant Cell 6:1177–1186

Kadlec J, Loureiro S, Abrescia NGA, Stuart DI, Jones IM (2008) The post-fusion structure of Baculovirus gp64 supports a unified view of viral fusion machines. Nat Struct Mol Biol 15:1024–1030

Kapitonov VV, Jurka J (1999) Molecular paleontology of transposable elements from *Arabidopsis thaliana*. Genetica 107:27–37

Kenward KD, Bai D, Ban MR, Brandle JE (1999) Isolation and characterization of Tnd-1, a retrotransposon marker linked to black root rot resistance in tobacco. Theor Appl Genet 98:387–395

Kikkert M, van Lent J, Storms M, Bodegom P, Kormelink R, Goldbach R (1999) Tomato spotted wilt virus particle morphogenesis in plant cells. J Virol 73:2288–2297

Kim A, Terzian C, Santamaria P, Pelisson A, Prud'homme N, Bucheton A (1994) Retroviruses in invertebrates: the gypsy retrotransposon is apparently an infectious retrovirus of *Drosophila melanogaster*. Proc Natl Acad Sci USA 91:1285–1289

Kim FJ, Battini JL, Manel N, Sitbon M (2004) Emergence of vertebrate retroviruses and envelope capture. Virology 318:183–191

King AMQ, Adams MJ, Carstens EB, Lefkowitz EJ (eds) (2012) Virus taxonomy: ninth report of the international committee on taxonomy of viruses. Elsevier, London

Kumar A (1998) The evolution of plant retroviruses: moving to green pastures. Trends Plant Sci 3:371–374

Laten HM (1999) Phylogenetic evidence for Ty1-copia-like endogenous retroviruses in plant genomes. Genetica 107:87–93

Laten HM, Bousios A (2012) Genus Sirevirus. In: Tidona C, Darai G (eds) The Springer index of viruses, 2nd edn. Springer, New York, NY, pp 1561–1564

Laten HM, Morris RO (1993) SIRE-1, a long interspersed repetitive DNA element from soybean with weak sequence similarity to retrotransposons: initial characterization and partial sequence. Gene 134:153–159

Laten HM, Majumdar A, Gaucher EA (1998) SIRE-1, a copia/Ty1-like retroelement from soybean, encodes a retroviral envelope-like protein. Proc Natl Acad Sci USA 95:6897–6902

Laten HM, Havecker ER, Farmer LM, Voytas DF (2003) SIRE1, an endogenous retrovirus family from *Glycine max*, is highly homogeneous and evolutionarily young. Mol Biol Evol 20:1222–1230

Laten HM, Mogil LS, Wright LN (2009) A shotgun approach to discovering and reconstructing consensus retrotransposons ex novo from dense contigs of short sequences derived from Genbank Genome Survey Sequence database records. Gene 448:168–173

Lavin M, Herendeen PS, Wojciechowski MF (2005) Evolutionary rates analysis of Leguminosae implicates a rapid diversification of lineages during the tertiary. Syst Biol 54:575–594

Lazarowitz SD (2007) Plant viruses. In: Knipe DM, Howley PM (eds) Fields virology. Lippincott Williams & Wilkins, Philadelphia, PA, pp 641–705

Lerat E, Capy P (1999) Retrotransposons and retroviruses: analysis of the envelope gene. Mol Biol Evol 16:1198–1207

Li F, Ding S-W (2006) Virus counter-defense: diverse strategies for evading the RNA-silencing immunity. Annu Rev Microbiol 60:503–531

Lin E (2001) Analysis of SIRE1 transcriptional activity. M.S. Thesis, Loyola University, Chicago

Llorens JV, Clark JB, Martinez-Garay I, Soriano S, de Frutos R, Martinez-Sebastian MJ (2008) Gypsy endogenous retrovirus maintains potential infectivity in several species of Drosophilids. BMC Evol Biol 8:302

Llorens C, Muñoz-Pomer A, Bernad L, Botella H, Moya A (2009) Network dynamics of eukaryotic LTR retroelements beyond phylogenetic trees. Biol Direct 4:41

Llorens C, Futami R, Covelli L, Domínguez-Escribá L, Viu JM, Tamarit D, Aguilar-Rodríguez J, Vicente-Ripolles M, Fuster G, Bernet GP, Maumus F, Munoz-Pomer A, Sempere JM, Latorre A, Moya A (2011) The Gypsy database (GyDB) of mobile genetic elements: release 2.0. Nucleic Acids Res 39:D70–D74

Macas J, Neumann P (2007) Ogre elements–a distinct group of plant Ty3/gypsy-like retrotransposons. Gene 390:108–116

Macas J, Neumann P, Navratilova A (2007) Repetitive DNA in the pea (*Pisum sativum* L.) genome: comprehensive characterization using 454 sequencing and comparison to soybean and Medicago truncatula. BMC Genomics 8:427

Maksakova IA, Romanish MT, Gagnier L, Dunn CA, van de Lagemaat LN, Mager DL (2006) Retroviral elements and their hosts: insertional mutagenesis in the mouse germ line. PLoS Genet 2:e2

Maksakova IA, Mager DL, Reiss D (2008) Keeping active endogenous retroviral-like elements in check: the epigenetic perspective. Cell Mol Life Sci 65:3329–3347

Malik HS, Henikoff S, Eickbush TH (2000) Poised for contagion: evolutionary origins of the infectious abilities of invertebrate retroviruses. Genome Res 10:1307–1318

McCarthy EM, Liu J, Lizhi G, McDonald JF (2002) Long terminal repeat retrotransposons of *Oryza sativa*. Genome Biol 3:0053

Mejlumian L, Pelisson A, Bucheton A, Terzian C (2002) Comparative and functional studies of Drosophila species invasion by the gypsy endogenous retrovirus. Genetics 160:201–209

Misseri Y, Labesse G, Bucheton A, Terzian C (2003) Comparative sequence analysis and predictions for the envelope glycoproteins of insect endogenous retroviruses. Trends Microbiol 11:253–256

Misseri Y, Cerutti M, Devauchelle G, Bucheton A, Terzian C (2004) Analysis of the Drosophila gypsy endogenous retrovirus envelope glycoprotein. J Gen Virol 85:11–31

Moyes D, Griffiths DJ, Venables PJ (2007) Insertional polymorphisms: a new lease of life for endogenous retroviruses in human disease. Trends Genet 23:326–333

Nagaki K, Song J, Stupar RM, Parokonny AS, Yuan Q, Ouyang S, Liu J, Hsiao J, Jones KM, Dawe RK, Buell CR, Jiang J (2003) Molecular and cytological analyses of large tracks of centromeric DNA reveal the structure and evolutionary dynamics of maize centromeres. Genetics 163:759–770

Nagata T, Inoue-Nagata AK, Prins M, Goldbach R, Peters D (2000) Impeded thrips transmission of defective tomato spotted wilt virus isolates. Phytopathology 90:454–459

Neumann P, Pozarkova D, Macas J (2003) Highly abundant pea LTR retrotransposon Ogre is constitutively transcribed and partially spliced. Plant Mol Biol 53:399–410

Neumann P, Pozarkova D, Koblizkova A, Macas J (2005) PIGY, a new plant envelope-class LTR retrotransposon. Mol Genet Genomics 273:43–53

Nielen S, Campos-Fonseca F, Leal-Bertioli S, Guimaraes P, Seijo G, Town C, Arrial R, Bertioli D (2010) FIDEL-a retrovirus-like retrotransposon and its distinct evolutionary histories in the A- and B-genome components of cultivated peanut. Chromosome Res 18:227–246

Nishimura MT, Dangl JL (2010) Arabidopsis and the plant immune system. Plant J 61:1053–1066

Obbard DJ, Gordon KHJ, Buck AH, Jiggins FM (2009) The evolution of RNAi as a defence against viruses and transposable elements. Philos Trans R Soc Lond B Biol Sci 364:99–115

Ozers MS, Friesen PD (1996) The Env-like open reading frame of the Baculovirus-integrated retrotransposon TED encodes a retrovirus-like envelope protein. Virology 226:252–259

Pearce SR (2007) SIRE-1, a putative plant retrovirus is closely related to a legume TY1-copia retrotransposon family. Cell Mol Biol Lett 12:120–126

Pearson MN, Rohrmann GF (2002) Transfer, incorporation, and substitution of envelope fusion proteins among members of the Baculoviridae, Orthomyxoviridae, and Metaviridae (insect retrovirus) families. J Virol 76:5301–5304

Pearson MN, Rohrmann GF (2004) Conservation of a proteinase cleavage site between an insect retrovirus (gypsy) Env protein and a Baculovirus envelope fusion protein. Virology 322:61–68

Pearson MN, Rohrmann GF (2006) Envelope gene capture and insect retrovirus evolution: the relationship between Errantivirus and Baculovirus envelope proteins. Virus Res 118:7–15

Pelisson A, Mejlumian L, Robert V, Terzian C, Bucheton A (2002) Drosophila germline invasion by the endogenous retrovirus gypsy: involvement of the viral env gene. Insect Biochem Mol Biol 32:1249–1256

Peterson-Burch BD, Voytas DF (2002) Genes of the Pseudoviridae (Ty1/copia Retrotransposons). Mol Biol Evol 19:1832–1845

Peterson-Burch BD, Wright DA, Laten HM, Voytas DF (2000) Retroviruses in plants? Trends Genet 16:151–152

Piedöel M, Gonçalves IR, Higuet D, Bonnivard E (2011) Eukaryotic DIRS1-like retrotransposons: an overview. BMC Genomics 12:621

Pinter A, Honnen WJ (1988) O-linked glycosylation of retroviral envelope gene products. J Virol 62:1016–1021

Poulter R, Goodwin T (2005) *DIRS-1* and the other tyrosine recombinase retrotransposons. Cytogenet Genome Res 110:575–588

Rabson AB, Graves BJ (1997) Synthesis and processing of viral RNA. In: Coffin JM, Hughes SH, Varmus HE (eds) Retroviruses. Cold Spring Harbor Laboratory Press, Plainview, NY, pp 205–261

Ramakrishna W, Dubcovsky J, Park YJ, Busso C, Emberton J, SanMiguel P, Bennetzen JL (2002) Different types and rates of genome evolution detected by comparative sequence analysis of orthologous segments from four cereal genomes. Genetics 162:1389–1400

Ribet D, Harper F, Dewannieux M, Pierron G, Heidmann T (2007) Murine MusD retrotransposon: structure and molecular evolution of an "intracellularized" retrovirus. J Virol 81:1888–1898

Ribet D, Harper F, Dupressoir A, Dewannieux M, Pierron G, Heidmann T (2008) An infectious progenitor for the murine IAP retrotransposon: emergence of an intracellular genetic parasite from an ancient retrovirus. Genome Res 18:597–609

Rohrmann GF, Karplus PA (2001) Relatedness of Baculovirus and gypsy retrotransposon envelope proteins. BMC Evol Biol 1:1

SanMiguel P, Vitte C (2009) The LTR-retrotransposons of maize. In: Bennetzen JL, Hake SC (eds) Handbook of maize: genetics and genomics. Springer, New York, NY, pp 307–327

SanMiguel P, Tikhonov A, Jin YK, Motchoulskaia N, Zakharov D, Melake-Berhan A, Springer PS, Edwards KJ, Lee M, Avramova Z, Bennetzen JL (1996) Nested retrotransposons in the intergenic regions of the maize genome. Science 274:765–768

Song SU, Gerasimova T, Kurkulos M, Boeke JD, Corces VG (1994) An env-like protein encoded by a Drosophila retroelement: evidence that gypsy is an infectious retrovirus. Genes Dev 8:2046–2057

Staginnus C, Richert-Poggeler KR (2006) Endogenous pararetroviruses: two-faced travelers in the plant genome. Trends Plant Sci 11:485–491

Steinbauerová V, Neumann P, Novák P, Macas J (2012) A widespread occurrence of extra open reading frames in plant Ty3/gypsy retrotransposons. Genetica. doi:10.1007/s10709-012-9654-9

Stocking C, Kozak CA (2008) Murine endogenous retroviruses. Cell Mol Life Sci 65:3383–3398

Terzian C, Pelisson A, Bucheton A (2001) Evolution and phylogeny of insect endogenous retroviruses. BMC Evol Biol 1:3

Teysset L, Burns JC, Shike H, Sullivan BL, Bucheton A, Terzian C (1998) A Moloney murine leukemia virus-based retroviral vector pseudotyped by the insect retroviral gypsy envelope can infect Drosophila cells. J Virol 72:853–856

Vicient CM, Kalendar R, Schulman AH (2001) Envelope-class retrovirus-like elements are widespread, transcribed and spliced, and insertionally polymorphic in plants. Genome Res 11:2041–2049

Vitte C, Bennetzen JL (2006) Analysis of retrotransposon structural diversity uncovers properties and propensities in angiosperm genome evolution. Proc Natl Acad Sci USA 103:17638–17643

Wang Y, Liska F, Gosele C, Sedova L, Kren V, Krenova D, Ivics Z, Hubner N, Izsvak Z (2010) A novel active endogenous retrovirus family contributes to genome variability in rat inbred strains. Genome Res 20:19–27

Wang X, Wang H, Wang J, Sun R, Wu J, Liu S, Bai Y, Mun JH, Bancroft I, Cheng F, Huang S, Li X, Hua W, Wang J, Wang X, Freeling M, Pires JC, Paterson AH, Chalhoub B, Wang B, Hayward A, Sharpe AG (2011) The genome of the mesopolyploid crop species *Brassica rapa*. Nat Genet 43:1035–1039

Weber B, Wenke T, Frommel U, Schmidt T, Heitkam T (2010) The Ty1-copia families SALIRE and Cotzilla populating the *Beta vulgaris* genome show remarkable differences in abundance, chromosomal distribution, and age. Chromosome Res 18:247–263

Whitfield AE, Ullman DE, German TL (2005) Tospovirus-thrips interactions. Annu Rev Phytopathol 43:459–489

Wicker T, Keller B (2007) Genome-wide comparative analysis of copia retrotransposons in Triticeae, rice, and Arabidopsis reveals conserved ancient evolutionary lineages and distinct dynamics of individual copia families. Genome Res 17:1072–1081

Wilson IB, Gavel Y, von Heijne G (1991) Amino acid distributions around O-linked glycosylation sites. Biochem J 275:529–534

Wright DA, Voytas DF (1998) Potential retroviruses in plants: Tat1 is related to a group of *Arabidopsis thaliana* Ty3/gypsy retrotransposons that encode envelope-like proteins. Genetics 149:703–715

Wright DA, Voytas DF (2002) Athila4 of Arabidopsis and Calypso of soybean define a lineage of endogenous plant retroviruses. Genome Res 12:122–131

Wu BW, Cannon PM, Gordon EM, Hall FL, Anderson WF (1998) Characterization of the proline-rich region of murine leukemia virus envelope protein. J Virol 72:5383–5391

Yang ZN, Ye XR, Molina J, Roose ML, Mirkov TE (2003) Sequence analysis of a 282-kilobase region surrounding the citrus Tristeza virus resistance gene (*Ctv*) locus in *Poncirus trifoliata* L. Raf. Plant Physiol 131:482–492

Yano ST, Panbehi B, Das A, Laten HM (2005) Diaspora, a large family of Ty3-gypsy retrotransposons in *Glycine max*, is an envelope-less member of an endogenous plant retrovirus lineage. BMC Evol Biol 5:30

Zhang Y, Maksakova IA, Gagnier L, van de Lagemaat LN, Mager DL (2008) Genome-wide assessments reveal extremely high levels of polymorphism of two active families of mouse endogenous retroviral elements. PLoS Genet 4:e1000007

Zhou Q, Haymer DS (1998) Molecular structure of yoyo, a gypsy-like retrotransposon from the Mediterranean fruit fly, *Ceratitis capitata*. Genetica 101:167–178

Chapter 7
MITEs, Miniature Elements with a Major Role in Plant Genome Evolution

Hélène Guermonprez, Elizabeth Hénaff, Marta Cifuentes, and Josep M. Casacuberta

Abstract Miniature Inverted-repeat Transposable Elements (MITEs) are a particular type of class II transposons found in genomes in high copy numbers. Most MITEs are deletion derivatives of class II transposons whose transposases have been shown to mobilize them by a typical cut-and-paste mechanism. However, unlike class II transposons, MITEs can amplify rapidly and dramatically and attain very high copy numbers, in particular, in plant genomes. This high copy number, together with their close association with genes, endows MITEs with a high potential to generate variability, and impact gene and genome evolution.

Keywords MITE-Class II transposons • Transposition mechanism • Impact of transposition • Amplification

Abbreviations

MITE	Miniature inverted-repeat transposable element
TE	Transposable element
TIRs	Terminal inverted repeats
TSD	Target site duplication

H. Guermonprez • E. Hénaff • J.M. Casacuberta (✉)
Department of Molecular Genetics, Center for Research in Agricultural Genomics, CRAG (CSIC-IRTA-UAB-UB), Campus UAB, Bellaterra (Cerdanyola del Vallés), 08193 Barcelona, Spain
e-mail: josep.casacuberta@cragenomica.es

M. Cifuentes
Institut Jean-Pierre Bourgin, UMR1318 INRA-AgroParisTech INRA Centre de Versailles-Grignon, Bâtiment 7, Route de St-Cyr (RD10), 78026 Versailles Cedex, France

M.-A. Grandbastien and J.M. Casacuberta (eds.), *Plant Transposable Elements*, Topics in Current Genetics 24, DOI 10.1007/978-3-642-31842-9_7,

7.1 Introduction

The term Miniature Inverted-repeat Transposable Elements (MITEs) was coined to designate different families of short mobile elements featuring Terminal Inverted Repeats (TIRs) and found in plant genomes in high copy number (Wessler et al. 1995). The first two families described were *Tourist* and the *Stowaway* from maize (Bureau and Wessler 1992, 1994). Sequence homology searches revealed their high similarity to transposons of *Mariner* and *PIF* families, respectively, suggesting that they could be deletion derivatives of class II transposons (Feschotte and Mouches 2000; Zhang et al. 2001). Since then, MITEs related to all major families of class II transposons have been reported (Benjak et al. 2009; Kuang et al. 2009; Yang and Hall 2003b), and MITE families have been described in both prokaryote and eukaryote genomes (Dufresne et al. 2007; Filee et al. 2007; Han et al. 2010; Piriyapongsa and Jordan 2007; Surzycki and Belknap 2000), including virtually all plant genomes analyzed (Benjak et al. 2009; Bergero et al. 2008; Bureau et al. 1996; Cantu et al. 2010; Casacuberta et al. 1998; Grzebelus et al. 2009; Kuang et al. 2009; Lyons et al. 2008; Momose et al. 2010; Sarilar et al. 2011; Schwarz-Sommer et al. 2010; Yang and Hall 2003b). However, while most MITEs seem to be deletion derivatives of autonomous elements, which probably mobilize them, in some cases the situation is less clear. Some MITEs cannot be related to long coding elements suggesting that in some cases MITEs may arise by the serendipitous juxtaposition of two inverted repeated sequences which may be recognized by an existing transposase (Feschotte and Pritham 2007). In other cases, like that of *mPing* in rice, the related long coding element has been identified but is absent from the varieties where *mPing* is active, suggesting that the element that gave rise to the MITE has been lost and that other transposases may catalyze its mobilization (Jiang et al. 2003). The emerging picture is thus a complex relationship between MITEs and their distantly related autonomous elements (Feschotte et al. 2005).

In addition to their small size and the presence of TIRs, a number of other characteristics have been associated with MITEs. The sequence of the first MITEs described was shown to be A/T-rich and to have the potential to form highly stable secondary structures (Bureau and Wessler 1992), and these characteristics seem to be shared by a high proportion of the MITEs described to date. However, during these years no evidence has demonstrated any relevance of these characteristics for MITEs' amplification dynamics.

MITEs are frequently found within or close to genes (Casacuberta and Santiago 2003), although this preference probably varies among different families (Mao et al. 2000). This trend, combined with their high copy number, endows MITEs with a great potential to modify gene expression upon mobilization (Deragon et al. 2008). In this chapter we summarize recent advances in the identification of MITEs, their mechanism of transposition, and their impact on genes and genomes. We also point out open questions regarding these miniature but highly complex elements.

7.2 MITE Identification

Due to their small size and absence of coding capacity, MITE identification and annotation is particularly difficult. As is the case for most TE (Transposable Element) families described to date, the first MITEs discovered were elements inserted in genes, causing a detectable mutation and phenotype. However, the availability of whole genome sequences together with the development of appropriate bioinformatic tools has enabled the discovery of the high prevalence of these elements in eukaryotic genomes.

7.2.1 Discovery by Insertional Mutagenesis

The first MITE, dubbed *Tourist*, was discovered in maize by insertional mutation in the *waxy* gene (Bureau and Wessler 1992). Its analysis revealed the presence of TIRs in the insert, which, combined with the fact that it was found in many copies in the available gene sequences of the same line, and the presence of a flanking duplicated sequence, led to the hypothesis that this was actually a mobile repeated element. Since then, other cases of insertional mutagenesis have led to the discovery of a few other MITEs such as *mPing* that was found inserted into the gene for *rice ubiquitin-related modifier-1* (*Rurml*) and whose excision resulted in the reversion of the "slender glume" phenotype (Nakazaki et al. 2003) and *dTstu1*, the source of a somaclonal variation inducing purple pigment synthesis in a usually red potato variety (Momose et al. 2010).

7.2.2 Discovery by Bioinformatic Methods

While MITEs as a new superfamily of transposable elements were stumbled upon by accident and studied using molecular biology techniques, the availability of genomic sequence data as well as sequence search tools has allowed the identification of MITE families by bioinformatic means. One category of methods is based on sequence similarity to a known MITE or autonomous class II transposon. The second is to identify MITE families de novo, exploiting their structural characteristics and the fact that they are found in large copy numbers.

Certain MITE families are shared among several species, as is the case for *Tourist* in cereals, and can be detected by sequence similarity to already defined MITEs. For example, elements similar to the consensus sequences of the MITEs first identified in maize and barley (Bureau and Wessler 1992) were found in rice and sorghum (Bureau and Wessler 1994).

Many MITEs arise as deletion derivatives of their autonomous counterparts, and thus display sequence similarity to class II transposons. Exploiting this sequence similarity, new MITEs can be discovered by searching with a full-length element as a query. However, while some MITEs are homologous to their autonomous counterparts in their entire length, others only share the TIR sequences and the rest of the internal sequence is unrelated, requiring different computational approaches for either case.

In the first case, MITEs can be identified by genome-wide similarity searches using the full-length TE as query, as was done in *Vitis vinifera* to identify MITEs related to known elements in the CACTA, hAT, and PIF superfamilies (Benjak et al. 2009).

The second case is more difficult as TIRs are short (10–20 nucleotides), and these can give many spurious hits. Various softwares have been developed to implement this search. A first example is TRANSPO that takes a TIR sequence and searches for inverted matches within a certain window and can be paired with the SPAT software, which performs a hierarchical clustering of the results, thus defining families of putative elements (Santiago et al. 2002). A second example is the MAK toolkit (Yang and Hall 2003a) that provides a suite of programs to identify MITE copies, or a related autonomous element, given a MITE query. This software implements various modes with different goals. The *Member Retriever* mode is designed to retrieve other MITEs similar to the supplied MITE query. The *Anchor* mode aims to identify autonomous elements that are related to a given MITE query, and the *Associator* mode reports gene annotations nearest to the hits.

With the recent proliferation of whole-genome sequencing data and comparative analyses, it becomes tempting to mine this wealth of information for entirely new MITEs using computational methods. Two different approaches for de novo MITE identification have been used to date, one based on comparative analyses of closely related organisms and the other exploiting the elements' structural characteristics and the fact that they are found in very high copy number.

The first approach is not specific to MITEs, but has lead to the identification of new MITE families in solanacae related to *hAT*, *Mutator*, *Stowaway* and *Tourist* elements by inspecting syntenic regions of resistance gene clusters in tomato, potato, and tobacco (Kuang et al. 2009). This method of searching for Related Empty Sites also provides indirect evidence for their mobilization, as discussed below.

The second approach is based on the fact that MITEs present very clear structural characteristics—exact TIRs and TSDs (target site duplication) upon insertion. However, these structures are very short and similar ones can arise by chance, leading to many false positives when the search criteria are limited to two inverted repeats flanked by direct ones. Thus, the true challenge of in silico MITE identification is eliminating false positives. Various programs have been developed for MITE identification in genomic sequences, the latest being MITE-hunter (Han and Wessler 2010). This software is the most sophisticated in that it provides several methods of eliminating false positives, at various steps of the algorithm. Similarly to others [FINDMITE (Tu 2001); MUST (Chen et al. 2009)], the first step is to identify candidate MITEs based on TIRs and TSDs. In a subsequent step, candidates are discriminated based on copy number by pairwise comparison—elements that do not align with any other are eliminated as false positives. Then a consensus sequence is generated for each family and the definition of its borders verified by multiple sequence alignment with its copies taken with flanking regions. This last step relies on the fact that within a certain family, the copies' terminal sequences (i.e., TIRs and TSDs) will be near identical and align well but the alignment will break down at the flanking regions as each element is inserted in a different genomic context.

The surge of available genomic sequence data is a wealth of information for studying transposons in general, and MITEs in particular. Whole genome sequences provide the possibility of mining for new elements, impossible until the advent of this data. Also, comparative analyses between genomes are a powerful tool for identification of new elements and following TE movement. Two major technological advances, besides the progress in sequencing technologies, permit this: the development of algorithms for accurate whole-genome alignments (Frith et al. 2010) and genome resequencing (Stratton 2008). Until now transposon discovery by comparative analysis has been limited to certain syntenic regions, but exploiting this type of data on a whole genome scale is a promising prospect. Resequencing of varieties or lines within a species has the advantage of providing highly comparable data of closely related organisms, giving a perspective of the variations of the transposon landscape at a small evolutionary scale. Recently, the resequencing of rice lines issued from cell culture led to the identification of 43 new insertions of 13 different TEs. Although the authors have not exploited this analysis to look for new elements, their approach could also be used for de novo identification. In conclusion, genomic data analysis has provided evidence for MITE mobility and enabled the discovery of new elements. Furthermore, we can expect that the level of detail and precision at which we can study mobile elements on the genomic scale will increase with progress in algorithms for sequence analysis and quantity of data available.

7.3 MITE Transposition Mechanisms

The analysis of *Tourist*, the first MITE family characterized (Bureau and Wessler 1992), allowed for a first description of the particular characteristics of MITEs. *Tourist* elements presented TIRs and subterminal repeated sequences, as well as TSDs flanking the elements, which make them similar to class II transposons. However, these elements were present at a higher copy number than typical class II elements, and their copies showed an unprecedented homogeneity in size and sequence. These characteristics, later shown to be shared by most MITEs, made it difficult at the time to classify them. Moreover, MITEs' transposition mechanism remained a mystery as no excision event had yet been observed (Wessler et al. 1995).

The first evidence of MITEs' capacity for excision came from the phylogenetic analysis of the *Stowaway* family in 30 *Triticae* species (Petersen and Seberg 2000) and was later confirmed by the analysis of a rice *slender glume* mutant, which carries an *mPing* MITE whose excision lead to the reversion of the mutant phenotype (Nakazaki et al. 2003). The confirmation of MITEs' potential for excision, together with the fact that some show high sequence similarity with class II transposons (Feschotte and Mouches 2000), strongly suggested that MITEs could be deletion derivatives of class II transposons, mobilized by transposases encoded by their related autonomous elements (Casacuberta and Santiago 2003; Feschotte et al. 2002; Zhang et al. 2001). This hypothesis gained further support from studies showing that the transposases encoded by class II transposons specifically bind the TIRs and subterminal sequences of related MITEs (Feschotte et al. 2005; Loot et al. 2006). The mobilization of MITEs by class II transposases was finally demonstrated in three independent reports in animals, plants, and fungi, which showed conclusive evidence that transposases from a related element were able to mobilize MITEs in vivo (Dufresne et al. 2007; Miskey et al. 2007; Yang et al. 2007). This mobilization has also been observed in heterologous systems (Hancock et al. 2010, 2011; Yang et al. 2007), suggesting that, as is the case for typical class II elements, the minimal requirements for MITEs transposition are a transposase and its binding sequences within the element. However, although MITEs' transposition seems in some respects very similar to that of typical class II elements, it also presents particular features that make MITEs a very unique type of defective class II elements.

First of all, MITEs seem to be particularly promiscuous with respect to the transposase they can use for mobilization. Phylogenetic analyses of rice *Mariner*-like elements and their related *Stowaway* MITEs suggested that homology restricted to the TIRs and subterminal sequences may be sufficient for cross-mobilization (Feschotte et al. 2003). This was confirmed by in vitro protein/DNA interaction studies showing that rice *Stowaway* MITEs can interact with transposases encoded by a panoply of *Mariner*-like *Osmar* elements (Feschotte et al. 2005). This promiscuity may explain the transposition of the rice *Tourist*-like element *mPing*, which is a deletion derivative of a class II element *Ping*, in rice cultivars that are devoid of active *Ping* elements but contain potentially active elements of the distantly related transposon *Pong* (Jiang et al. 2003). Indeed, recent experiments have demonstrated that *mPing* can be mobilized in vivo by both *Ping* and *Pong*'s transposases (Hancock et al. 2010). Based on these observations a model of MITE dynamics has been proposed in which MITEs would be generated through a deletion in an autonomous transposon, then amplification would take place maybe long afterwards, catalyzed by the element's encoded transposase or that of a distantly related element, as the former may even have disappeared (Jiang et al. 2004).

Second, some reports suggest that MITEs may be mobilized more efficiently than typical class II transposons. It has been shown that some transposases bind with higher affinity to the MITE sequence than to the transposase-encoding element, either because the MITE contains additional transposase binding sites in the subterminal repeated regions (Loot et al. 2006) or because it lacks repressive sequences present in the original autonomous element (Yang et al. 2009). Both MITEs' promiscuity and their higher transposase binding affinity could account for an increased transposition efficiency with respect to typical class II transposons. However, this does not seem to explain the third and most striking particularity of MITEs: their high copy number. Indeed, although the transposition process may in some cases lead to a moderate increase in copy number (as is the case for typical class II transposons), it is hard to imagine that the very high copy numbers MITEs can attain in very short evolutionary timescales (see below) are the result of an increased number of normal cut-and-paste transposition events. Moreover, while

MITEs do excise, excision events seem to be rare, as most MITE insertions are relatively stable even to the point of being used as genetic markers (Feschotte et al. 2002), suggesting that excisions do not correlate with MITE amplification.

What it is known to date explains how MITEs transpose but not how they amplify to the elevated copy numbers they usually reach in genomes. MITE transposition and amplification may be two different and uncoupled processes (Casacuberta and Santiago 2003; Feschotte et al. 2002) with the standard cut-and-paste transposition generating a moderate or no increase in copy number and amplification occurring rarely. Alternatively, amplification may result from transposition in particular cell types or conditions with higher DNA replication with respect to cell division, such as endoreduplicating cells.

A structural particularity of most MITEs for which a function has not yet been determined is their capacity to form highly stable single strand secondary structures. While it does not seem to be required for MITE cut-and-paste transposition (Sinzelle et al. 2008), it could affect MITEs amplification. It is tempting to hypothesize that the formation of single-strand hairpin structures, with double stranded TIRs, could allow transposase binding and single-stranded excision. It is interesting to note that the bacterial transposons of the IS200/IS605 family move by the excision and reintegration of only one of the strands of the transposon leaving the complementary strand behind. This mechanism is catalyzed by a very particular type of transposase and linked to replication (Guynet et al. 2008; Ton-Hoang et al. 2010). This particular mode of transposition could easily explain an increase of transposon copies. In plants, where endoreduplication or re-replication processes are commonplace, such a mechanism could be particularly relevant.

Irrespective of the mechanism responsible for MITEs amplification, their high copy number suggests that these elements are particularly successful in avoiding genome control. Interestingly MITEs are present at a much higher copy number than the elements coding for the transposase, which mobilize them and from which they frequently derive from. As silencing is the most general and efficient mechanism to control transposons (Lisch 2009), the separation of the transposase encoding element, which can be maintained at a low copy number and thus will not attract silencing, from the transposing unit, the MITE, more difficult to control as it does not need to be transcribed, could in part explain their success in invading genomes (Casacuberta and Santiago 2003; Feschotte and Pritham 2007). In accordance with this, it has been shown that the number of sequences related to the *Mariner*-like element *Lemi*1 is low in *Medicago truncatula*, where it has not given rise to MITEs, while it is much higher in *Arabidopsis* where it has given rise to the *Emigrant* MITE (Guermonprez et al. 2008).

7.4 Prevalence of MITEs and Their Impact in Plant Genomes

One of the characteristics that make MITEs a singular type of defective class II transposons is their capacity to reach high copy numbers in genomes (Casacuberta and Santiago 2003). MITEs are present in virtually all plant genomes, where their copy number can vary but usually exceeds that of typical class II transposons. For example, more than 90,000 MITEs grouped into approximately 100 different families are present in the rice genome (Feschotte et al. 2003; Jiang et al. 2004; Juretic et al. 2004). Individual families such as the *Tourist* and *Stowaway* families are found in more than 33,000 and 24,000 copies, respectively, in rice, and some 7,200 and 28,000 copies, respectively, in sorghum (Paterson et al. 2009). Even though these families are very large, the overall genome fraction MITEs occupy is relatively small, due to the diminutive size of these elements. Indeed, *Tourist* and *Stowaway* elements combined only occupy 3.24 % and 1.12 % of the rice and sorghum genomes, respectively, (Paterson et al. 2009). The size of a particular MITE family may vary greatly among closely related species and even between landraces. Indeed, it has been reported that while most rice strains only contain 1–50 copies of the *mPing* MITE, the EG4 strain and related landraces contain up to 1,000 (Naito et al. 2006). These data highlight these elements' capacity to multiply rapidly by bursts of amplification, which endows them with the capability to have an impact in genomes in spite of the low fraction they occupy.

Most MITEs are closely associated with genes in plant genomes. The first MITE described, *Tourist*, was shown to be closely associated to maize genes (Bureau and Wessler 1992), and this characteristic was found to be shared by most MITEs (Casacuberta and Santiago 2003; Wessler et al. 1995). For example, in rice and *Arabidopsis*, the majority of MITEs are located in the euchromatin (Feng et al. 2002; Santiago et al. 2002; Wright et al. 2003). This close association with genes could be the result of an insertion site preference or, alternatively, the effect of selection, as it seems to be the case for some *Arabidopsis* MITEs (Santiago et al. 2002). MITEs are not only located close to genes in plants but can also insert within genes, providing new promoter regulatory sequences (Naito et al. 2009; Sarilar et al. 2011), transcription termination elements (Kuang et al. 2009; Santiago et al. 2002), or even new alternative exons. Indeed, a recent report shows that the insertion of a MITE provides a functionally indispensable alternative exon in the tobacco mosaic virus N resistance gene (Kuang et al. 2009). While there are only a limited number of reports showing an unambiguous implication of MITE in creating new gene functions, there are many more examples of MITE insertions generating variability in gene sequences. A paradigmatic case is that of MITE insertions within resistance genes, which have been reported in rice (Song et al. 1998), barley (Wei et al. 2002), and potato (Huang et al. 2005). MITEs are also an important target of siRNAs, and their silencing may affect the expression of neighboring genes. The siRNAs that target MITEs can be of 24 nt (Kuang et al. 2009) or 21 nt (Cantu et al. 2010), suggesting that MITEs are targets of both transcriptional and posttranscriptional gene silencing. Thus, a MITE insertion within a gene promoter may attract heterochromatin and silence it transcriptionally, as it has been shown for other transposons (Lisch 2009), and an insertion within a transcribed region may make it prone to posttranscriptional gene silencing and mRNA degradation.

This close association with genes, together with their capability of reaching high copy numbers in short periods of time, makes MITEs a potent motor of gene evolution. MITE insertions polymorphic among accessions cultivars or lines have been reported in pea, sugar beet, grapevine, potato, and *Medicago truncatul*a (Benjak et al. 2009; Grzebelus et al. 2009; Macas et al. 2005; Menzel et al. 2006; Momose et al. 2010), and occasionally this variability correlates with phenotypic differences (Momose et al. 2010). The analysis of a recent burst of amplification of the *mPing* element in rice shows an important number of insertions into the 5′ region of rice genes, which in some cases result in their transcriptional upregulation (Naito et al. 2009). The simultaneous insertion of different copies of the same MITE into different gene promoters may result in the coordinated regulation of multiple genes creating a so-called regulatory network (Feschotte 2008), as it has been proposed for *mPing* insertions in rice (Naito et al. 2009). However MITEs can also contribute to the coordinated expression of genes in a more subtle way. It has been shown that MITEs can encode miRNAs and siRNAs in plants (Kuang et al. 2009; Piriyapongsa and Jordan 2008). The frequent insertion of MITEs within transcribed regions of genes (Benjak et al. 2009; Kuang et al. 2009), and their capacity to form stable single strand secondary structures, may facilitate the production of siRNAs from the transcribed elements. Interestingly, it has been recently shown that MITE-derived siRNAs regulate ABA signaling and stress responses in rice (Yan et al. 2011). In this context, the insertion of multiple copies of the siRNA-producing MITE within different genes may also generate a regulatory network, as created by *mPing* MITE in rice (Naito et al. 2009).

7.5 Concluding Remarks

MITEs have been particularly successful in colonizing complex genomes. This is in part due to the difficulty of silencing them by homology-dependent pathways, as they are frequently mobilized by transposases to which they are only distantly related. Their success is probably also a consequence of their capacity to generate more subtle mutations than most other transposons. Indeed, MITEs are very short elements and their insertion within the non-translated regions of genes may be easier to tolerate. Their frequent association with genes, which seems more pronounced that that of their related DNA transposons, suggests that MITE insertions near or within genes have been selected for during evolution. The last few years have seen many reports highlighting the impact of these elements on plant genes' function and regulation, attesting to the role MITEs have played in the evolution of plant genomes.

References

Benjak A, Boue S, Forneck A, Casacuberta JM (2009) Recent amplification and impact of MITEs on the genome of grapevine (*Vitis vinifera* L.). Gen Biol Evol 1:75–84

Bergero R, Forrest A, Charlesworth D (2008) Active miniature transposons from a plant genome and its nonrecombining Y chromosome. Genetics 178:1085–1092

Bureau TE, Wessler SR (1992) Tourist: a large family of small inverted repeat elements frequently associated with maize genes. Plant Cell 4:1283–1294

Bureau TE, Wessler SR (1994) Stowaway: a new family of inverted repeat elements associated with the genes of both monocotyledonous and dicotyledonous plants. Plant Cell 6:907–916

Bureau TE, Ronald PC, Wessler SR (1996) A computer-based systematic survey reveals the predominance of small inverted-repeat elements in wild-type rice genes. Proc Natl Acad Sci USA 93:8524–8529

Cantu D, Vanzetti LS, Sumner A, Dubcovsky M, Matvienko M, Distelfeld A, Michelmore RW, Dubcovsky J (2010) Small RNAs, DNA methylation and transposable elements in wheat. BMC Genomics 11:408

Casacuberta JM, Santiago N (2003) Plant LTR-retrotransposons and MITEs: control of transposition and impact on the evolution of plant genes and genomes. Gene 311:1–11

Casacuberta E, Casacuberta JM, Puigdomenech P, Monfort A (1998) Presence of miniature inverted-repeat transposable elements (MITEs) in the genome of *Arabidopsis thaliana*: characterisation of the Emigrant family of elements. Plant J 16:79–85

Chen Y, Zhou FF, Li GJ, Xu Y (2009) MUST: A system for identification of miniature inverted-repeat transposable elements and applications to *Anabaena variabilis* and *Haloquadratum walsbyi*. Gene 436:1–7

Deragon JM, Casacuberta JM, Panaud O (2008) Plant transposable elements. Genome Dyn 4:69–82

Dufresne M, Hua-Van A, El Wahab HA, Ben M'Barek S, Vasnier C, Teysset L, Kema GH, Daboussi MJ (2007) Transposition of a fungal miniature inverted-repeat transposable element through the action of a Tc1-like transposase. Genetics 175:441–452

Feng Q, Zhang Y, Hao P, Wang S, Fu G, Huang Y, Li Y, Zhu J, Liu Y, Hu X, Jia P, Zhao Q, Ying K, Yu S, Tang Y, Weng Q, Zhang L, Lu Y, Mu J, Zhang LS, Yu Z, Fan D, Liu X, Lu T, Li C, Wu Y, Sun T, Lei H, Li T, Hu H, Guan J, Wu M, Zhang R, Zhou B, Chen Z, Chen L, Jin Z, Wang R, Yin H, Cai Z, Ren S, Lv G, Gu W, Zhu G, Tu Y, Jia J, Chen J, Kang H, Chen X, Shao C, Sun Y, Hu Q, Zhang X, Zhang W, Wang L, Ding C, Sheng H, Gu J, Chen S, Ni L, Zhu F, Chen W, Lan L, Lai Y, Cheng Z, Gu M, Jiang J, Li J, Hong G, Xue Y, Han B (2002) Sequence and analysis of rice chromosome 4. Nature 420:316–320

Feschotte C (2008) Transposable elements and the evolution of regulatory networks. Nat Rev Genet 9:397–405

Feschotte C, Mouches C (2000) Evidence that a family of miniature inverted-repeat transposable elements (MITEs) from the *Arabidopsis thaliana* genome has arisen from a pogo-like DNA transposon. Mol Biol Evol 17:730–737

Feschotte C, Pritham EJ (2007) DNA transposons and the evolution of eukaryotic genomes. Annu Rev Genet 41:331–368

Feschotte C, Jiang N, Wessler SR (2002) Plant transposable elements: where genetics meets genomics. Nat Rev Genet 3:329–341

Feschotte C, Swamy L, Wessler SR (2003) Genome-wide analysis of mariner-like transposable elements in rice reveals complex relationships with stowaway miniature inverted repeat transposable elements (MITEs). Genetics 163:747–758

Feschotte C, Osterlund MT, Peeler R, Wessler SR (2005) DNA-binding specificity of rice mariner-like transposases and interactions with Stowaway MITEs. Nucleic Acids Res 33:2153–2165

Filee J, Siguier P, Chandler M (2007) Insertion sequence diversity in archaea. Microbiol Mol Biol Rev 71:121–157

Frith MC, Hamada M, Horton P (2010) Parameters for accurate genome alignment. BMC Bioinformatics 11:80

Grzebelus D, Gladysz M, Macko-Podgorni A, Gambin T, Golis B, Rakoczy R, Gambin A (2009) Population dynamics of miniature inverted-repeat transposable elements (MITEs) in *Medicago truncatula*. Gene 448:214–220

Guermonprez H, Loot C, Casacuberta JM (2008) Different strategies to persist: the pogo-like Lemi1 transposon produces miniature inverted-repeat transposable elements or typical defective elements in different plant genomes. Genetics 180:83–92

Guynet C, Hickman AB, Barabas O, Dyda F, Chandler M, Ton-Hoang B (2008) *In vitro* reconstitution of a single-stranded transposition mechanism of IS608. Mol Cell 29:302–312

Han Y, Wessler SR (2010) MITE-Hunter: a program for discovering miniature inverted-repeat transposable elements from genomic sequences. Nucleic Acids Res 38:e199

Han MJ, Shen YH, Gao YH, Chen LY, Xiang ZH, Zhang Z (2010) Burst expansion, distribution and diversification of MITEs in the silkworm genome. BMC Genomics 11:520

Hancock CN, Zhang F, Wessler SR (2010) Transposition of the Tourist-MITE mPing in yeast: an assay that retains key features of catalysis by the class 2 PIF/Harbinger superfamily. Mob DNA 1:5

Hancock CN, Zhang F, Floyd K, Richardson AO, Lafayette P, Tucker D, Wessler SR, Parrott WA (2011) The rice miniature inverted repeat transposable element mPing is an effective insertional mutagen in soybean. Plant Physiol 157:552–562

Huang S, van der Vossen E, Kuang H, Vleeshouwers V, Zhang N, Borm T, van Eck H, Baker B, Jacobsen E, Visser R (2005) Comparative genomics enabled the isolation of the R3a late blight resistance gene in potato. Plant J 42:251–261

Jiang N, Bao Z, Zhang X, Hirochika H, Eddy SR, McCouch SR, Wessler SR (2003) An active DNA transposon family in rice. Nature 421:163–167

Jiang N, Feschotte C, Zhang X, Wessler SR (2004) Using rice to understand the origin and amplification of miniature inverted repeat transposable elements (MITEs). Curr Opin Plant Biol 7:115–119

Juretic N, Bureau TE, Bruskiewich RM (2004) Transposable element annotation of the rice genome. Bioinformatics 20:155–160

Kuang H, Padmanabhan C, Li F, Kamei A, Bhaskar PB, Ouyang S, Jiang J, Buell CR, Baker B (2009) Identification of miniature inverted-repeat transposable elements (MITEs) and biogenesis of their siRNAs in the Solanaceae: New functional implications for MITEs. Genome Res 19:42–56

Lisch D (2009) Epigenetic regulation of transposable elements in plants. Annu Rev Plant Biol 60:43–66

Loot C, Santiago N, Sanz A, Casacuberta JM (2006) The proteins encoded by the pogo-like Lemi1 element bind the TIRs and subterminal repeated motifs of the Arabidopsis Emigrant MITE: consequences for the transposition mechanism of MITEs. Nucleic Acids Res 34:5238–5246

Lyons M, Cardle L, Rostoks N, Waugh R, Flavell AJ (2008) Isolation, analysis and marker utility of novel miniature inverted repeat transposable elements from the barley genome. Mol Genet Genomics 280:275–285

Macas J, Koblizkova A, Neumann P (2005) Characterization of Stowaway MITEs in pea (*Pisum sativum* L.) and identification of their potential master elements. Genome 48:831–839

Mao L, Wood TC, Yu Y, Budiman MA, Tomkins J, Woo S, Sasinowski M, Presting G, Frisch D, Goff S, Dean RA, Wing RA (2000) Rice transposable elements: a survey of 73,000 sequence-tagged-connectors. Genome Res 10:982–990

Menzel G, Dechyeva D, Keller H, Lange C, Himmelbauer H, Schmidt T (2006) Mobilization and evolutionary history of miniature inverted-repeat transposable elements (MITEs) in *Beta vulgaris* L. Chromosome Res 14:831–844

Miskey C, Papp B, Mates L, Sinzelle L, Keller H, Izsvak Z, Ivics Z (2007) The ancient mariner sails again: transposition of the human Hsmar1 element by a reconstructed transposase and activities of the SETMAR protein on transposon ends. Mol Cell Biol 27:4589–4600

Momose M, Abe Y, Ozeki Y (2010) Miniature inverted-repeat transposable elements of Stowaway are active in potato. Genetics 186:59–66

Naito K, Cho E, Yang G, Campbell MA, Yano K, Okumoto Y, Tanisaka T, Wessler SR (2006) Dramatic amplification of a rice transposable element during recent domestication. Proc Natl Acad Sci USA 103:17620–17625

Naito K, Zhang F, Tsukiyama T, Saito H, Hancock CN, Richardson AO, Okumoto Y, Tanisaka T, Wessler SR (2009) Unexpected consequences of a sudden and massive transposon amplification on rice gene expression. Nature 461:1130–1134

Nakazaki T, Okumoto Y, Horibata A, Yamahira S, Teraishi M, Nishida H, Inoue H, Tanisaka T (2003) Mobilization of a transposon in the rice genome. Nature 421:170–172

Paterson AH, Bowers JE, Bruggmann R, Dubchak I, Grimwood J, Gundlach H, Haberer G, Hellsten U, Mitros T, Poliakov A, Schmutz J, Spannagl M, Tang H, Wang X, Wicker T, Bharti AK, Chapman J, Feltus FA, Gowik U, Grigoriev IV, Lyons E, Maher CA, Martis M, Narechania A, Otillar RP, Penning BW, Salamov AA, Wang Y, Zhang L, Carpita NC, Freeling M, Gingle AR, Hash CT, Keller B, Klein P, Kresovich S, McCann MC, Ming R, Peterson DG, Mehboob ur R, Ware D, Westhoff P, Mayer KF, Messing J, Rokhsar DS (2009) The *Sorghum bicolor* genome and the diversification of grasses. Nature 457:551–556

Petersen G, Seberg O (2000) Phylogenetic evidence for excision of Stowaway miniature inverted-repeat transposable elements in triticeae (Poaceae). Mol Biol Evol 17:1589–1596

Piriyapongsa J, Jordan IK (2007) A family of human microRNA genes from miniature inverted-repeat transposable elements. PLoS One 2:e203

Piriyapongsa J, Jordan IK (2008) Dual coding of siRNAs and miRNAs by plant transposable elements. RNA 14:814–821

Santiago N, Herraiz C, Goni JR, Messeguer X, Casacuberta JM (2002) Genome-wide analysis of the Emigrant family of MITEs of *Arabidopsis thaliana*. Mol Biol Evol 19:2285–2293

Sarilar V, Marmagne A, Brabant P, Joets J, Alix K (2011) BraSto, a Stowaway MITE from Brassica: recently active copies preferentially accumulate in the gene space. Plant Mol Biol 77:59–75

Schwarz-Sommer Z, Gubitz T, Weiss J, Gomez-di-Marco P, Delgado-Benarroch L, Hudson A, Egea-Cortines M (2010) A molecular recombination map of *Antirrhinum majus*. BMC Plant Biol 10:275

Sinzelle L, Jegot G, Brillet B, Rouleux-Bonnin F, Bigot Y, Auge-Gouillou C (2008) Factors acting on Mos1 transposition efficiency. BMC Mol Biol 9:106

Song WY, Pi LY, Bureau TE, Ronald PC (1998) Identification and characterization of 14 transposon-like elements in the noncoding regions of members of the Xa21 family of disease resistance genes in rice. Mol Gen Genet 258:449–456

Stratton M (2008) Genome resequencing and genetic variation. Nat Biotechnol 26:65–66

Surzycki SA, Belknap WR (2000) Repetitive-DNA elements are similarly distributed on *Caenorhabditis elegans* autosomes. Proc Natl Acad Sci USA 97:245–249

Ton-Hoang B, Pasternak C, Siguier P, Guynet C, Hickman AB, Dyda F, Sommer S, Chandler M (2010) Single-stranded DNA transposition is coupled to host replication. Cell 142:398–408

Tu ZJ (2001) Eight novel families of miniature inverted repeat transposable elements in the African malaria mosquito, *Anopheles gambiae*. Proc Natl Acad Sci USA 98:1699–1704

Wei F, Wing RA, Wise RP (2002) Genome dynamics and evolution of the Mla (powdery mildew) resistance locus in barley. Plant Cell 14:1903–1917

Wessler SR, Bureau TE, White SE (1995) LTR-retrotransposons and MITEs: important players in the evolution of plant genomes. Curr Opin Genet Dev 5:814–821

Wright SI, Agrawal N, Bureau TE (2003) Effects of recombination rate and gene density on transposable element distributions in *Arabidopsis thaliana*. Genome Res 13:1897–1903

Yan Y, Zhang Y, Yang K, Sun Z, Fu Y, Chen X, Fang R (2011) Small RNAs from MITE-derived stem-loop precursors regulate abscisic acid signaling and abiotic stress responses in rice. Plant J 65:820–828

Yang G, Hall TC (2003a) MAK, a computational tool kit for automated MITE analysis. Nucleic Acids Res 31:3659–3665

Yang G, Hall TC (2003b) MDM-1 and MDM-2: two mutator-derived MITE families in rice. J Mol Evol 56:255–264

Yang G, Zhang F, Hancock CN, Wessler SR (2007) Transposition of the rice miniature inverted repeat transposable element mPing in *Arabidopsis thaliana*. Proc Natl Acad Sci USA 104:10962–10967

Yang G, Nagel DH, Feschotte C, Hancock CN, Wessler SR (2009) Tuned for transposition: molecular determinants underlying the hyperactivity of a Stowaway MITE. Science 325:1391–1394

Zhang X, Feschotte C, Zhang Q, Jiang N, Eggleston WB, Wessler SR (2001) P instability factor: an active maize transposon system associated with the amplification of Tourist-like MITEs and a new superfamily of transposases. Proc Natl Acad Sci USA 98:12572–12577

Chapter 8
Glue for Jumping Elements: Epigenetic Means for Controlling Transposable Elements in Plants

Thierry Pélissier and Olivier Mathieu

Abstract Transposable elements (TEs) and their derivatives are highly abundant in plant genomes. The potential mobilization of TEs poses a constant threat to genome integrity; this hazardous situation may explain why epigenetic regulation initially emerged. Plants use different epigenetic silencing mechanisms to restrain TE mobility during different stages of their life cycle. DNA methylation, posttranslational modification of histone tails and small RNA-based pathways contribute to restraining TE activity. The frontier between these mechanisms is sometimes blurry, and their exact contributions are complicated to delineate. The availability of several silencing mechanisms provides versatility that has allowed the hosts' genomes to individualize the silencing of particular TEs. There is recent evidence, particularly in *Arabidopsis thaliana*, that the silencing of TEs is much more dynamic than had been previously thought and can be relieved in certain cell lineages or under adverse environmental conditions.

Keywords Transposable elements • Epigenetics • Silencing • Arabidopsis • DNA methylation • Histone modification • Stress

8.1 Introduction

The rapid development of new sequencing methodologies and their wide-scale implementation over the past 10 years has resulted in the compiling of genomic sequences from a variety of organisms. Contrary to previous assumptions, genomes, particularly plant genomes, are loaded with transposable element (TE) sequences, and TEs can be the major constituent of a genome (Wessler 2006; Tenaillon et al. 2010). For example, the maize genome contains about six times

T. Pélissier • O. Mathieu (✉)
Centre National de la Recherche Scientifique (CNRS), UMR 6293 – GReD – INSERM U, 1103, Clermont Université, 24 av. des landais, BP 80026 Aubière, France
e-mail: olivier.mathieu@univ-bpclermont.fr

M.-A. Grandbastien and J.M. Casacuberta (eds.), *Plant Transposable Elements*, Topics in Current Genetics 24, DOI 10.1007/978-3-642-31842-9_8,

more transposable element genes than genes that are not encoded for by TEs (Baucom et al. 2009). TEs are mobile pieces of DNA that have been viewed as genomic parasites that make additional copies of themselves and insert them at new genomic positions. Class I TEs, known as retrotransposons, move through an RNA intermediate, while class II elements transpose via a DNA excision ("cut-and-paste") mechanism or, in the case of the more recently identified Helitrons, an apparent rolling-circle mechanism (Wicker et al. 2007).

In most eukaryotic genomes, including plants, most TEs are clustered around centromeres; however, a large number of TEs are present on the euchromatic arms of chromosomes. The vast majority of TEs are defective because of mutations or deletions, but some full-length elements still retain an intact code and the potential to transpose. This presents a constant threat to genomic integrity, which led to their discovery in maize by Barbara McClintock in the late 1940s (McClintock 1948). Although TE activity has been beneficial to host genomes in some instances, it more often generates selectively disadvantageous outcomes such as chromosome breakage or disruption of gene function. To cope with the harmful potential of TE activity, host genomes have evolved sophisticated mechanisms that counteract TE mobilization and maintain TEs in a silent, quiescent state. Nonetheless, even silent TEs are likely to play an important role in the evolution of animal and plant genomes, because TEs and the mechanisms that regulate their activity have been co-opted for a wide variety of major cellular processes, ranging from gene regulation to centromere and telomere function, genomic imprinting, and X-chromosome inactivation (Slotkin and Martienssen 2007; Lisch 2009; Chow et al. 2010).

The mechanisms that silence TEs are epigenetic; they heritably silence the expression of TEs but do not alter their coding potential. Most often, these mechanisms generate a repressive chromatin environment that is associated with specific small RNA signatures and a range of epigenetic marks that affect DNA and histone proteins. Alternatively, some mechanisms employ small RNAs derived from TEs to actively target TE mRNA for degradation.

In this chapter, we present these epigenetic processes and how they contribute to the silencing of TEs with a focus on recent studies that have highlighted the plasticity of these mechanisms under certain developmental or environmental cues.

8.2 DNA Methylation of Cytosine Residues

8.2.1 Propagation of DNA Methylation Patterns

Although still controversial in animals (Suzuki and Bird 2008), plants incontestably use DNA methylation to defend against TEs. The cytosine residues of most eukaryotic genomes can be modified through the addition of a methyl group to position 5 of the pyridine ring. There are a few exceptions among the commonly used laboratory model organisms, including yeast (*Schizosaccharomyces pombe* and *Saccharomyces cerevisiae*), worms (*Caenorhabditis elegans*), and fruit flies (*Drosophila melanogaster*); the genomes of all other organisms analyzed to date contain detectable levels of DNA methylation, attesting to the ancient evolutionary origin of this epigenetic modification (Feng et al. 2010; Zemach et al. 2010). Defects in maintaining DNA methylation lead to a wide range of developmental abnormalities in plants and to embryonic lethality in mammals. In mammals, DNA methylation occurs almost exclusively at symmetric CGs. In plants, non-CG methylation is common, and DNA methylation affects cytosines in all sequence contexts, including the symmetric CG and CHG contexts (H = A, T or C) and the asymmetric CHH context.

Mammalian DNA methyltransferase 1 (DNMT1) and its counterpart in plants, METHYLTRANSFERASE1 (MET1), propagate CG methylation patterns following each round of DNA replication. Hemimethylated CG sites that are generated during replication are recognized by the chromatin-associated proteins UHRF1 in mammals and VARIANT IN METHYLATION (VIM) in plants, which likely recruit DNMT1 and MET1 to hemimethylated DNA (Bostick et al. 2007; Liu et al. 2007; Woo et al. 2008). A plant-specific protein, CHROMOMETHYLASE3 (CMT3), maintains DNA methylation in the CHG context, while DOMAINS REARRANGED METHYLTRANSFERASE2 (DRM2) largely ensures the persistence of CHH methylation. It is thought that CMT3 is recruited to chromatin via its chromodomain that binds to methylated histone H3 tails (Lindroth et al. 2004). DRM2, an ortholog of the mammalian DNMT3 methyltransferases, functions in a pathway known as RNA-directed DNA methylation (RdDM), which is related to the canonical RNA interference pathway and is targeted to DNA by 24-nucleotide (nt) small interfering RNAs (siRNAs) (Cao and Jacobsen 2002; Cao et al. 2003; Matzke et al. 2009). This pathway requires the activity of the plant-specific RNA polymerases IV and V (Zhang and Zhu 2011). In addition, maintenance of DNA methylation in mammals and plants also requires the chromatin remodeling factors LSH1 and DECREASE IN DNA METHYLATION1 (DDM1), respectively; however, the mechanism through which these two factors affect DNA methylation remains unknown (Vongs et al. 1993; Jeddeloh et al. 1998, 1999; Dennis et al. 2001).

8.2.2 DNA Methylation and Defense Against TE Activity

Studies in maize were among the first to reveal the importance of DNA methylation in TE silencing. Roughly 40 years after the discovery of TEs by B. McClintock, studies of the *Activator* (*Ac*), *Suppressor–mutator* (*Spm*), and *Mutator* (*Mu*) elements revealed that inactivation of these elements was correlated with the methylation of their DNA (Chandler and Walbot 1986; Chomet et al. 1987; Banks et al. 1988). Indeed, active transcription of the *Ac* and *Spm* elements requires hypomethylation of their transposase promoter, while DNA methylation of the autonomous *Mu*-family elements *MuDR* is associated with transcriptional silencing (Hershberger et al. 1991; Rudenko and Walbot 2001). Additional work on

Arabidopsis later confirmed that the global genomic hypomethylation induced by mutations of *DDM1* results in the transcriptional activation of several classes of TEs (Miura et al. 2001; Singer et al. 2001; Lippman et al. 2003, 2004). Moreover, mobilization was detected for the MULE-like element *AtMu1* and the *CACTA* DNA transposons (Miura et al. 2001; Singer et al. 2001).

More recent analyses of *ddm1* mutants have expanded the list of TEs that can be mobilized in this mutant background based on increases in the copy number of several families of retrotransposons and DNA transposons (Tsukahara et al. 2009). In mice, mutations in *LSH1* result in hypomethylation and transcriptional reactivation of TEs (Yan et al. 2003; Huang et al. 2004). These and other experimental data have led to the conclusion that DNA methylation is used as part of a genomic immune system against TE activity.

Perhaps the strongest evidence for such a function of DNA methylation can be found in recent studies that mapped the distribution of DNA methylation along the entire *Arabidopsis* genome. These studies highlighted the ubiquitous methylation of TEs and showed that TEs and other types of repeated sequences are the most highly methylated sequences in the genome (Zhang et al. 2006; Zilberman and Henikoff 2007; Cokus et al. 2008; Lister et al. 2008). This observation is not specific to *Arabidopsis* and can likely be extended to all plant types based on two recent, large-scale studies that quantified DNA methylation in an additional 21 eukaryotic genomes (Feng et al. 2010; Zemach et al. 2010).

8.2.3 Selectivity in the Use of DNA Methylation

TEs are not only modified by CG methylation; consistent with their preferential association with H3K9me2 and their large contribution to the pool of small RNAs, they are also preferential targets of both CHG and asymmetrical methylation (Zhang et al. 2006, 2007; Bernatavichute et al. 2008; Lister et al. 2008). Several lines of evidence suggest that all three types of methylation are used in TE silencing. In *Arabidopsis*, it was shown that mutations in *MET1* or *CMT3* activate the transcription of *CACTA* elements. However, the high-frequency mobility of this TE is only seen in the *met1 cmt3* double mutant (Kato et al. 2003). Therefore, the presence of both CG and CHG methylation is necessary to maintain the silencing of class II *CACTA* elements, and only in the absence of both is the element actively transposed. Similarly, class I *ATGP3* retroelements are transcribed and transposed when the maintenance of both CG and non-CG methylation is compromised in the *met1 cmt3* double mutant and the *ddm1* mutant backgrounds. Yet, in contrast to *CACTA, ATGP3* elements remain transcriptionally silent in *met1* or *cmt3* single mutants (Tsukahara et al. 2009), showing that CG and CHG methylation act in a redundant manner to silence transcription of these TEs. These examples highlight that host genomes use combinations of CG and CHG methylation to curb TE activity. This implies that DNA methylation has distinctive impacts on different types of TEs. This assumption can be further illustrated by an analysis of the epigenetic control of the *copia*-like retrotransposon, *Evadé* (*EVD*) (Mirouze et al. 2009). The promoter region of this element is methylated at CG dinucleotides and contains low levels of CHH methylation in wild-type plants. Loss of CG methylation is sufficient to allow transcription of *EVD* and even mobilization in late generation *met1* mutants. Mutations known to impair CHH methylation, including *drm2* and *pol IV/V*, do not interfere with *EVD* transcriptional silencing; however, when mutated in combination with *met1*, they lead to a synergistic increase in transcription, indicating that the RdDM pathway is largely used to reinforce the existing silenced state of *EVD*. In contrast, *ATGP3* silencing is efficiently maintained even in backgrounds that are defective for both CG and CHH methylation (Tsukahara et al. 2009). These findings underscore that host genomes have evolved selectivity in the use of DNA methylation to control TE activity. Noticeably, in the absence of CG methylation defects in NPRE2, the common subunit of pol IV and V, a burst of *EVD* transposition occurs; however, other transcriptionally activated TEs, such as *CACTA*, remain immobile [(Mirouze et al. 2009); see Sect. 8.4].

In *Arabidopsis*, loci that lose DNA methylation and its associated epigenetic silencing in *met1* and *ddm1* mutants, including TEs, retain activity and hypomethylation for several generations following removal of the mutations (Vongs et al. 1993; Kakutani et al. 1999; Soppe et al. 2002; Lippman et al. 2003). In contrast, TEs are not heritably activated when the RdDM pathway is compromised (Chan et al. 2006). Because mutations in *DDM1* and *MET1* both have dramatic effects on CG methylation but the RdDM pathway mostly affects CHG and CHH methylation, this suggests that CG methylation is the primary platform for heritable silencing information. CG hypomethylation-induced reactivation of TEs is not necessarily irreversible. Progressive remethylation over successive generations can occur at a subset of TEs when DDM1 function is restored in a *ddm1* hypomethylated background. TEs that become remethylated are characterized by high amounts of corresponding siRNAs and the retention of a certain level of CHH methylation in *ddm1* (Teixeira et al. 2009), illustrating that CHH methylation is maintained by distinct mechanisms at various TEs. The progressive remethylation suggests that the RdDM pathway may act as a backup system against transgenerational loss of DNA methylation at TEs (Mathieu et al. 2007; Teixeira et al. 2009). As mentioned above, "remethylatable" TEs appear less demethylated in *ddm1* than "nonremethylatable" TEs, suggesting that only incomplete demethylation can be corrected and restored to wild-type patterns. Whether activated TEs can be resilenced following a complete loss of all types of DNA methylation remains an open question.

8.3 Post-translational Histone Modifications

8.3.1 Methylation of Histone H3 at Lysine 9 and 27

The repressive environment typical at TE chromatin is not only characterized by the presence of DNA methylation but also by a variety of post-translational modifications of histone proteins; modifications affecting histone H3 are the most accurately described. Modifications of histone amino-terminal tails enable or inhibit the binding of various proteins that directly or indirectly impact transcription. In plants, nucleosomes that are associated with TEs located in pericentric heterochromatin are enriched for H3K9me2 and H3K27me1, which are signals of transcriptionally repressive chromatin (Lindroth et al. 2004; Mathieu et al. 2005; Bernatavichute et al. 2008; Jacob et al. 2009). In *Arabidopsis*, the propagation of H3K9me2 is ensured by the partly redundant activity of three histone methyltransferases, namely KRYPTONITE/SUVH4 (KYP), SUVH5 and SUVH6 (Jackson et al. 2002; Malagnac et al. 2002; Ebbs et al. 2005; Ebbs and Bender 2006; Johnson et al. 2007). However, deposition of H3K27me1 is catalyzed by the histone methyltransferases, *ARABIDOPSIS* TRITHORAX-RELATED PROTEIN 5 (ATXR5) and ATXR6, which also exhibit some functionally redundant activity (Jacob et al. 2009).

Mutations in genes encoding these histone methyltransferases or their homologs in other organisms often lead to TE reactivation, although the upregulation is more modest (in terms of intensity or spectrum) than that induced by the loss of DNA methylation. In *Arabidopsis*, TE reactivation was observed in *kyp suvh5 suvh6* triple mutants (Ebbs and Bender 2006) and *atxr5 atxr6* double mutants (Jacob et al. 2009). Defects in the histone H3K9 methyltransferases genes *SGD714* and *SDG728* also activate TE transcription in rice (Ding et al. 2007; Qin et al. 2010). The role of H3K27me1 in transcriptional silencing of TEs is independent of DNA methylation and H3K9me2 (Jacob et al. 2009). TEs reactivated in *atxr5 atxr6* mutants retain high levels of these two repressive marks, and conversely, H3K27me1 levels are not changed in *met1* and *ddm1* mutants, which show reduced DNA methylation, and the *kyp* mutant, which shows reduced H3K9me2 levels (Lindroth et al. 2004; Mathieu et al. 2005). How H3K27me1 represses TE transcription is still unknown. H3K27me1 has been shown to prevent DNA replication from occurring more than once per cell cycle preferentially at TE-rich heterochromatic regions, which led to speculation that this mark may have evolved to restrain excess heterochromatic DNA replication and reactivation of TEs (Jacob et al. 2010).

Unlike H3K27me1, H3K9me2 is tightly interwoven with DNA methylation, in particular, CHG methylation. Consequently, the exact contribution of H3K9me2 to TE silencing is more difficult to identify. In the *Arabidopsis* genome, approximately 90% of CHG methylation overlaps with H3K9me2-enriched regions (Bernatavichute et al. 2008), and TEs represent preferential targets for CHG methylation (Tompa et al. 2002; Lippman et al. 2004; Zhang et al. 2006; Zilberman et al. 2007; Cokus et al. 2008). Mutants for H3K9me2 histone methyltransferases show not only reduced H3K9me2 levels at reactivated TEs but also a significant reduction in DNA methylation at CHG sites (Jackson et al. 2002; Malagnac et al. 2002; Ebbs et al. 2005; Ebbs and Bender 2006; Johnson et al. 2007). Reciprocally, TEs activated in mutants for *CMT3*, the maintenance DNA methyltransferase for CHG sites, also exhibit reduced H3K9me2 levels (Johnson et al. 2002). Biochemical studies have provided a molecular explanation for these genetic observations. KYP, SUVH5, and SUVH6 contain SET and RING associated (SRA) domains, which have been shown to bind to DNA at methylated cytosines, and KYP and SUVH6 display a preference for CHG methylation over CG methylation (Johnson et al. 2007; Rajakumara et al. 2011). This binding affinity is thought to recruit these histone methyltransferases to their genomic targets. CMT3 contains a chromodomain that likely interacts with H3K9me2, thereby recruiting the DNA methyltransferase to chromatin (Feng and Jacobsen 2011). Therefore, H3K9me2 and CHG methylation are maintained through a self-reinforcing feedback loop such that influencing one modification likely impacts the other. Further blurring the line between DNA methylation and H3K9me2, an additional pathway that is dependent on CG methylation maintains H3K9me2 at a subset of TEs (Inagaki et al. 2010). Recently, genetic studies have identified the *INCREASE IN BONSAI METHYLATION1* (*IBM1*) gene that encodes a putative H3K9me2 demethylase. Interestingly, *ibm1* mutations result in ectopic H3K9me2 and CHG DNA hypermethylation in a large number of genes, while TEs are unaffected (Saze et al. 2008; Miura et al. 2009; Inagaki et al. 2010). Thus, IBM1 protects genes but not TEs from H3K9me2 and CHG methylation. Along with multiple pathways that serve to maintain H3K9me2, this situation contributes to the perpetuation of robustly silenced TEs.

8.3.2 Impact of Other Histone Modifications on TE Silencing

The presence or absence of other modifications at histone tails have been implicated in TE silencing; however, these modifications also appear linked to DNA methylation and their exact contribution to silencing is difficult to identify. For instance, histone deacetylation appears to be important for the maintenance of TE silencing. Mutations of the *Arabidopsis* histone deacetylase gene, *HDA6*, or downregulation of the histone deacetylase gene, *OsSRT1*, in rice result in transcriptional activation of several classes of TEs (Lippman et al. 2003; Huang et al. 2007). Reactivation of these TEs is also correlated with a reduction in H3K9me2, the appearance of H3K4me2, and, occasionally, a reduction in DNA methylation. Similar to histone acetylation, ubiquitination of histone H2B is required to maintain chromatin in an open state (Zhang 2003). Mutations in an *Arabidopsis* deubiquitination enzyme, UBP36, release transcriptional silencing of several different classes of TEs. Histone H2B is mostly non-ubiquitinated at silent TEs, and the deubiquitination of H2B seems to be required for H3K9me2 deposition and subsequent non-CG DNA methylation (Sridhar et al. 2007). Interestingly, this suggests that TEs may be associated with active histone modification marks by default and that faithful maintenance of a repressive chromatin state depends on the continuous removal of these activating marks.

Histone methyltransferases may function not only in blocking TE transcription but also in inhibiting the later stages of the TE life cycle. It was recently shown that the KYP histone methyltransferase plays a role in restraining the mobilization of the *EVD* retroelement, specifically at the posttranscriptional level (Mirouze et al. 2009). Analogous to mammalian lysine methyltransferases, which also methylate nonhistone proteins (Chuikov et al. 2004; Kouskouti et al. 2004; Sampath et al. 2007; Esteve et al. 2009), KYP has been proposed to inactivate a TE-encoded protein required for the translation and/or reverse transcription of *EVD* transcripts through methylation (Mirouze et al. 2009). Further investigations are needed to clarify the molecular basis and the extent of this silencing pathway.

8.4 Small RNAs and TE Silencing

8.4.1 Various Small RNA Pathways

Over the last two decades, small RNAs have emerged as key regulators of gene expression in eukaryotes and have been shown to be involved in a wide range of biological processes, including developmental timing, cell differentiation, metabolic control, antiviral defense, genome rearrangement, and TE silencing. Various classes of ~20–35 nt small RNAs have been described that can guide silencing at a variety of points, including posttranscriptionally through mRNA degradation or translation inhibition, transcriptionally through DNA methylation and/or chromatin modifications, and cotranscriptionally through inhibiting transcription elongation (Brodersen et al. 2008; Guang et al. 2010; Vazquez et al. 2010; Ketting 2011; Zhang and Zhu 2011).

In animal systems, two classes of small RNAs, known as endogenous siRNAs (endo-siRNAs) and PIWI-associated RNAs (piRNAs), are more targeted at TE silencing. The pi-RNA pathway appears to primarily operate in the germ line, while endo-siRNAs are also produced and active in somatic tissues (Kim et al. 2009). The ~21-nt long endo-siRNAs are processed by Dicer from transposon-derived long double-stranded RNAs (dsRNAs), whereas ~24–31-nt long piRNAs are derived from single-stranded transcripts originating from particular transposon-containing genomic clusters in a Dicer-independent mechanism. Both pathways target TEs for posttranscriptional silencing (PTGS) and can induce transcriptional silencing (TGS) through DNA methylation or heterochromatin formation in mice and Drosophila, respectively (Pal-Bhadra et al. 2004; Aravin et al. 2008; Kuramochi-Miyagawa et al. 2008; Okamura and Lai 2008; Fagegaltier et al. 2009; Kim et al. 2009; Bourc'his and Voinnet 2010; Siomi et al. 2011).

Plants produce several classes of siRNAs, but no piRNAs have been identified (Vazquez et al. 2010). The vast majority of siRNAs in *Arabidopsis* consists of ~24-nt long siRNAs that are known to guide the DRM2 DNA methyltransferase for de novo DNA methylation and silencing of complementary genomic sequences in the RdDM pathway (Matzke et al. 2009). These siRNAs are predominantly derived from various types of repeats, including TEs, which are enriched for DNA methylation and H3K9me2 (Zhang et al. 2006; Kasschau et al. 2007; Bernatavichute et al. 2008; Cokus et al. 2008; Lister et al. 2008). Paradoxically, siRNA-mediated silencing requires transcription of silent loci, which depends on the activity of two plant-specific RNA polymerase, Pol IV and Pol V, which are homologs of the DNA-dependent RNA polymerase II. It is still unclear how Pol IV is recruited to these loci; however, current models propose a Pol IV affinity for methylated DNA templates (Fig. 8.1) (Vazquez et al. 2010; Zhang and Zhu 2011). Pol IV is thought to generate single stranded RNAs (ssRNAs) that are then converted into dsRNAs by the RNA-dependent RNA polymerase, RDR2, and cleaved into 24-nt siRNAs by the DICER-LIKE3 (DCL3) endonuclease. The siRNAs are then loaded onto RNA-binding ARGONAUTE proteins (AGO4 or AGO6), and these

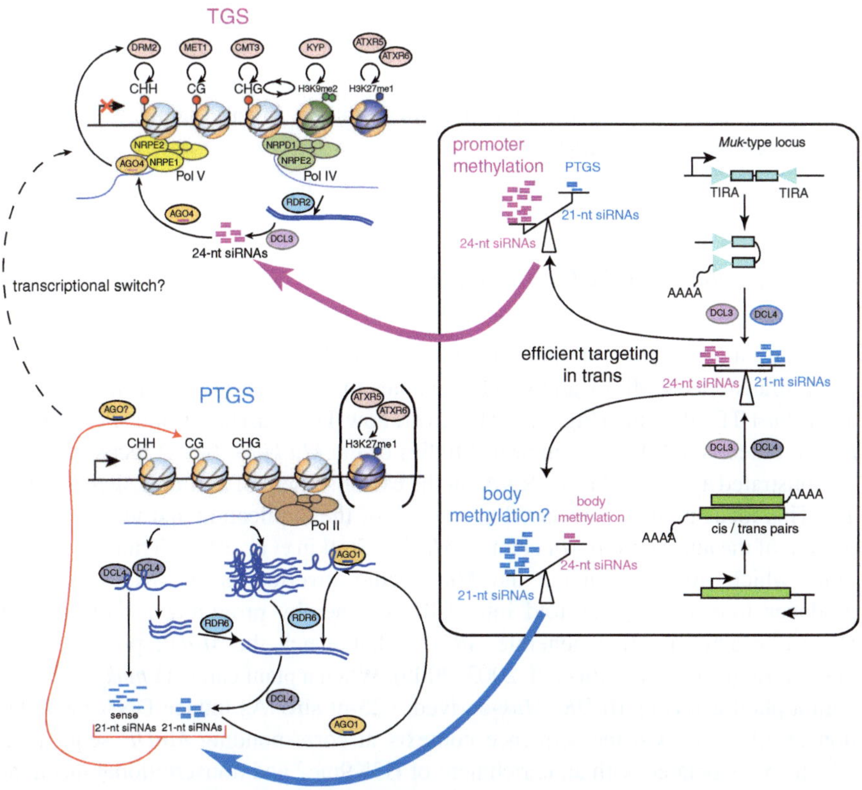

Fig. 8.1 Schematic representation of posttranscriptional silencing (PTGS) and transcriptional silencing (TGS) pathways during the initiation and maintenance steps of transposable element (TE) silencing. The *box* describes the chromosomal rearrangements that would encourage the induction of TGS and/or PTGS. The former is exemplified by the transposon rearrangements at the *Muk* locus (*upper portion*). An inverted repeat (IR) of a *MuDr* portion downstream of an endogeneous promoter leads to the production of dsRNA templates that can then be processed by DCL3 and/or DCL4 to produce variable levels of 21- and 24-nt siRNAs. The 21-nt siRNAs could be targeted and result in PTGS of ectopic copies and amplify. The presence of the *mudrA*-promoter region, TIRA (Terminal Inverted Repeats *mudrA*), would allow 24-nt siRNAs to actively target RdDM in *cis* and *trans*, leading to DNA methylation and transcriptional repression of ectopic copies of *MuDR*, correlating with an increased accumulation of 24-nt siRNA and a loss of most 21-nt siRNAs. Following the induction of TGS, the *Muk* locus is dispensable and can be lost without affecting TGS maintenance. Robust TGS maintenance depends on several overlapping pathways that are of variable importance in each class of TEs (TGS section); MET1, CMT3, and siRNA-targeted DRM2 methyltransferases, in combination with H3K9me2, contribute to the perpetuation of DNA methylation patterns. Maintenance of H3K27me1 is DNA methylation-independent and requires the activity of ATXR5 and ATXR6. Other rearrangements that do not contain transposon promoter sequences, including IR or sense–antisense pairs derived from TEs, are primarily channeled into PTGS processes (*box-lower part*), and AGO1-containing complexes target complementary transcripts for cleavage. To some extent, the direct cleavage of highly structured TE RNA regions by DCL4 can also restrict TE RNA accumulation. Additionally, 21-nt-mediated imprinting in TE bodies and/or a high level of transcription may generate TE transcripts that may be sensed as "aberrant." In all cases, an RDR6-dependent amplification loop allows for

ARGONAUTE/siRNA complexes bind nascent scaffold RNAs generated from intergenic regions by Pol V and/or Pol II (Zhang and Zhu 2011). This binding appears to be required for the recruitment of downstream effectors, including DRM2, which can induce cytosine methylation in all sequence contexts (Wierzbicki et al. 2008, 2009).

8.4.2 Initiation of TE Silencing

The continuous action of the RdDM pathway, together with the maintenance methyltransferases MET1 and CMT3, enables the propagation of TE methylation and robust TE silencing (Fig. 8.1). However, how TE methylation and silencing are first initiated is still largely unclear. Studies of the *Mu killer* (*Muk*) locus in maize demonstrated a role for 24-nt siRNAs in the establishment of heritable silencing at a TE. This locus naturally occurred as a result of the duplication and inversion of a portion of the autonomous *Mu* element *MuDR* (Slotkin et al. 2005). Transcription of *Muk*, which initiates outside the *Mu*-specific inverted-repeat (IR) sequences, produces long RNAs that fold into dsRNAs, thereby providing a template for Dicer-like activities that generate siRNAs that match the *MuDR* promoter and internal sequences (Slotkin et al. 2003, 2005). When a plant carrying *muk* is crossed with a plant carrying *MuDR*, *Muk*-derived, ~25-nt siRNAs induce DNA methylation in all three cytosine sequence contexts at corresponding *MuDR* sequences, which are associated with an enrichment of H3K9me2 and transcriptional inactivation of *MuDR* (Fig. 8.1) (Slotkin et al. 2003, 2005; Li et al. 2010). Although transposons are subject to frequent rearrangements, the prevalence of a similar process to initiate silencing of a particular family or class of TEs in plant genomes is difficult to evaluate.

When a new TE invades a naïve genome, it is likely actively transcribed. This may parallel the burst of TE transcription observed in specific cell lineages, in plant cell cultures, and in *ddm1* or *met1* mutants (see the following section for further information). In these cases, TE activation is accompanied by the appearance of TE-matching 21-nt siRNAs (Lister et al. 2008; Tanurdzic et al. 2008; Mirouze et al. 2009; Slotkin et al. 2009; Teixeira et al. 2009). As shown for *Athila* and *EVD* retrotransposons, the 21-nt siRNAs typically originate from limited internal regions

---

Fig. 8.1 (continued) efficient PTGS of these elements (PTGS section). At specific genomic loci, the DNA methylation that is often detected within the transcribed regions of PTGS targets may spread, leading to a potential switch to TGS. Conversely, an alteration of DNA methylation patterns can alleviate transcriptional silencing. In such cases, posttranscriptional regulation may immediately be reengaged to restrain TE expression and/or transposition as described for the EVD retrotransposon (see the text for details). The importance of epigenetic marks that can persist upon DNA hypomethylation-associated transcriptional reactivation (e.g., H3K27me1) is unknown. For clarity, only some of the key protein activities of the complexes involved in the silencing pathways are depicted

of the elements, in contrast to the relatively dispersed distribution of 24-nt siRNAs along the entire length of retrotransposon sequences [(Mirouze et al. 2009; Slotkin 2010); Pélissier and Mathieu, unpublished results]. Whether this pattern of 21-nt siRNAs accumulation reflects direct recognition and processing of secondary structures within TE transcripts (Molnár et al. 2005; Jakubiec et al. 2012) and/or the release of "aberrant" RNA transcripts is currently unknown. Whatever their origin, 21-nt siRNAs generally target PTGS, potentially preventing the accumulation of TE transcripts. However, PTGS often correlates with genomic DNA methylation within the transcribed regions (Vaucheret 2006), and evidence revealing considerable connections between the PTGS and TGS pathways is emerging (Eamens et al. 2008; Daxinger et al. 2009; Bourc'his and Voinnet 2010). Therefore, a role for these 21-nt siRNAs in the initiation of transcriptional TE taming during genome colonization may be plausible (Fig. 8.1). TE gene-body methylation, initially triggered by the 21-nt siRNAs, could then spread into the TE promoter sequences (Daxinger et al. 2009), thereby inducing robust TE silencing at the transcriptional level.

8.4.3 Maintenance of TE Silencing

Once established, the maintenance of DNA methylation and silencing involves several partially overlapping pathways (Fig. 8.1). RdDM-mediated non-CG methylation appears to be used largely to reinforce the preexisting silencing at TEs, and defects in the RdDM pathway in *Arabidopsis* only result in selective transcription reactivation of a subset of TEs (Kanno et al. 2004, 2005, 2008, 2010; Herr et al. 2005; Onodera et al. 2005; Pontier et al. 2005; Huettel et al. 2006; Gao et al. 2010). Further illustrating the connection between TGS and PTGS silencing, it has been shown recently that NRPE2 (the common subunit of Pol IV and Pol V and a basal component of the RdDM pathway) also restricts the mobilization of the *EVD* retroelement at the posttranscriptional level after it has been transcriptionally activated (Mirouze et al. 2009). Similarly, as mentioned above, the efficient mobilization of *EVD* has been observed in *met1 kyp* double mutants (Mirouze et al. 2009). Whether the 21-/24-nt siRNAs that accumulate after *EVD* transcriptional activation are involved in posttranscriptional regulation by KYP and NRPE2 and whether KYP and NRPE2 function in the same or in different mechanisms are still unclear.

The analysis of *ddm1*-derived epiRILs has highlighted the potential importance of small RNAs and RdDM pathways in reimposing TE silencing following an alteration in the epigenetic pattern (Teixeira et al. 2009). In another study, Olmedo-Monfil et al. (2010) demonstrated that 24-nt TE siRNAs are also essential for maintaining TE silencing in the *Arabidopsis* female gametophyte (the egg and neighboring cells), in contrast to their modest impact in somatic tissues. Interestingly, this control pathway requires the function of another ARGONAUTE protein, AGO9, which belongs to the same clade as AGO4 and 6. AGO9 is not expressed in the female gametophyte itself but is expressed in the surrounding somatic companion cells. The inactivation of TEs in the cells surrounding the female gametophyte appears to be necessary in order to maintain their cellular identity and to prevent them from abnormally differentiating into gametic cells. AGO9-mediated TE silencing requires several known components of the RdDM pathway (i.e., *RDR2*, *DCL3*, Pol IV, and/or Pol V) in addition to factors known to be involved in distinct small RNA pathways (Olmedo-Monfil et al. 2010). Therefore, AGO9 appears to be involved in an unorthodox, and potentially specific, siRNA-silencing pathway that is necessary for maintaining TE silencing in female gametes. Interestingly, AGO9 is also highly expressed in anthers (Olmedo-Monfil et al. 2010), where it may play a similar role.

8.5 Dynamics of TE Silencing

Although TE silencing can be inherited over multiple generations, this state can be reversed during specific developmental windows and in response to a wide range of stress conditions.

8.5.1 Developmental Reprogramming of TE Silencing

A drastic reprogramming of DNA methylation has been recently reported in both male and female germinal lineages from *Arabidopsis*. This takes place in "dead-end" cells that do not contribute to the next generation and correlates to some extent with TE reactivation. Paradoxically, this reactivation could contribute to keeping TEs quiescent in the egg and sperm cells and later the embryo, thereby protecting the genomic integrity of the offspring. In angiosperms, female gametogenesis leads to the formation of a haploid egg cell and a homodiploid central cell that are surrounded by several accessory cells. The male gametophyte, or pollen grain, contains three haploid cells—two sperm cells that are embedded inside the cytoplasm of a larger vegetative cell. In the double fertilization process common to angiosperms, one sperm cell fertilizes the haploid egg cell, giving rise to a diploid embryo, while the second sperm cell fertilizes the homodiploid central cell, producing a triploid endosperm that provides a nurturing tissue for embryo development. Recent studies in *Arabidopsis* have revealed widespread reductions in DNA methylation at TEs and other repeats in the endosperm (Gehring et al. 2009; Hsieh et al. 2009). This loss of methylation is probably initiated in the central cell prior to fertilization as a result of specific induction of DEMETER, a DNA glycosylase that excises methylcytosines in all sequence contexts, in this cell type in conjunction with the reduced expression of MET1 during gametogenesis (Choi et al. 2002; Gehring et al. 2006, 2009; Morales-Ruiz et al. 2006; Jullien et al. 2008; Hsieh et al. 2009). Hypomethylation occurs primarily at CG sites and is accompanied by increased methylation at CHH sites at repeated sequences, which are consistent with previous observations from *met1* mutants that have lost CG methylation (Mathieu et al. 2007; Hsieh et al. 2009). Methylation at CHH sites is an hallmark of RdDM, and consistent with this observation, these methylation patterns correlate with a massive production of 24-nt siRNAs that initiates in the central cell and persists in the endosperm (Mosher et al. 2009). Strikingly, the methylation level in the embryo is higher than in the aerial tissues in all of the sequence contexts, and CHH hypermethylation is particularly extensive (Hsieh et al. 2009). As 21-nt and 24-nt siRNAs can exert their silencing functions over a distance (Dunoyer et al. 2010; Molnar et al. 2010; Melnyk et al. 2011), it has been proposed that some of the 24-nt siRNAs produced in the central cell/endosperm migrate into the egg cell/ embryo to reinforce epigenetic marks that silence transposons (Hsieh et al. 2009). Whether reduced DNA methylation in the endosperm correlates with a reduction in TE silencing has not been tested. On the male side, a related process occurs in the vegetative cell; DNA hypomethylation is detected at TEs in association with transcriptional activity and, for some of them, with transposition (Slotkin et al. 2009). In this accessory cell, DNA hypomethylation likely results from a downregulation of genes, such as *MET1* and *DDM1*, in conjunction with active demethylation by an unidentified DNA demethylase (Borges et al. 2008; Jullien et al. 2008; Slotkin et al. 2009). TE reactivation is associated with decreased accumulation of 24-nt siRNAs and with a dramatic gain of 21-nt siRNAs, easily detected for the high copy number *Athila* retrotransposon (Slotkin et al. 2009). These mobile 21-nt siRNAs accumulate in sperm cells, where they could provide an additional layer of TE silencing. They may also do this later in the embryo through posttranscriptional degradation of TEs that would escape TGS and/or by reinforcing some of the preexisting chromatin imprints associated with TGS. Interestingly, the epigenetic reprogramming of TE silencing has also been documented in maize, where *MuDR* silencing is reversed upon the change from the juvenile to adult phase in a tissue adjacent to the one that will produce the germ line (Li et al. 2010).

8.5.2 Environmental Changes and TE Silencing

In plants, the germ line is established during late sporophyte development. The optimum window for a TE to mobilize and successfully invade a host genome is between the differentiation of the gametophyte precursors and the formation of the early embryo. There is growing evidence that TEs contribute to the structure, evolution, and (epi)genetic control architecture of plant genomes (Deragon et al. 2008; Lisch and Bennetzen 2011). To some extent, a "programmed" loss of silencing of these elements may be potentially beneficial to the host genome. Nonetheless, the accumulation of extra layers of TE silencing in these cells likely reflects an evolutionary need to secure the genome's integrity, preventing wayward

TE activation. What could trigger an "accidental" relaxation of TE silencing? As demonstrated experimentally, mutations in epigenetic regulators involved in the maintenance of CG methylation, such as MET1 or DDM1, may trigger TE mobilization (Miura et al. 2001; Singer et al. 2001; Mirouze et al. 2009; Tsukahara et al. 2009). Such mutations are expected to occur relatively infrequently, but a number of studies have shown that sudden changes in environmental conditions can efficiently interfere with TE silencing. Two recent reports have demonstrated that heat stress can overcome TE silencing in *Arabidopsis*, at least at the transcriptional level (Pecinka et al. 2010; Tittel-Elmer et al. 2010). Transcription of several classes of TEs and other heterochromatic targets, which are silent at ambient temperature, is significantly upregulated upon prolonged exposure to 37 °C (15–30 h). This release from silencing occurs across the genome and is mainly transient; most of the targets return to the silenced state within 2–7 days. Interestingly, stress-induced reactivation appears to occur without altering the common repressive epigenetic marks, including DNA methylation and H3K9me2 (Pecinka et al. 2010; Tittel-Elmer et al. 2010), indicating that these marks are not sufficient for efficient transcriptional silencing under these conditions. The molecular mechanisms in action are unknown. One possibility is that heat stress produces an activating signal that overcomes the presence of these silencing marks or interferes with their readout. Alternatively, heat stress may induce the removal of an additional repressive mark that has yet to be identified. It remains to be tested whether stresses have a similar impact on the gametes, gamete precursors, and early embryo. It would be interesting to determine if the "extra layers" of silencing present in these cell lineages, which are represented by high levels of specific siRNAs and the possible action of an unconventional siRNA pathway involving AGO9, moderate the stress sensitivity of these cells.

Heat stress-induced transcriptional stimulation can be associated with active transposition, as was recently shown for the *ONSEN* LTR retroelement (Ito et al. 2011). Stress conditions similar to the one described above (Pecinka et al. 2010; Tittel-Elmer et al. 2010) were applied to young seedlings, either wild type or ones mutant for components of the RdDM pathway. In all cases, a transient induction of *ONSEN* transcription was observed for a few days, correlating with the accumulation of transposition intermediates that were undetectable 20 days after stress. Remarkably, while no somatic transposition events could be detected in stressed plants of any genotype, transposition was revealed in the progeny of RdDM mutants. The authors demonstrated that in the absence of a functional RdDM pathway, the "memory" of the heat-stress persists during plant development, resulting in *ONSEN* mobilization. Interestingly, this mobilization occurred prior to gametophyte development (Ito et al. 2011), illustrating yet another crucial function of the RdDM pathway, which is used in the plant for other tasks in addition to only reinforcing or restoring preexisting silenced states.

8.6 Conclusion

A combination of genomic and genetic studies in plants have revealed that TEs are the main targets of epigenetic silencing pathways and that TEs are controlled by several layers of silencing. The precise contribution of each layer to maintain the silenced state appears highly variable. Although we have begun to dissect the molecular components of each layer, our understanding of their connection is still fragmentary.

TEs represent highly dynamic genomic components. Because there are potentially deleterious consequences to TE mobilization, evolutionary forces have likely driven the production of multiple epigenetic mechanisms to ensure TE silencing. Certain silencing layers appear to be devoted to specific cell types or developmental stages; however, the purpose and evolutionary origin of this specificity remain elusive. Various intrinsic factors may dictate this specificity, including the TE size, copy number, transposition competence, genomic location, and local chromatin environment. Accumulating evidence has demonstrated that a variety of environmental stresses can challenge TE silencing and possibly lead to bursts of transposition. Interestingly, distinct stresses affect TEs differentially, suggesting that selectivity and complexity in TE silencing may also have evolved as a consequence of stress exposure. Additional studies will be needed to reveal how the environment interferes with silencing and influences TE dynamics.

References

Aravin AA, Sachidanandam R, Bourc'his D, Schaefer C, Pezic D, Toth KF, Bestor T, Hannon GJ (2008) A piRNA pathway primed by individual transposons is linked to *de novo* DNA methylation in mice. Mol Cell 31:785–799

Banks JA, Masson P, Fedoroff N (1988) Molecular mechanisms in the developmental regulation of the maize suppressor-mutator transposable element. Genes Dev 2:1364–1380

Baucom RS, Estill JC, Chaparro C, Upshaw N, Jogi A, Deragon JM, Westerman RP, Sanmiguel PJ, Bennetzen JL (2009) Exceptional diversity, non-random distribution, and rapid evolution of retroelements in the B73 maize genome. PLoS Genet 5:e1000732

Bernatavichute YV, Zhang X, Cokus S, Pellegrini M, Jacobsen SE (2008) Genome-wide association of histone H3 lysine nine methylation with CHG DNA methylation in *Arabidopsis thaliana*. PLoS One 3:e3156

Borges F, Gomes G, Gardner R, Moreno N, McCormick S, Feijo JA, Becker JD (2008) Comparative transcriptomics of Arabidopsis sperm cells. Plant Physiol 148:1168–1181

Bostick M, Kim JK, Esteve PO, Clark A, Pradhan S, Jacobsen SE (2007) UHRF1 plays a role in maintaining DNA methylation in mammalian cells. Science 317:1760–1764

Bourc'his D, Voinnet O (2010) A small-RNA perspective on gametogenesis, fertilization, and early zygotic development. Science 330:617–622

Brodersen P, Sakvarelidze-Achard L, Bruun-Rasmussen M, Dunoyer P, Yamamoto YY, Sieburth L, Voinnet O (2008) Widespread translational inhibition by plant miRNAs and siRNAs. Science 320:1185–1190

Cao X, Jacobsen SE (2002) Role of the arabidopsis DRM methyltransferases in *de novo* DNA methylation and gene silencing. Curr Biol 12:1138–1144

Cao X, Aufsatz W, Zilberman D, Mette MF, Huang MS, Matzke M, Jacobsen SE (2003) Role of the DRM and CMT3 methyltransferases in RNA-directed DNA methylation. Curr Biol 13:2212–2217

Chan SW, Henderson IR, Zhang X, Shah G, Chien JS, Jacobsen SE (2006) RNAi, DRD1, and histone methylation actively target developmentally important non-CG DNA methylation in Arabidopsis. PLoS Genet 2:e83

Chandler VL, Walbot V (1986) DNA modification of a maize transposable element correlates with loss of activity. Proc Natl Acad Sci USA 83:1767–1771

Choi Y, Gehring M, Johnson L, Hannon M, Harada JJ, Goldberg RB, Jacobsen SE, Fischer RL (2002) DEMETER, a DNA glycosylase domain protein, is required for endosperm gene imprinting and seed viability in arabidopsis. Cell 110:33–42

Chomet PS, Wessler S, Dellaporta SL (1987) Inactivation of the maize transposable element Activator (Ac) is associated with its DNA modification. EMBO J 6:295–302

Chow JC, Ciaudo C, Fazzari MJ, Mise N, Servant N, Glass JL, Attreed M, Avner P, Wutz A, Barillot E, Greally JM, Voinnet O, Heard E (2010) LINE-1 activity in facultative heterochromatin formation during X chromosome inactivation. Cell 141:956–969

Chuikov S, Kurash JK, Wilson JR, Xiao B, Justin N, Ivanov GS, McKinney K, Tempst P, Prives C, Gamblin SJ, Barlev NA, Reinberg D (2004) Regulation of p53 activity through lysine methylation. Nature 432:353–360

Cokus SJ, Feng S, Zhang X, Chen Z, Merriman B, Haudenschild CD, Pradhan S, Nelson SF, Pellegrini M, Jacobsen SE (2008) Shotgun bisulphite sequencing of the Arabidopsis genome reveals DNA methylation patterning. Nature 452:215–219

Daxinger L, Kanno T, Bucher E, Van Der Winden J, Naumann U, Matzke AJM, Matzke M (2009) A stepwise pathway for biogenesis of 24-nt secondary siRNAs and spreading of DNA methylation. EMBO J 28:48–57

Dennis K, Fan T, Geiman T, Yan Q, Muegge K (2001) Lsh, a member of the SNF2 family, is required for genome-wide methylation. Genes Dev 15:2940–2944

Deragon JM, Casacuberta JM, Panaud O (2008) Plant transposable elements. Genome Dyn 4:69–82

Ding Y, Wang X, Su L, Zhai J, Cao S, Zhang D, Liu C, Bi Y, Qian Q, Cheng Z, Chu C, Cao X (2007) SDG714, a histone H3K9 methyltransferase, is involved in Tos17 DNA methylation and transposition in rice. Plant Cell 19:9–22

Dunoyer P, Brosnan CA, Schott G, Wang Y, Jay F, Alioua A, Himber C, Voinnet O (2010) An endogenous, systemic RNAi pathway in plants. EMBO J 29:1699–1712

Eamens A, Vaistij FNE, Jones L (2008) NRPD1a and NRPD1b are required to maintain post-transcriptional RNA silencing and RNA-directed DNA methylation in Arabidopsis. Plant J 55:596–606

Ebbs ML, Bender J (2006) Locus-specific control of DNA methylation by the Arabidopsis SUVH5 histone methyltransferase. Plant Cell 18:1166–1176

Ebbs ML, Bartee L, Bender J (2005) H3 lysine 9 methylation is maintained on a transcribed inverted repeat by combined action of SUVH6 and SUVH4 methyltransferases. Mol Cell Biol 25:10507–10515

Esteve PO, Chin HG, Benner J, Feehery GR, Samaranayake M, Horwitz GA, Jacobsen SE, Pradhan S (2009) Regulation of DNMT1 stability through SET7-mediated lysine methylation in mammalian cells. Proc Natl Acad Sci USA 106:5076–5081

Fagegaltier D, Bouge AL, Berry B, Poisot E, Sismeiro O, Coppee JY, Theodore L, Voinnet O, Antoniewski C (2009) The endogenous siRNA pathway is involved in heterochromatin formation in Drosophila. Proc Natl Acad Sci USA 106:21258–21263

Feng S, Jacobsen SE (2011) Epigenetic modifications in plants: an evolutionary perspective. Curr Opin Plant Biol 14:179–186

Feng S, Cokus SJ, Zhang X, Chen PY, Bostick M, Goll MG, Hetzel J, Jain J, Strauss SH, Halpern ME, Ukomadu C, Sadler KC, Pradhan S, Pellegrini M, Jacobsen SE (2010) Conservation and divergence of methylation patterning in plants and animals. Proc Natl Acad Sci USA 107:8689–8694

Gao Z, Liu HL, Daxinger L, Pontes O, He X, Qian W, Lin H, Xie M, Lorkovic ZJ, Zhang S, Miki D, Zhan X, Pontier D, Lagrange T, Jin H, Matzke AJ, Matzke M, Pikaard CS, Zhu JK (2010) An RNA polymerase II- and AGO4-associated protein acts in RNA-directed DNA methylation. Nature 465:106–109

Gehring M, Huh JH, Hsieh TF, Penterman J, Choi Y, Harada JJ, Goldberg RB, Fischer RL (2006) DEMETER DNA glycosylase establishes MEDEA polycomb gene self-imprinting by allele-specific demethylation. Cell 124:495–506

Gehring M, Bubb KL, Henikoff S (2009) Extensive demethylation of repetitive elements during seed development underlies gene imprinting. Science 324:1447–1451

Guang S, Bochner AF, Burkhart KB, Burton N, Pavelec DM, Kennedy S (2010) Small regulatory RNAs inhibit RNA polymerase II during the elongation phase of transcription. Nature 465:1097–1101

Herr AJ, Jensen MB, Dalmay T, Baulcombe DC (2005) RNA polymerase IV directs silencing of endogenous DNA. Science 308:118–120

Hershberger RJ, Warren CA, Walbot V (1991) Mutator activity in maize correlates with the presence and expression of the Mu transposable element Mu9. Proc Natl Acad Sci USA 88:10198–10202

Hsieh TF, Ibarra CA, Silva P, Zemach A, Eshed-Williams L, Fischer RL, Zilberman D (2009) Genome-wide demethylation of Arabidopsis endosperm. Science 324:1451–1454

Huang J, Fan T, Yan Q, Zhu H, Fox S, Issaq HJ, Best L, Gangi L, Munroe D, Muegge K (2004) Lsh, an epigenetic guardian of repetitive elements. Nucleic Acids Res 32:5019–5028

Huang L, Sun Q, Qin F, Li C, Zhao Y, Zhou DX (2007) Down-regulation of a silent information regulator2-related histone deacetylase gene, OsSRT1, induces DNA fragmentation and cell death in rice. Plant Physiol 144:1508–1519

Huettel B, Kanno T, Daxinger L, Aufsatz W, Matzke AJ, Matzke M (2006) Endogenous targets of RNA-directed DNA methylation and Pol IV in Arabidopsis. EMBO J 25:2828–2836

Inagaki S, Miura-Kamio A, Nakamura Y, Lu F, Cui X, Cao X, Kimura H, Saze H, Kakutani T (2010) Autocatalytic differentiation of epigenetic modifications within the Arabidopsis genome. EMBO J 29:3496–3506

Ito H, Gaubert H, Bucher E, Mirouze M, Vaillant I, Paszkowski J (2011) An siRNA pathway prevents transgenerational retrotransposition in plants subjected to stress. Nature 472:115–119

Jackson JP, Lindroth AM, Cao X, Jacobsen SE (2002) Control of CpNpG DNA methylation by the KRYPTONITE histone H3 methyltransferase. Nature 416:556–560

Jacob Y, Feng S, LeBlanc CA, Bernatavichute YV, Stroud H, Cokus S, Johnson LM, Pellegrini M, Jacobsen SE, Michaels SD (2009) ATXR5 and ATXR6 are H3K27 monomethyltransferases required for chromatin structure and gene silencing. Nat Struct Mol Biol 16:763–768

Jacob Y, Stroud H, Leblanc C, Feng S, Zhuo L, Caro E, Hassel C, Gutierrez C, Michaels SD, Jacobsen SE (2010) Regulation of heterochromatic DNA replication by histone H3 lysine 27 methyltransferases. Nature 466:987–991

Jakubiec A, Yang SW, Chua N-H (2012) Arabidopsis DRB4 protein in antiviral defense against Turnip yellow mosaic virus infection. Plant J 69:14–25

Jeddeloh JA, Bender J, Richards EJ (1998) The DNA methylation locus DDM1 is required for maintenance of gene silencing in Arabidopsis. Genes Dev 12:1714–1725

Jeddeloh JA, Stokes TL, Richards EJ (1999) Maintenance of genomic methylation requires a SWI2/SNF2-like protein. Nat Genet 22:94–97

Johnson L, Cao X, Jacobsen S (2002) Interplay between two epigenetic marks. DNA methylation and histone H3 lysine 9 methylation. Curr Biol 12:1360–1367

Johnson LM, Bostick M, Zhang X, Kraft E, Henderson I, Callis J, Jacobsen SE (2007) The SRA methyl-cytosine-binding domain links DNA and histone methylation. Curr Biol 17:379–384

Jullien PE, Mosquna A, Ingouff M, Sakata T, Ohad N, Berger F (2008) Retinoblastoma and its binding partner MSI1 control imprinting in Arabidopsis. PLoS Biol 6:e194

Kakutani T, Munakata K, Richards EJ, Hirochika H (1999) Meiotically and mitotically stable inheritance of DNA hypomethylation induced by ddm1 mutation of *Arabidopsis thaliana*. Genetics 151:831–838

Kanno T, Mette MF, Kreil DP, Aufsatz W, Matzke M, Matzke AJ (2004) Involvement of putative SNF2 chromatin remodeling protein DRD1 in RNA-directed DNA methylation. Curr Biol 14:801–805

Kanno T, Huettel B, Mette MF, Aufsatz W, Jaligot E, Daxinger L, Kreil DP, Matzke M, Matzke AJ (2005) Atypical RNA polymerase subunits required for RNA-directed DNA methylation. Nat Genet 37:761–765

Kanno T, Bucher E, Daxinger L, Huettel B, Bohmdorfer G, Gregor W, Kreil DP, Matzke M, Matzke AJ (2008) A structural-maintenance-of-chromosomes hinge domain-containing protein is required for RNA-directed DNA methylation. Nat Genet 40:670–675

Kanno T, Bucher E, Daxinger L, Huettel B, Kreil DP, Breinig F, Lind M, Schmitt MJ, Simon SA, Gurazada SG, Meyers BC, Lorkovic ZJ, Matzke AJ, Matzke M (2010) RNA-directed DNA methylation and plant development require an IWR1-type transcription factor. EMBO Rep 11:65–71

Kasschau KD, Fahlgren N, Chapman EJ, Sullivan CM, Cumbie JS, Givan SA, Carrington JC (2007) Genome-wide profiling and analysis of Arabidopsis siRNAs. PLoS Biol 5:e57

Kato M, Miura A, Bender J, Jacobsen SE, Kakutani T (2003) Role of CG and non-CG methylation in immobilization of transposons in Arabidopsis. Curr Biol 13:421–426

Ketting RF (2011) The many faces of RNAi. Dev Cell 20:148–161

Kim VN, Han J, Siomi MC (2009) Biogenesis of small RNAs in animals. Nat Rev Mol Cell Biol 10:126–139

Kouskouti A, Scheer E, Staub A, Tora L, Talianidis I (2004) Gene-specific modulation of TAF10 function by SET9-mediated methylation. Mol Cell 14:175–182

Kuramochi-Miyagawa S, Watanabe T, Gotoh K, Totoki Y, Toyoda A, Ikawa M, Asada N, Kojima K, Yamaguchi Y, Ijiri TW, Hata K, Li E, Matsuda Y, Kimura T, Okabe M, Sakaki Y, Sasaki H, Nakano T (2008) DNA methylation of retrotransposon genes is regulated by Piwi family members MILI and MIWI2 in murine fetal testes. Genes Dev 22:908–917

Li H, Freeling M, Lisch D (2010) Epigenetic reprogramming during vegetative phase change in maize. Proc Natl Acad Sci USA 107:22184–22189

Lindroth AM, Shultis D, Jasencakova Z, Fuchs J, Johnson L, Schubert D, Patnaik D, Pradhan S, Goodrich J, Schubert I, Jenuwein T, Khorasanizadeh S, Jacobsen SE (2004) Dual histone H3 methylation marks at lysines 9 and 27 required for interaction with CHROMOMETHYLASE3. EMBO J 23:4286–4296

Lippman Z, May B, Yordan C, Singer T, Martienssen R (2003) Distinct mechanisms determine transposon inheritance and methylation via small interfering RNA and histone modification. PLoS Biol 1:E67

Lippman Z, Gendrel AV, Black M, Vaughn MW, Dedhia N, McCombie WR, Lavine K, Mittal V, May B, Kasschau KD, Carrington JC, Doerge RW, Colot V, Martienssen R (2004) Role of transposable elements in heterochromatin and epigenetic control. Nature 430:471–476

Lisch D (2009) Epigenetic regulation of transposable elements in plants. Annu Rev Plant Biol 60:43–66

Lisch D, Bennetzen JL (2011) Transposable element origins of epigenetic gene regulation. Curr Opin Plant Biol 14:156–161

Lister R, O'Malley RC, Tonti-Filippini J, Gregory BD, Berry CC, Millar AH, Ecker JR (2008) Highly integrated single-base resolution maps of the epigenome in Arabidopsis. Cell 133:523–536

Liu S, Yu Y, Ruan Y, Meyer D, Wolff M, Xu L, Wang N, Steinmetz A, Shen WH (2007) Plant SET- and RING-associated domain proteins in heterochromatinization. Plant J 52:914–926

Malagnac F, Bartee L, Bender J (2002) An Arabidopsis SET domain protein required for maintenance but not establishment of DNA methylation. EMBO J 21:6842–6852

Mathieu O, Probst AV, Paszkowski J (2005) Distinct regulation of histone H3 methylation at lysines 27 and 9 by CpG methylation in Arabidopsis. EMBO J 24:2783–2791

Mathieu O, Reinders J, Caikovski M, Smathajitt C, Paszkowski J (2007) Transgenerational stability of the Arabidopsis epigenome is coordinated by CG methylation. Cell 130:851–862
Matzke M, Kanno T, Daxinger L, Huettel B, Matzke AJ (2009) RNA-mediated chromatin-based silencing in plants. Curr Opin Cell Biol 21:367–376
McClintock B (1948) Mutable loci in maize. Carnegie Inst Wash Yearbook 47:155–169
Melnyk CW, Molnar A, Baulcombe DC (2011) Intercellular and systemic movement of RNA silencing signals. EMBO J 30:3553–3563
Mirouze M, Reinders J, Bucher E, Nishimura T, Schneeberger K, Ossowski S, Cao J, Weigel D, Paszkowski J, Mathieu O (2009) Selective epigenetic control of retrotransposition in Arabidopsis. Nature 461:427–430
Miura A, Yonebayashi S, Watanabe K, Toyama T, Shimada H, Kakutani T (2001) Mobilization of transposons by a mutation abolishing full DNA methylation in Arabidopsis. Nature 411:212–214
Miura A, Nakamura M, Inagaki S, Kobayashi A, Saze H, Kakutani T (2009) An Arabidopsis jmjC domain protein protects transcribed genes from DNA methylation at CHG sites. EMBO J 28:1078–1086
Molnár A, Csorba T, Lakatos L, Várallyay E, Lacomme C, Burgyán J (2005) Plant virus-derived small interfering RNAs originate predominantly from highly structured single-stranded viral RNAs. J Virol 79:7812–7818
Molnar A, Melnyk CW, Bassett A, Hardcastle TJ, Dunn R, Baulcombe DC (2010) Small silencing RNAs in plants are mobile and direct epigenetic modification in recipient cells. Science 328:872–875
Morales-Ruiz T, Ortega-Galisteo AP, Ponferrada-Marin MI, Martinez-Macias MI, Ariza RR, Roldan-Arjona T (2006) Demeter and repressor of silencing 1 encode 5-methylcytosine DNA glycosylases. Proc Natl Acad Sci USA 103:6853–6858
Mosher RA, Melnyk CW, Kelly KA, Dunn RM, Studholme DJ, Baulcombe DC (2009) Uniparental expression of PolIV-dependent siRNAs in developing endosperm of Arabidopsis. Nature 460:283–286
Okamura K, Lai EC (2008) Endogenous small interfering RNAs in animals. Nat Rev Mol Cell Biol 9:673–678
Olmedo-Monfil V, Duran-Figueroa N, Arteaga-Vazquez M, Demesa-Arevalo E, Autran D, Grimanelli D, Slotkin RK, Martienssen RA, Vielle-Calzada JP (2010) Control of female gamete formation by a small RNA pathway in Arabidopsis. Nature 464:628–632
Onodera Y, Haag JR, Ream T, Nunes PC, Pontes O, Pikaard CS (2005) Plant nuclear RNA polymerase IV mediates siRNA and DNA methylation-dependent heterochromatin formation. Cell 120:613–622
Pal-Bhadra M, Leibovitch BA, Gandhi SG, Rao M, Bhadra U, Birchler JA, Elgin SC (2004) Heterochromatic silencing and HP1 localization in Drosophila are dependent on the RNAi machinery. Science 303:669–672
Pecinka A, Dinh HQ, Baubec T, Rosa M, Lettner N, Mittelsten Scheid O (2010) Epigenetic regulation of repetitive elements is attenuated by prolonged heat stress in Arabidopsis. Plant Cell 22:3118–3129
Pontier D, Yahubyan G, Vega D, Bulski A, Saez-Vasquez J, Hakimi MA, Lerbs-Mache S, Colot V, Lagrange T (2005) Reinforcement of silencing at transposons and highly repeated sequences requires the concerted action of two distinct RNA polymerases IV in Arabidopsis. Genes Dev 19:2030–2040
Qin FJ, Sun QW, Huang LM, Chen XS, Zhou DX (2010) Rice SUVH histone methyltransferase genes display specific functions in chromatin modification and retrotransposon repression. Mol Plant 3:773–782
Rajakumara E, Law JA, Simanshu DK, Voigt P, Johnson LM, Reinberg D, Patel DJ, Jacobsen SE (2011) A dual flip-out mechanism for 5mC recognition by the Arabidopsis SUVH5 SRA domain and its impact on DNA methylation and H3K9 dimethylation in vivo. Genes Dev 25:137–152

Rudenko GN, Walbot V (2001) Expression and post-transcriptional regulation of maize transposable element MuDR and its derivatives. Plant Cell 13:553–570

Sampath SC, Marazzi I, Yap KL, Krutchinsky AN, Mecklenbrauker I, Viale A, Rudensky E, Zhou MM, Chait BT, Tarakhovsky A (2007) Methylation of a histone mimic within the histone methyltransferase G9a regulates protein complex assembly. Mol Cell 27:596–608

Saze H, Shiraishi A, Miura A, Kakutani T (2008) Control of genic DNA methylation by a jmjC domain-containing protein in *Arabidopsis thaliana*. Science 319:462–465

Singer T, Yordan C, Martienssen RA (2001) Robertson's Mutator transposons in *A. thaliana* are regulated by the chromatin-remodeling gene Decrease in DNA Methylation (DDM1). Genes Dev 15:591–602

Siomi MC, Sato K, Pezic D, Aravin AA (2011) PIWI-interacting small RNAs: the vanguard of genome defence. Nat Rev Mol Cell Biol 12:246–258

Slotkin RK (2010) The epigenetic control of the Athila family of retrotransposons in Arabidopsis. Epigenetics 5:483–490

Slotkin RK, Martienssen R (2007) Transposable elements and the epigenetic regulation of the genome. Nat Rev Genet 8:272–285

Slotkin RK, Freeling M, Lisch D (2003) Mu killer causes the heritable inactivation of the Mutator family of transposable elements in *Zea mays*. Genetics 165:781–797

Slotkin RK, Freeling M, Lisch D (2005) Heritable transposon silencing initiated by a naturally occurring transposon inverted duplication. Nat Genet 37:641–644

Slotkin RK, Vaughn M, Borges F, Tanurdzic M, Becker JD, Feijo JA, Martienssen RA (2009) Epigenetic reprogramming and small RNA silencing of transposable elements in pollen. Cell 136:461–472

Soppe WJ, Jasencakova Z, Houben A, Kakutani T, Meister A, Huang MS, Jacobsen SE, Schubert I, Fransz PF (2002) DNA methylation controls histone H3 lysine 9 methylation and heterochromatin assembly in Arabidopsis. EMBO J 21:6549–6559

Sridhar VV, Kapoor A, Zhang K, Zhu J, Zhou T, Hasegawa PM, Bressan RA, Zhu JK (2007) Control of DNA methylation and heterochromatic silencing by histone H2B deubiquitination. Nature 447:735–738

Suzuki MM, Bird A (2008) DNA methylation landscapes: provocative insights from epigenomics. Nat Rev Genet 9:465–476

Tanurdzic M, Vaughn MW, Jiang H, Lee TJ, Slotkin RK, Sosinski B, Thompson WF, Doerge RW, Martienssen RA (2008) Epigenomic consequences of immortalized plant cell suspension culture. PLoS Biol 6:2880–2895

Teixeira FK, Heredia F, Sarazin A, Roudier F, Boccara M, Ciaudo C, Cruaud C, Poulain J, Berdasco M, Fraga MF, Voinnet O, Wincker P, Esteller M, Colot V (2009) A role for RNAi in the selective correction of DNA methylation defects. Science 323:1600–1604

Tenaillon MI, Hollister JD, Gaut BS (2010) A triptych of the evolution of plant transposable elements. Trends Plant Sci 15:471–478

Tittel-Elmer M, Bucher E, Broger L, Mathieu O, Paszkowski J, Vaillant I (2010) Stress-induced activation of heterochromatic transcription. PLoS Genet 6:e1001175

Tompa R, McCallum CM, Delrow J, Henikoff JG, van Steensel B, Henikoff S (2002) Genome-wide profiling of DNA methylation reveals transposon targets of CHROMOMETHYLASE3. Curr Biol 12:65–68

Tsukahara S, Kobayashi A, Kawabe A, Mathieu O, Miura A, Kakutani T (2009) Bursts of retrotransposition reproduced in Arabidopsis. Nature 461:423–426

Vaucheret H (2006) Post-transcriptional small RNA pathways in plants: mechanisms and regulations. Genes Dev 20:759–771

Vazquez F, Legrand S, Windels D (2010) The biosynthetic pathways and biological scopes of plant small RNAs. Trends Plant Sci 15:337–345

Vongs A, Kakutani T, Martienssen RA, Richards EJ (1993) *Arabidopsis thaliana* DNA methylation mutants. Science 260:1926–1928

Wessler SR (2006) Transposable elements and the evolution of eukaryotic genomes. Proc Natl Acad Sci USA 103:17600–17601

Wicker T, Sabot F, Hua-Van A, Bennetzen JL, Capy P, Chalhoub B, Flavell A, Leroy P, Morgante M, Panaud O, Paux E, SanMiguel P, Schulman AH (2007) A unified classification system for eukaryotic transposable elements. Nat Rev Genet 8:973–982

Wierzbicki AT, Haag JR, Pikaard CS (2008) Noncoding transcription by RNA polymerase Pol IVb/Pol V mediates transcriptional silencing of overlapping and adjacent genes. Cell 135:635–648

Wierzbicki AT, Ream TS, Haag JR, Pikaard CS (2009) RNA polymerase V transcription guides ARGONAUTE4 to chromatin. Nat Genet 41:630–634

Woo HR, Dittmer TA, Richards EJ (2008) Three SRA-domain methylcytosine-binding proteins cooperate to maintain global CpG methylation and epigenetic silencing in Arabidopsis. PLoS Genet 4:e1000156

Yan Q, Cho E, Lockett S, Muegge K (2003) Association of Lsh, a regulator of DNA methylation, with pericentromeric heterochromatin is dependent on intact heterochromatin. Mol Cell Biol 23:8416–8428

Zemach A, McDaniel IE, Silva P, Zilberman D (2010) Genome-wide evolutionary analysis of eukaryotic DNA methylation. Science 328:916–919

Zhang Y (2003) Transcriptional regulation by histone ubiquitination and deubiquitination. Genes Dev 17:2733–2740

Zhang H, Zhu J-K (2011) RNA-directed DNA methylation. Curr Opin Plant Biol 14:142–147

Zhang X, Yazaki J, Sundaresan A, Cokus S, Chan SW, Chen H, Henderson IR, Shinn P, Pellegrini M, Jacobsen SE, Ecker JR (2006) Genome-wide high-resolution mapping and functional analysis of DNA methylation in arabidopsis. Cell 126:1189–1201

Zhang X, Henderson IR, Lu C, Green PJ, Jacobsen SE (2007) Role of RNA polymerase IV in plant small RNA metabolism. Proc Natl Acad Sci USA 104:4536–4541

Zilberman D, Henikoff S (2007) Genome-wide analysis of DNA methylation patterns. Development 134:3959–3965

Zilberman D, Gehring M, Tran RK, Ballinger T, Henikoff S (2007) Genome-wide analysis of *Arabidopsis thaliana* DNA methylation uncovers an interdependence between methylation and transcription. Nat Genet 39:61–69

Chapter 9
Responses of Transposable Elements to Polyploidy

Christian Parisod and Natacha Senerchia

Abstract Polyploidy (i.e., hybridization between more or less divergent genomes, associated with whole genome duplication) has been shown to result in drastic genome reorganization. Such changes involved major restructuring and epigenetic repatterning, mainly in transposable element (TE) fractions. Polyploidy thus is an adequate model to explore the mechanisms generating genome variation and their impact on evolution. In this chapter, we will review available evidence on the importance of TEs in the short-term and the long-term changes in polyploid genomes. We will argue that the study of polyploid systems not only offers the opportunity to highlight specific mechanisms controlling the activity of TEs but also the evolutionary impact of TE-induced genome reorganization.

Keywords Epigenetic changes • Genome reorganization • Genome shock • Hybridization • Restructuring • si-RNA • Speciation • Whole genome doubling

9.1 Polyploidy, a Prominent Evolutionary Process

Polyploidy is a recurrent process in the evolutionary history of most organisms and can be understood as a major speciation mechanism (Wood et al. 2009). It is prominent in plants, but also commonly occurs in several animal taxa (Otto 2007; Mable et al. 2011). In particular, all angiosperms have been demonstrated as having gone through one or more rounds of whole genome duplication (Jiao et al. 2011), and plant genomes thus contain considerable genetic redundancy (Fig. 9.1). Two main types of polyploids, representing extreme cases of a continuum, have been traditionally recognized (Stebbins 1971). Autopolyploids are polyploids with chromosomes derived

C. Parisod (✉) • N. Senerchia
Laboratory of Evolutionary Botany, Institute of Biology, University of Neuchâtel, Rue Emile-Argand 11, CH-2000 Neuchâtel, Switzerland
e-mail: christian.parisod@unine.ch

M.-A. Grandbastien and J.M. Casacuberta (eds.), *Plant Transposable Elements*, Topics in Current Genetics 24, DOI 10.1007/978-3-642-31842-9_9,

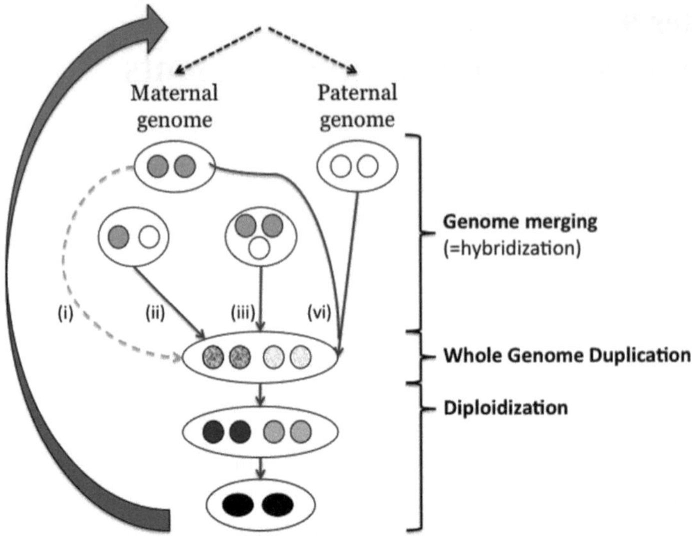

Fig. 9.1 Evolution of natural polyploids. The merging (i.e., hybridization) of more or less diverged parental genomes associated with whole genome duplication leads to the formation of a nascent polyploid lineage. Autopolyploidy involves hybridization between closely related (i.e., homologous) genomes, while allopolyploidy is the merging of widely divergent parental genomes (i.e., homeologous). Genome changes occurring after the origin of the polyploid are referred to as diploidization, restoring a diploid-like genetic system. Seed plant genomes have evolved through successive rounds of polyploidy. The most common natural pathways to polyploidy are depicted: (i) spontaneous genome doubling, which is extremely rare under natural conditions; (ii) homoploid hybrid intermediate; (iii) triploid bridge through the union of an unreduced gamete with a reduced one, and (iv) one-step formation through the union of two unreduced gametes

from two homologous genomes (AAAA) and are characterized by predominant polysomic inheritence at meiosis (Parisod et al. 2010b). Allopolyploids present chromosomes resulting from the merging of divergent (i.e., homeologous) genomes (AABB) and mostly show disomic inheritance (Leitch and Leitch 2008). The distinction between homologous and homeologous genomes is hardly clear-cut and there is a continuum between auto- and allopolyploidy. It is thus important to realize that the evolutionary origin of all natural polyploids (i.e., both auto- and allopolyploids) involves hybridization between variously related genomes.

9.2 Reorganization of Polyploid Genomes

Polyploid genomes are expected to be the addition of parental genomes, and departure from this additivity highlights genome reorganization. Recent studies revealed drastic polyploidy-induced genome reorganization, including reproducible structural and epigenetic alteration (Soltis and Soltis 1999; Comai 2000, 2005; Comai et al. 2000; Wendel 2000; Levin 2002; Adams and Wendel 2005; Chen 2007;

Doyle et al. 2008; Feldman and Levy 2009). Such processes restore a secondary diploid-like genetics in polyploids and are commonly referred to as diploidization. Following Levy and Feldman (2004), genome reorganization after polyploidization can be conveniently classified as (1) short-term changes (or revolutionary changes), acting immediately after polyploidization, and (2) long-term changes (or evolutionary changes), occurring during the lifetime of the polyploid lineage (Fig. 9.1).

Genome reorganization is commonly observed in the first few generations following polyploidy and sometimes as early as in F1 hybrids (Parisod et al. 2009). Both intra- and intergenomic structural rearrangements have been reported and include (1) elimination of DNA sequences from hom(e)ologous chromosomes and gene loss (Ozkan et al. 2001; Chantret et al. 2005), (2) amplification or reduction of repetitive sequences (Zhao et al. 1998; Petit et al. 2010), and (3) chromosomal repatterning (Pires et al. 2004; Udall et al. 2005). Genome downsizing after polyploidization appears to be a general trend (Leitch and Bennett 2004). In addition to restructuring, drastic epigenetic changes have been commonly reported in allopolyploids (Liu and Wendel 2003). These changes include (1) alteration of gene expression through alterations of cytosine methylation (Kashkush et al. 2002; Salmon et al. 2005) and through transcriptional activation of retroelements (Kashkush et al. 2003; Kashkush and Khasdan 2007), and (2) chromatin remodeling due to modification of DNA methylation and acetylation (e.g., Wang et al. 2006). Polyploidy-induced epigenetic variation is certainly linked to intergenomic interactions and dosage compensation among subgenomes (Riddle and Birchler 2003). Methylation repatterning sometimes affects subgenomes equally (e.g., Song et al. 1995), but most often differentially affects the paternal (e.g., Shaked et al. 2001) or the maternal (e.g., Ainouche et al. 2009). Epigenetic changes were further associated with organ-specific silencing of coding genes in allopolyploids (Adams et al. 2003; Adams and Wendel 2005; Chen 2007). As a whole, diploidization could be a foster for new phenotypes that could potentially be linked to the evolutionary outcomes of polyploidy (e.g., Levy and Feldman 2004; Doyle et al. 2008; Leitch and Leitch 2008; Parisod 2012). It could indeed be that genome reorganization in nascent polyploids leads to novel properties as compared to the addition of the parental genomes and may support the emergence of new species. Our knowledge on the causes and consequences of polyploidy-induced genome reorganization, however, remains elusive.

The confinement of divergent genomes in the single nucleus of nascent polyploids can induce troubles such as inaccurate pairing between hom(e)ologous sequences or dosage-dependent interactions (Doyle et al. 2008). Accordingly, quick sequence rearrangement (including DNA insertion/deletion) and epigenetic modifications could increase the divergence between subgenomes. Such changes could futher impede the pairing of homeologous chromosome and thus indirectly facilitating proper homologous pairing at meiosis (Levy and Feldman 2002; Eilam et al. 2008), or could participate in the regulation of gene dosage, promoting intergenomic coordination (Rieseberg 2001). Reorganization targeted toward one of the parental subgenome is commonly interpreted as evidence that cytoplasmic–nuclear interactions represent crucial incompatibilities to be overcome after genome merging, but it has been noted that nuclear–nuclear interactions may be important as well

(Josefsson et al. 2006). Although the exact cause of immediate genome reorganization after polyploidy deserve further work, a greater rate of genome reorganization is expected to be necessary to resolve conflicts in hybrids derived from genetically divergent parents. Accordingly, we can predict more changes to occur in allopolyploids than in autopolyploids. Evidence accumulated so far is coherent with this hypothesis (Parisod et al. 2010b), but we almost completely lack knowledge about genome reorganization after autopolyploidy. Additional studies involving hybridization between closely related genomes may help to shed light on the mechanisms inducing immediate diploidization.

9.3 Reorganization of TE Genome Fractions After Polyploidy

For those used to see TEs as major supporters of natural genetic engineering, it might be already clear that the plethora of mechanisms occurring after polyploidy can be related to TEs. In the formulation of the "Genome Shock" hypothesis, Barbara McClintock (1984) stated that challenges such as species cross may induce transposition bursts. This hypothesis, stating that transpositions should play a critical role in polyploidy-induced genome reorganization, has been repeatedly put forward (Matzke and Matzke 1998; Soltis and Soltis 1999; Comai et al. 2000; Wendel 2000). Although data showing an activation of TEs after hybridization and polyploidy have recently accumulated, conclusive evidence is still scarce and we are still far from understanding the mechanisms and the consequences of polyploid genome evolution under the influence of TEs.

Due to their prevalence in eukaryote genomes (Gaut and Ross-Ibarra 2008), it can be expected that TEs play a major role in the molecular events leading to the establishment of a viable polyploid genome. Furthermore, TEs can have a dual role in genome reorganization, affecting both structural features and epigenetic states of sequences throughout the host genome (Teixeira et al. 2009). In case of transposition, new TE insertions can promote proper pairing at meiosis by triggering structural divergence between subgenomes through microchromosomal rearrangements. Transposition can also promote intergenomic coordination by disrupting genes or altering the epigenetic state of neighboring sequences, impacting on genome function by affecting chromatin structure and/or gene expression (Hollister et al. 2011). On the other hand, dispersed TE insertions might represent homologous substrate sustaining illegitimate recombination and fostering reorganization of TE fractions. Such changes without transposition can have similar consequences for subgenomes divergence and/or coordination.

The commonly anticipated proliferation of TEs in polyploid genomes can be explained by three non-mutually exclusive hypotheses. (1) Whole genome duplication may relax purifying selection against deleterious TE insertions (Matzke and Matzke 1998). In other words, gene redundancy may lead to an overall increase in the number of neutral sites available for TEs to insert and fix without strong selective constraints (the Redundancy hypothesis). Accordingly, under a constant transposition rate, TE insertions would accumulate neutrally in polyploids until such sites are all occupied. (2) The origin of a polyploid lineage represents a transient period with a low population size (i.e., bottleneck; Lynch 2007). As selection efficiency decreases when population size decreases, moderately deleterious TE insertions could be fixed in nascent polyploid genomes with a higher probability (the Bottleneck hypothesis; Parisod et al. 2010a). Accordingly, under a constant transposition rate, TE insertions would accumulate in nascent polyploids until the establishment of a large population. (3) The merging of divergent genomes into a single nucleus would generate conflicts between the TEs and the host repressors (Box 9.1; Figs. 9.2 and 9.3), inducing a genome shock promoting TE activation and ultimately transposition (Genome Shock hypothesis; Comai et al. 2003). Accordingly, polyploidy would induce a change in the activity of TEs.

Mechanisms behind these three hypotheses are expected to result in different patterns of TE proliferation and may thus be distinguished by assessing TE activity, rate of accumulation during and after polyploidization, and the parental genome divergence. Under the Redundancy hypothesis, no discrete burst of TE activity is expected, and the rate of TE accumulation should be continuous until full diploidization is reached. A bona fide change in TE activity (transcriptional and, to a certain extent, transpositional) is postulated immediately after polyploidy under the Genome Shock hypothesis. Accordingly, both the Genome Shock and the Bottleneck hypothesis are expected to result in the accumulation of transposed TE copies during the first generations after polyploidy. However, genome merging should reveal genetic conflicts between specific TE families (see Box 9.1), and only these TEs should be affected under the Genome Shock hypothesis, while a bottleneck would change the frequency of all polymorphic TE insertions. Noticeably, the Redundancy and the Bottleneck hypothesis could explain TE dynamics in both auto- and allopolyploids, while a genome shock is expected to result in reorganization of fewer TEs in hybrids between closely related genomes (i.e., autopolyploids) than in allopolyploids. While theory can help to predict the impact of polyploidy on TE activity, empirical data are still too scarce to test the different hypotheses. Accordingly, what follows remains a narrative review of the levels and timing of reorganization in TE genome fractions of polyploids.

9.4 Short-Term Reorganization of TE Fractions

Short-term genome reorganization related to TEs can be straightforwardly evaluated by comparing the genome of experimental (i.e., resynthetized) or recent (i.e., less than a few hundred years old) polyploids to the expected addition of their parents (Fig. 9.4). Several PCR-based fingerprint techniques can be used to assess reorganization throughout the genomes of both autopolyploids and allopolyploids (Parisod et al. 2010a; Kalendar et al. 2011). As different molecular methods allow

Box 9.1 Dynamics of TE-Repressing Mechanisms During Polyploidy

As a majority of transposition events are expected to have a deleterious effect, host genomes have evolved sophisticated mechanisms repressing the activity of functional TEs (Fig. 9.2a). Recent studies have considerably improved our understanding of the various epigenetic pathways controlling TEs, but much remains to be done in order to decipher these overlapping mechanisms (Feng et al. 2010). Two main mechanisms are responsible for silencing of TEs: DNA methylation (in CG, CHG, and CHH sequences contexts; H = C, T, or A) and histone methylation (H3K9 dimethylation and H3K27 monomethylation). These pathways are triggered by repeat-derived small interfering RNA (siRNA) that target TE insertions through sequence homology and recruit the enzyme machinery responsible for DNA methylation and heterochromatinization (Martienssen 2010). Genomes typically contain specific TE sequences inducing the production of specific siRNAs silencing corresponding TEs and thus assure genome stability during plant development.

While plants do not show a proper demethylated germ line, it seems that both female (Fig. 9.2b) and male (Fig. 9.2c) gametogenesis relaxes the repression of TEs in accessory cells, ensuring the massive production of siRNAs and reinforcing the silencing of TEs in the germ cells (i.e., consolidation; Bourc'his and Voinnet 2010). During male gametogenesis, post-meiosis microspores develop into a vegetative cell and two sperm cells. Epigenetic pathways responsible for the maintenance of methylation are downregulated, and TEs are reactivated in pollen grains. It seems, however, that hypomethylation is exclusive to the vegetative cell and would serve the production of 21 nucleotide siRNAs mediating the repression of TEs in the adjacent sperm cells through CHG methylation. During female gametogenesis, post-meiosis megaspore gives rise to one egg cell (participating to the zygote), one central cell with two nuclei (participating the endosperm), and other accessory cells. The genome of the central cell is specifically demethylated, leading to the expression of maternal alleles and TEs in the endosperm (i.e. imprinting; Fig. 9.2d). Such soft reactivation of TEs in the central cell and the endosperm may serve the massive production of 24 nucleotides siRNAs to reinforce the silencing of TEs in the egg cell and maybe the endosperm and the zygote.

The confrontation of paternal and maternal genomes (Fig. 9.2e) presenting qualitative and/or quantitative mismatch in their respective TEs and siRNAs may result in the failure of the TE-siRNA system to reach equilibrium at fertilization (Fig. 9.3). In other words, hybridization between lineages with incompatible TE loads is expected to results in conflicts between TEs and siRNAs. If siRNAs in the central cell do not match TE insertions in pollen, then corresponding TEs could be transcribed and could possibly transpose in the endosperm. Note that a massive proliferation of TEs in the endosperm is expected to have deleterious consequences such as seed failure. Similarly, if siRNAs in the egg cytoplasm do not match with TE insertions from the sperm cells, corresponding TEs could be activated and may proliferate in the zygote. The outcome of a cross thus depends on both the copy number of TEs and of siRNAs, but also on the dose of paternal and maternal genomes. Such a reactivation of TEs in F1 hybrids is similar to hybrid dysgenesis as described in *Drosophila* and may lead to strong incompatibility between gene pools (i.e., intrinsic postzygotic isolation; Josefsson et al. 2006; Martienssen 2010; Parisod et al. 2010b).

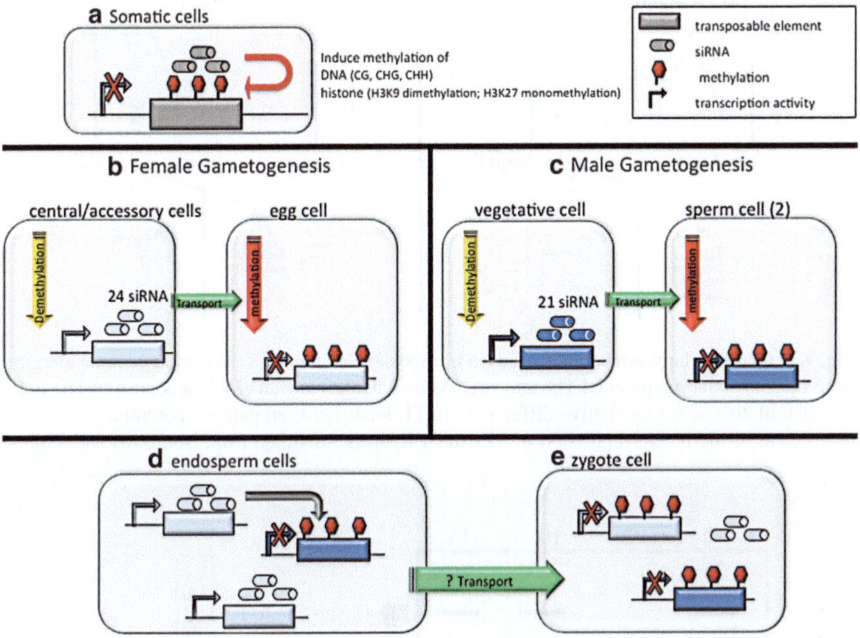

Fig. 9.2 Transposable element (TE) silencing by siRNA during plant development and reproduction (after Feng et al. 2010; Bourc'his and Voinnet 2010). In somatic cells (**a**) siRNA derived from TEs recruit the methylation machinery in order to maintain the repression of TE transcription through methylated DNA or histones. During both female (**b**) and male (**c**) gametogenesis, 24-nucleotide-long and 21- nucleotide-long siRNAs are produced by the demethylated genomes of the central/accessory cells and from the vegetative cells, respectively. Those siRNAs maintain or reinforce TE repression in the egg and sperm cells. During fecundation (**d**) the endosperm is demethylated and further produce siRNAs. Putative transport of siRNAs from the endosperm to the zygote might help to sustain TE methylation in the zygote. See Box 9.1 for details

focusing on either genome restructuring or methylation changes in TE fractions vs. random sequences (Box 9.2), it is possible to assess the reorganization of TE genome fractions as compared to genome-wide changes (Table 9.1).

Recent hybrids are rare in nature and/or difficult to identify, and most studies used experimentally resynthetized hybrids. Massive reorganization in TE genome fractions has been documented during the first generations after polyploidization.

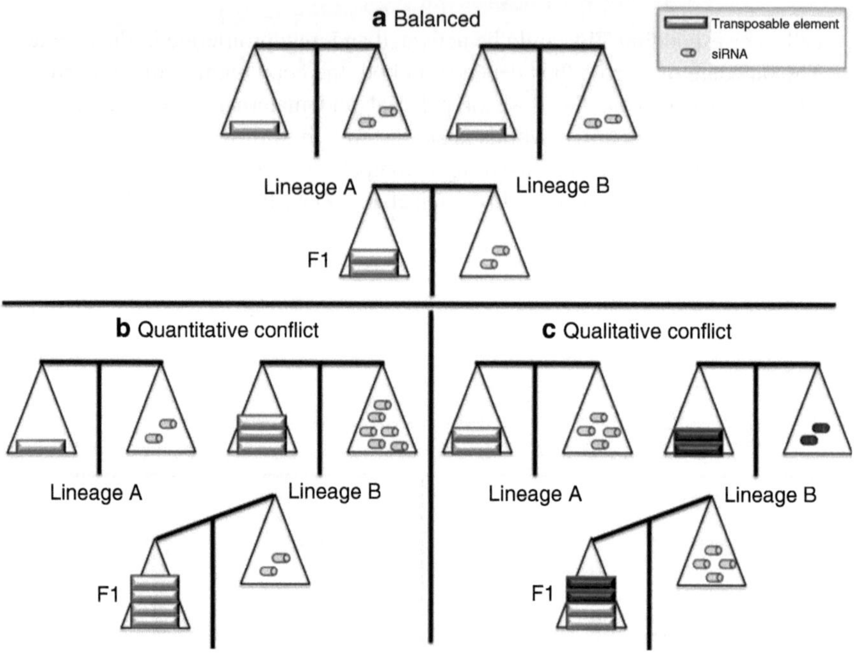

Fig. 9.3 Conflicts between parental loads in transposable elements (TEs) during genome merging. (**a**) Balanced situation: parental TEs and siRNAs match, allowing an efficient control of TEs in F1. (**b**) Quantitative or (**c**) qualitative differences in TE loads between parents, potentially leading to insufficient or inefficient repression of TEs in F1 hybrids (modified from Bourc'his and Voinnet 2010)

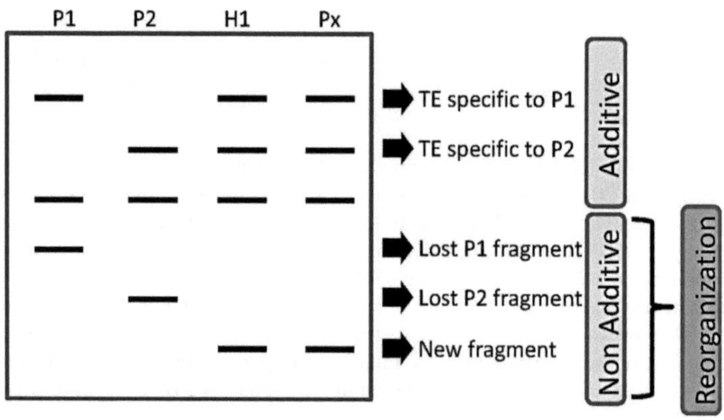

Fig. 9.4 Principle of fingerprint analyses in polyploids. Genetic profiles in the hybrid (H1) and the polyploid (Px) are expected to be the addition of the parents (P1 and P2). Deviations from this additivity indicate genome reorganization in contrasted genome fractions. See Box 9.2 for detailed explanations

Box 9.2. Molecular Fingerprint Techniques to Assess Genomes Reorganization in Nonmodel Polyploid Species

Amplified Fragment Length Polymorphism (AFLP) is a high-resolution fingerprint technique generating markers following the digestion of genomic DNA with restriction enzymes, the ligation of adaptors and PCR amplifications of fragments. The resulting markers are dominant and anonymous, but are widely distributed throughout the genome (Meudt and Clarke 2007) and thus assess genome-wide variation in random sequences. Sequence-Specific Amplified Polymorphism (SSAP) is similar to AFLP, except that it is a TE-anchored PCR strategy (i.e., Transposon Display) allowing the simultaneous detection of multiple insertions (Waugh et al. 1997; Syed and Flavell 2006). Briefly, the amplification of digested genomic DNA, specifically targeting TEs insertions, generates a pool of labeled fragments containing the termini of inserted copies of a given TE and its flanking genomic region. As retrotransposons do not excise, particular insights concerning the molecular mechanisms underlying SSAP polymorphisms can be gathered: new bands are indicative of new TE insertions (i.e., transpositions), while lost bands point to restructuring in TE sequences (comprehensively described in Parisod et al. 2010b). Note that new SSAP bands should be cautiously interpreted as new transposition events, because they can result from other molecular events changing the band size of inserted TEs (Petit et al. 2010). As a whole, the comparison of AFLP versus SSAP profiles generated from the same individuals assesses the respective variation in random sequences versus specific TE fractions.

Methyl-sensitive derivative of multilocus fingerprint techniques can be exploited by using restriction enzymes with differential sensitivity to DNA methylation on the same samples. Methyl-sensitive AFLP is named Methyl-Sensitive Amplified Polymorphism (MSAP), while methyl-sensitive SSAP has been termed Methyl-Sensitive Transposon Display (MSTD; Parisod et al. 2009). The isoschizomers *Msp*I and *Hpa*II recognize the same tetranucleotide sequence (5′-CCGG-3′), but *Hpa*II is sensitive to methylation of any cytosine at both strands (i.e., cuts 5′-CCGG-3′), while *Msp*I cuts methylated internal cytosine (i.e., cuts 5′-C5mCGG-3′). These enzymes thus assess the methylation status of internal cytosine at restriction sites (CG methylated fractions of the genome). As a whole, comparing MSAP versus MSTD profiles, respectively, can assess CG methylation changes in random sequences versus TE fractions.

In a few cases, the impact of genome merging (i.e., hybridization) vs. genome doubling has been experimentally contrasted, and hybridization seems to induce most genomic changes (reviewed in Parisod et al. 2010a), but additional studies on autopolyploids are required before reaching conclusions.

Table 9.1 Summary of reviewed studies reporting evidence on the reorganization of the transposable element (TE) genome fraction after polyploidy

Model species	TEs	Restructuring[a]	Transcription	Methylation changes[b]	References
Short term reorganization					
Arabidopsis thaliana × *A. lyrata* (F1/S0, S1, S2)	CAC, Ac-III	0/–	.	+	Beaulieu et al. (2009)
Arabidopsis thaliana × *A. arenosa* (F4)	En-Spm like	./.	+	+	Madlung et al. (2005)
Arabidopsis thaliana × *A. arenosa* (F1/F7)	Various TEs	./.	.	+ (si)	Ha et al. (2009)
Spartina alterniflora × *S. maritima* (F1)	Ins2, Cassandra, Wis-like	+/– –	.	++	Parisod et al. (2009)
Nicotiana sylvestris × *N. tomentosiformis* (F1/S0)	Tnt1	0/0	.	.	Petit et al. (2010)
Nicotiana sylvestris × *N. tomentosiformis* (S4)	Tnt1	++/– –	.	.	Petit et al. (2010)
Aegilops sharonensis × *Triticum monoccocum* (F1/S1)	Wis2-1A	0/.	+	+	Kashkush et al. (2002)
Triticum turgidum × *Aegilops tauschi* (F1/S0)	Retrotransposons, *CACTA*	0/0	.	.	Mestiri et al. (2010)
Triticum turgidum × *Aegilops tauschi* (S1-S4)	Balduin, Apollo and Thalos	+/– –	++	++	Yaakov and Kashkush (2011)
Triticum turgidum × *Aegilops tauschi* (S1-S5)	Veju	+/–	+	+	Kraitshtein et al. 2010
Triticum turgidum × *Aegilops tauschi* (F1)	Veju and Wis2-1A	./.	+	+ (si)	Kenan Eichler et al. (2010)
Long term reorganization					
Arabidopsis arenosa	Ac-like	+/–	.	.	Hazzouri et al. (2008)
Arabidopsis suecica	Ac-like	0/0	.	.	Hazzouri et al. (2008)
Nicotiana tabacum	Tnt1, Tnt2 and Tto1	+/–	.	.	Petit et al. (2007)
Nicotiana tabacum	*Gypsy* elements	./–	.	.	Renny-Byfield et al. (2011)
Brassica napus, *B.carinata*, *B.juncea*	Retrotransposons	0/.	.	.	Alix and Heslop Harrison (2004)

Brassica napus	MITE BraSto	+/0	.	.	Sarilar et al. (2011)
Brassica napus	*CACTA* Bot1	0/.	.	.	Alix et al. (2008)
Gossypium hirsutum	VariousTEs	+/−−	.	.	Grover et al. (2008)
Gossypium hirsutum	Retrotransposon	0/.	.	.	Hu et al. (2010)
Gossypium hirsutum	LINE	+/.	.	.	Hu et al. (2010)
Oryza ssp.	VariousTEs	+/−	.	.	Lu et al. (2009)
Triticum aestivum	Fatima	+/.	.		Salina et al. (2011)
Triticum aestivum	Various TEs	+/−	.	.	Charles et al. (2008)
Triticum aestivum	Various TEs	./.	.	. (si)	Cantu et al. (2010)
Triticum aestivum	Athila-like, *gypsy* and *copia* elements	0/.	.	+	Bento et al. (2008)
Sacharum ssp.	Various TEs	./.	+	.	Garsmeur et al. (2011)
Zea mays	CRM1	./−	+	.	Sharma et al. (2008)
Zea mays	Various TEs	./−	.	.	Schnable et al. (2011)
Arachis monticola	AhMITE1	./−	+	.	Gowda et al. (2011)
Coffea arabica	*Copia* elements	+/.	.	.	Yu et al. (2011)

0, no evidence; + evidence of transposition (++, > 10%); − evidence of sequence loss in TE fractions (−−, >10 %); . not evaluated
[a]Transposition/loss of TE sequences
[b](si) Accounts for changes in siRNAs of the corresponding TEs

9.4.1 Structural Changes in TE Fractions

Beaulieu et al. (2009) analyzed genome reorganization in synthetic allotetraploids between *Arabidopsis thaliana* and *A. lyrata* subsp. *petrea* and identified substantial restructuring. Changes assessed through various fingerprint techniques were mostly sequence deletions and no burst was revealed for the two DNA transposons surveyed (CAC and *Ac-III*). Another study on resynthesized *A. suecica* allopolyploids (*A. thaliana* × *A. arenosa*) used genomic microarray and fingerprint techniques to examine a region of the chromosome 4 (Madlung et al. 2005). This work highlighted transcriptional activation of En-Spm-like transposon in the allopolyploids and also identified chromosome abnormalities, suggesting possible polyploidy-induced restructuring at specific loci. These events may be related, but the exact role of TEs remains unknown. Similarly, the long terminal repeat (LTR) retrotransposon WIS2-A was transcriptionally activated in experimental polyploids between *Aegilops sharonensis* and *Triticum monococcum* (Kashkush et al. 2002). However, new TE transcripts apparently did not increase the transposition rate. Accordingly, experimental F1 hexaploids of wheat were shown to be the addition of parental *T. turgidum* and *Ae. tauschii* at hundreds of loci (Mestiri et al. 2010). As many markers were targeting specific TE insertions, this work further indicates limited restructuring in TE fractions. In the 150-year-old *Spartina* allopolyploids, Parisod et al. (2009) found limited evidence of immediate TE proliferation, with very few new SSAP bands revealed for Ins2 (*hAT* DNA transposon), Cassandra (Terminal-repeat Retrotransposon In Miniature, TRIM), and Wis-like (*copia* LTR retrotransposon) as compared to the addition of the parents. Moreover, the level of structural changes in TE fractions was comparable to random sequences, indicating no specific restructuring of TE fractions after genome merging or genome doubling. Noticeably, most structural changes occurred in F1 hybrids, suggesting that genome merging is inducing genome reorganization.

Contrasting with studies indicating limited transposition, young populations of the Tnt1 retrotransposon showed a transposition burst in early generations of synthetic allopolyploid tobacco (Petit et al. 2010). While newly synthesized polyploids were the addition of the parents, new insertion sites were detected at the fourth generation. Although the causes of Tnt1 transposition remain unclear, this work suggests that polyploidy may induce transposition of specific TEs in some cases.

While systematic and immediate transposition bursts seem to occur in specific polyploids only, the study of polyploidy-induced restructuring of TE genome fractions highlighted sequence elimination to a large extent. Studies on *Triticaceae* species (Feldman et al. 1997; Ozkan et al. 2001) showed that synthesized allopolyploids between *Triticum* and *Aegilops* have rapidly eliminated high-copy, low-copy, coding and noncoding DNA sequences. Allopolyploidy in *Spartina* was associated with a predominant loss of bands, principally from maternal origin, suggesting DNA elimination within or including TE insertions (Parisod et al. 2009). Petit et al. (2010) identified losses and indels around insertions of paternal Tnt1 sites in synthesized allotetraploid tobacco.

As a whole, the study of different polyploid systems revealed no evidence of immediate and systematic TE bursts after polyploidy, but suggest that TE genome fraction are affected by elimination of DNA sequences in the first generations after allopolyploidization.

9.4.2 Epigenetic Modification in TE Genome Fraction

In synthetic allotetraploids (*Arabidopsis thaliana* × *A. lyrata* subsp. *petrea*), methylation changes at 25 % of the genome-wide loci surveyed was assessed by MSAP (Beaulieu et al. 2009). Another study on newly synthetized allotetraploids *Arabidopsis suecica* identified that TE activation was correlated with sequence demethylation, but this was not associated with significantly higher rate of transposition (Madlung et al. 2005). Comparing reorganization of CG methylation in the whole genome vs. TE genome fractions, Parisod et al. (2009) revealed that most methylation changes occurred in the TE fraction of recent *Spartina* polyploid. The investigation of methylation changes around insertion sites of three DNA transposons (Balduin, Apollo and Thalos) during the first four generations of newly formed allohexaploid wheats revealed that 54 % of the sites have undergone CG methylation changes (Yaakov and Kashkush 2011). Noticeably, these epigenetic modifications were hypermethylation to a large extent and occurred mainly during the first two generations. Recently, study on newly formed wheat allohexaploids demonstrated substantial methylation changes around the TRIM *Veju* during the first four generations. Interestingly, hypomethylation was predominant in the first generation and quickly followed by hypermethylation (Kraitshtein et al. 2010). The study of 3,072 transcripts in wheat allotetraploids [genome SSAA: *Aegilops sharonensis* (SS) × *Triticum monococcum* ssp. *aegilopoides* (AA)] showed that 12 transcripts, including retrotransposons, were activated at early stage after polyploidization probably in correlation with methylation changes (Kashkush et al. 2002). Such activation of TEs was shown to influence the expression of adjacent genes through methylation changes (Kashkush et al. 2003).

Twenty-four-nucleotide-long small interfering RNAs (siRNA) maintain DNA methylation and are enriched in and around TEs, suggesting that they play a major role in controlling transposition (Slotkin and Martienssen 2007; Teixeira et al. 2009; Bourc'his and Voinnet 2010). Comparisons of F1 and F7 generations of synthetic allotetraploids *Arabidopsis suecica* with the two parental diploids *A. thaliana* and *A. arenosa* showed that methylation changes were associated with variation in siRNAs (Ha et al. 2009). The expression of siRNAs in the hybrids deviated from the additivity of the parents and presented drastic changes during the first generation (F1) before stabilizing in later generations (F7). Accordingly, siRNAs produced during interspecific hybridization seem to support a greater stability of the allopolyploid genome and may "serve as a buffer against the genome shock." Correspondingly, in a synthetic hexaploid wheat, the massive sequencing of siRNAs revealed that the proportion of siRNAs related to TEs decreased in allopolyploids compared to the parental lines or F1 hybrids, suggesting that TE regulation was destabilized in polyploids (Kenan Eichler et al. 2011). Detailed investigations of two *copia* LTR retrotransposons (Veju and Wis2-1A) indicated that their transcription rate was higher in the polyploids, but no formal link was established between the levels of siRNA and transcription.

As a whole, polyploidy induces considerable reshuffling of epigenetic marks, mainly in TE fractions. This may change TE dynamics, but the formal link between these processes remains to be clarified. As genetic and epigenetic variation sit on top of each other, it is crucial to further understand the fuelling role of TEs on restructuring and epigenetic repatterning across the genome. Polyploidy seem to induce the transcriptional activation of specific TEs (although not necessarily transposition) and may help to shed light on the mechanisms underlying the control of such elements.

9.5 Long-Term Restructuring in TE Genome Fraction

Long-term genome reorganization underlying evolutionary changes during the species lifespan includes mutations, exchanges of chromosome sections, evolution of TE families in subgenomes and introgression between polyploids (Comai 2005; Doyle et al. 2008; Leitch and Leitch 2008; Feldman and Levy 2009). The properties of polyploid genomes as compared to diploids are not fully clear yet, but it seems that genetic redundancy might allow higher accumulation of mutations, which may be recruited by adaptive processes to improve the success of polyploids in nature (Feldman and Levy 2005; Otto 2007; Parisod et al. 2010b). Our knowledge of the causes and consequences of polyploid genomes evolution over thousands of years is still limited, because it is experimentally impossible to reproduce and thus can only be indirectly analyzed (Table 9.1). Genome changes are indeed investigated by comparing established polyploids to extent diploids and, since both diploids and polyploids may have evolved since the polyploidy event, it remains hard to distinguish between changes due to allopolyploidy and those that occurred during the polyploid species lifespan. As the turnover of TE insertions is relatively high (Vitte and Panaud 2005), the study of TE dynamics in millions-year old polyploids is challenging.

Several studies on the polyploid wheats (*Triticum durum*; genome BA and *T. aestivum*; genome BAD) investigated the TEs by sequencing large genomic regions and identified waves of TE insertions proliferation at different time and in different genomes. A detailed survey of parts of the chromosome 3B of hexaploid wheat highlighted more than 3,000 TEs that evolved through several waves of transposition within the last four million years (Choulet et al. 2010). While fluorescent in situ hybridization revealed that the retrotransposon Fatima contributed to B-genome-specific patterns (Salina et al. 2011), Charles et al. (2008) used BAC sequencing and assessed that 90 % of the divergence between the A and B subgenomes was due to restructuring of TE fractions. However, the inferred timing of transposition for athila-like, other *gypsy* and *copia* retrotransposons was not matching the polyploidy events, indicating that their proliferation was related to the divergence of parental genomes before merging more than to genome merging. While significant transposition seems to rarely occur in polyploid wheats, evidence from transposon displays (Bento et al. 2008) and from the comparison of the hardness locus (Chantret et al. 2005) in diploid, tetraploid, and hexaploid wheat species showed major rearrangements in repetitive fractions of polyploid genomes. TE insertions were indeed often truncated and/or presented large indels in the polyploids, suggesting that TEs sustain unequal or illegitimate recombination in response to polyploidy. Moreover, a recent study investigating the dynamics of siRNAs in natural hexaploids wheat confirmed their important role in repressing TE activation through methylation in the short term, but also noticed an increased mutation rate in heavily methylated TEs (Cantu et al. 2010). Interestingly, this suggests that short-term repression might turn into a long-term mechanism of TE inactivation and genome evolution.

The sequencing of partial reverse transcriptase from six diploid and related allotetraploids of *Brassica* showed that most *copia* and *gypsy* sequences are shared by all species (Alix and Heslop Harrison 2004). No evidence of specific amplification in polyploids was revealed based on sequence similarity. More recently, Alix et al. (2008) provided evidence for several waves of amplification of a specific *CACTA* transposon (BOT1) in the diploid *Brassica oleracea* as compared to the allopolyploid *Brassica napa*. Accordingly, the transposition of BOT1 was responsible for the divergence between diploid species but no recent transposition activation was assessed in polyploids. While the BraSto MITE apparently amplified in the two parental genomes (*B. rapa* and *B. oleracea*) and their allotetraploid (*B. napus*), no specific burst at allopolyploidization was inferred (Sarilar et al. 2011). Based on the sequencing of reverse transcriptase in the allopolyploid cotton (*Gossypium hirsutum*) and its parental diploids (*Gossypium arboreum* and *G. raimondii*), different activity of *copia*, of *gypsy Gorge3* LTR retrotransposon, and of long interspersed nuclear elements (LINEs) was highlighted (Hu et al. 2010). While various proliferation periods were identified for the different TEs in the different species, bursts were apparently TE specific and hardly related to polyploidy. The comparison of sequences around the cellulose synthase locus (Grover et al. 2004) and the alcohol dehydrogenate locus (Grover et al. 2007) in the diploid progenitors and tetraploid cottons revealed a similar rate of TE activity, but a higher turnover in the polyploid TE fraction (Grover et al. 2008). Small deletions in TEs were indeed found to be extremely frequent in the polyploid, underlying genome contraction as compared to diploids. Corresponding conclusions were reached by comparing the MONOCULM1 region in diploids and tetraploids *Oryza* species (Lu et al. 2009). While different TEs amplified in divergent species and were associated with different genome size, polyploid TE fractions were characterized by sequence elimination and, mostly, TE truncation.

A few studies provided circumstantial evidence of significant TE proliferation in polyploid genomes. BAC sequencing in diploid progenitors and allopolyploid coffeas (*Coffea caneophora*, *C. eugenioides*, and the polyploid *C. arabica*) revealed differential transposition of specific TEs in the polyploid (Yu et al. 2011). In particular, a recent proliferation of *copia* retrotransposons was highlighted in *C. arabica* and participated to size variation of the corresponding subgenome as compared to its diploid state. Similarly, confronting hom(eo)ologous sequences of modern sugarcane (*Saccharum* spp.), breakdown of colinearity was specifically observed in the TE fraction, suggesting a dynamic of expansion of TEs (Garsmeur et al. 2011). Focusing on evolutionary dynamic of several *copia* retrotransposons (Tnt1, Tnt2 and Tto1) in allotetraploids *Nicotiana tabacum* and its two parental species (*N. sylvestris* and *N. tomentosiformis*) with SSAP, Petit et al. (2007) inferred considerable turnover in TEs sequences, including several new bands suggestive of transposition as well as sequence loss. Recently, Renny-Byfield et al. (2011) used low-coverage 454 sequencing to investigate the dynamics of transposable elements in *N. tabacum* and the its progenitors. The high degree of similarity between *gypsy* sequences indicated a potential TE expansion in *N. sylvestris*, but not in *N. tomentosiformis* or in the allopolyploid *N. tabacum*. The characterization of a large number of TE insertions in a single analysis strongly suggests the observed pattern to be explained by TE expansion in *N. sylvestris* after the polyploidization, but cannot entirely rule out massive TE deletions in polyploids. Associated with rigorous statistical treatment still to be developed, new sequencing techniques will offer decisive insights on the impact of TEs on long-term polyploid genome evolution, because they enable the investigation of whole genome reorganization.

Some of the difficulties inherent to the inference of long-term evolutionary processes can be circumvented by population approaches surveying genome diversity and interpreting patterns within a reliable population genetics framework. Little work adopted this promising method. Investigation of a stress-inducible MITE (AhMITE1) transposon in polyploid peanuts showed that a specific insertion at the FST-1 locus was segregating within the allopolyploid lineages (Gowda et al. 2011). As the AhMITE1 insertion was absent from the primitive allopolyploids (*Arachis monticola*), but present in derived *Arachis hypogaea*, this may suggest TE activation after polyploidy. Hazzouri et al. (2008) compared the distribution of insertions of Ac-like transposon in populations of the allopolyploid *A. suecica* and of the autopolyploid *A. arenosa*. In stark contrast with expectations raised under the hypothesis of a polyploidy-induced burst of transposition, the allopolyploids had mostly fixed insertions (i.e., non-polymorphic and mainly inherited from the parents). Autopolyploids showed significant segregation of polymorphic insertions, indicating that some TEs recently transposed and were not removed by selection. A similar approach was used in the 4.5 million-year-old polyploids of the monophyletic *Nicotiana* section *Repandae* and highlighted considerable restructuring in TE fractions (Parisod et al. 2012; Lim et al. 2007). Although the exact timing of restructuring events was hardly assessed, most new and lost SSAP bands were shared by all polyploid species, suggesting that substantial genome changes occurred shortly after the polyploidy event. Noticeably, the different TEs showed contrasted segregation patterns in the different polyploid species, indicating that long-term genome turnover may depend not only on intrinsic properties of TE populations but also on constraints imposed by host populations.

As a whole, insights on the impact of polyploidy on TEs and on the long-term genome evolution of polyploid genomes remain hardly conclusive. Available evidence seems to suggest that polyploidy per se did seldom influence transposition rate on the long term. However, the evolution of TEs after genome merging has not been extensively addressed yet. Interestingly, Sharma et al. (2008) noticed recombination between centromeric TE family (CRM1) from the two parental subgenomes of maize and suggested that such novel recombinant TE might proliferate in relation to polyploidy. Although massive TE proliferation long after the polyploidy event seems not to be the rule, TE fractions show considerable restructuring and apparently foster genome evolution in the long term. Polyploid TE insertions indeed reveal indels and truncation to a large extent, suggesting that TEs represent opportune substrate for recombination to actively shape genome architecture (Devos et al. 2002).

9.6 Conclusion

Polyploidy is a major evolutionary process leading to massive restructuring events and/or epigenetic modifications throughout the genome. Evidence is accumulating that TEs play a central role in fuelling such genome reorganization (Table 9.1). In contrast to a common belief, recent studies on several polyploids systems indicate that polyploidy-induced transposition bursts are far from being a general rule. Only few studies assessed an important burst of transposition from young and specific TE families (Parisod et al. 2010a). Available evidence however indicates that restructuring events associated with polyploidy are more frequent in TE genome fractions than in random sequences, but predominantly involve DNA sequence deletion rather than transposition. It suggests TE-specific mechanisms, but untargeted DNA lesions affecting the predominant fraction of genomes (i.e., TEs) cannot be ruled out.

Available data suggest that genome reorganization generally occurs in the first generations following the polyploidy event and involves epigenetic changes in the vicinity of TEs to a large extent. Such evidence matches the expectations of the Genome Shock hypothesis and suggests that hybridization reveals TE-specific incompatibilities. Genome merging is indeed prone to alter the balance between TEs and siRNAs and such conflict might thus induce the activation of TEs during polyploidy (Box 9.1). It should, however, be noted that a massive transpositional activation of TEs could be strongly deleterious to the nascent hybrid genome. Accordingly, it is tempting to speculate that only polyploids having controlled transposition through substantial repatterning of epigenetic marks and/or having lost TE fragments could be viable.

9.7 Perspectives

Despite a growing number of examples illustrating the central role of transposable elements during genome evolution, many crucial issues remain unanswered. We are indeed far from understanding the molecular mechanisms or the evolutionary forces underlying genome reorganization. The race between host genomes and highly mutagenic TEs deserves additional work (Blumenstiel 2011), and polyploidy seem to represent a convenient process to further explore the mechanisms activating and repressing TEs in both the short and the long term.

Future studies shall address whether the necessary genome changes related to TEs could turn beneficial by improving the viability and fertility of the nascent polyploid genome. Although some cases of adaptive evolution through TE insertion have been assessed (Bennetzen 2005), the frequency of beneficial vs. neutral vs. deleterious insertions is still largely unknown. As polyploids often see the expression pattern of duplicated genes modified, such system may help to assess to what extent TEs trigger phenotypic evolution through non-functionalization, subfunctionalization, or neo-functionalization (Walsh 2003). Moreover, nascent polyploids have to establish populations and form reproductively isolated lineages to persist in nature. Accordingly, it remains to be assessed to what extent (TE-induced) genome reorganization sustains ecological shifts associated with polyploid speciation (Parisod 2012).

Most studies reviewed here relied on allopolyploid species originating from the merging of widely divergent genomes. Accordingly, further comparison of autopolyploids vs. allopolyploids could be fruitful in order to better understand the impact of genome merging vs. genome doubling on the control of TEs and the evolutionary forces acting on the resulting variation. Furthermore, conflicts between subgenomes as put forward here to explain TE-induced reorganization after polyploidy is a process occurring at the fundamental level of the genome, while evolutionary forces such as selection or genetic drift act at the level of populations. Accordingly, the Genome Shock, the Redundancy, and the Bottleneck hypotheses are not mutually exclusive. Future work addressing the causes and consequences of TE activation on (polyploid) genome evolution shall integrate this full hierarchy (Tenaillon et al. 2010).

Acknowledgment This work was funded by the National Centre of Competence in Research "Plant Survival" and a grant (PZ00P3-131950 to CP), both from the Swiss National Science Foundation.

References

Adams KL, Wendel JF (2005) Novel patterns of gene expression in polyploid plants. Trends Genet 21:539–543

Adams KL, Cronn R, Percifield R, Wendel JF (2003) Genes duplicated by polyploidy show unequal contributions to the transcriptome and organ-specific reciprocal silencing. Proc Natl Acad Sci USA 100:4649–4654

Ainouche ML, Fortune PM, Salmon A, Parisod C (2009) Hybridization, polyploidy and invasion: lessons from *Spartina* (*Poaceae*). Biol Invasions 11:1159–1173

Alix K, Heslop Harrison JS (2004) The diversity of retroelements in diploid and allotetraploid *Brassica* species. Plant Mol Biol 54:895–909

Alix K, Joets J, Ryder C, Moore J, Barker G (2008) The CACTA transposon Bot1 played a major role in *Brassica* genome divergence and gene proliferation. Plant J 56:1030–1044

Beaulieu J, Jean M, Belzile F (2009) The allotetraploid *Arabidopsis thaliana-Arabidopsis lyrata* subsp *petraea* as an alternative model system for the study of polyploidy in plants. Mol Genet Genomics 281:421–435

Bennetzen JL (2005) Transposable elements, gene creation and genome rearrangement in flowering plants. Curr Opin Genet Dev 15:621–627

Bento M, Pereira HS, Rocheta M, Gustafson P, Viegas W (2008) Polyploidization as a retraction force in plant genome evolution: sequence rearrangements in Triticale. PLoS One 3:e1402

Blumenstiel J (2011) Evolutionary dynamics of transposable elements in a small RNA world. Trends Genet 27:23–31

Bourc'his D, Voinnet O (2010) A Small-RNA perspective on gametogenesis, fertilization, and early zygotic development. Science 330:617–622

Cantu D, Vanzetti L, Sumner A, Dubcovsky M, Matvienko M (2010) Small RNAs, DNA methylation and transposable elements in wheat. BMC Genomics 11:408

Chantret N, Salse J, Sabot F, Rahman S, Bellec A (2005) Molecular basis of evolutionary events that shaped the hardness locus in diploid and polyploid wheat species (*Triticum* and *Aegilops*). Plant Cell 17:1033–1045

Charles M, Belcram H, Just J, Huneau C, Viollet A (2008) Dynamics and differential proliferation of transposable elements during the evolution of the B and A genomes of wheat. Genetics 180:1071–1086

Chen ZJ (2007) Genetic and epigenetic mechanisms for gene expression and phenotypic variation in plant polyploids. Annu Rev Plant Biol 58:377–406

Choulet F, Wicker T, Rustenholz C, Paux E, Salse J (2010) Megabase level sequencing reveals contrasted organization and evolution patterns of the wheat gene and transposable element spaces. Plant Cell 22:1686–1701

Comai L (2000) Genetic and epigenetic interactions in allopolyploid plants. Plant Mol Biol 43:387–399

Comai L (2005) The advantages and disadvantages of being polyploid. Nat Rev Genet 6:836–846

Comai L, Tyagi AP, Winter K, Holmes Davis R, Reynolds SH (2000) Phenotypic instability and rapid gene silencing in newly formed *Arabidopsis* allotetraploids. Plant Cell 12:1551–1567

Comai L, Madlung A, Josefsson C, Tyagu A (2003) Do the different parental 'heteronomes' cause genomic shock in newly formed allopolyploids? Philos Trans R Soc Lond B Biol Sci 358:1149–1155

Devos KM, Brown JKM, Bennetzen JL (2002) Genome size reduction through illegitimate recombination counteracts genome expansion in *Arabidopsis*. Genome Res 12:1075–1079

Doyle J, Flagel L, Paterson A, Rapp R, Soltis D (2008) Evolutionary genetics of genome merger and doubling in plants. Annu Rev Genet 42:443–461

Eilam T, Anikster Y, Millet E, Manisterski J, Feldman M (2008) Nuclear DNA amount and genome downsizing in natural and synthetic allopolyploids of the genera *Aegilops* and *Triticum*. Genome 51:616–627

Feldman M, Levy AA (2005) Allopolyploidy - a shaping force in the evolution of wheat genomes. Cytogenet Genome Res 109:250–258

Feldman M, Levy A (2009) Genome evolution in allopolyploid wheat—a revolutionary reprogramming followed by gradual changes. J Genet Genomics 36:511–518

Feldman M, Liu B, Segal G, Abbo S, Levy AA (1997) Rapid elimination of low-copy DNA sequences in polyploid wheat: A possible mechanism for differentiation of homoeologous chromosomes. Genetics 147:1381–1387

Feng SH, Jacobsen SE, Reik W (2010) Epigenetic reprogramming in plant and animal development. Science 330:622–627

Garsmeur O, Charron C, Bocs S, Jouffe V, Samain S, Couloux A, Droc G, Zini C, Glaszmann JC, Van Sluys M-A, D'Hont A (2011) High homologous gene conservation despite extreme autopolyploid redundancy in sugarcane. New Phytol 189:629–642

Gaut BS, Ross-Ibarra J (2008) Selection on major components of angiosperm genomes. Science 320:484–486

Gowda MVC, Bhat RS, Sujay V, Kusuma P (2011) Characterization of AhMITE1 transposition and its association with the mutational and evolutionary origin of botanical types in peanut (*Arachis* spp.). Plant Syst Evol 291:153–158

Grover CE, Kim HR, Wing RA, Paterson AH, Wendel JF (2004) Incongruent patterns of local and global genome size evolution in cotton. Genome Res 14:1474–1482

Grover C, Kim H, Wing R, Paterson A, Wendel J (2007) Microcolinearity and genome evolution in the AdhA region of diploid and polyploid cotton (*Gossypium*). Plant J 50:995–1006

Grover C, Yu Y, Wing R, Paterson A, Wendel J (2008) A phylogenetic analysis of indel dynamics in the cotton genus. Mol Biol Evol 25:1415–1428

Ha M, Lu J, Tian L, Ramachandran V, Kasschau K (2009) Small RNAs serve as a genetic buffer against genomic shock in *Arabidopsis* interspecific hybrids and allopolyploids. Proc Natl Acad Sci USA 106:17835–17840

Hazzouri R, Mohajer A, Dejak S, Otto S, Wright S (2008) Contrasting patterns of transposable-element insertion polymorphism and nucleotide diversity in autotetraploid and allotetraploid *Arabidopsis* species. Genetics 179:581–592

Hollister J, Smith L, Guo Y-L, Ott F, Weigel D (2011) Transposable elements and small RNAs contribute to gene expression divergence between *Arabidopsis thaliana* and *Arabidopsis lyrata*. Proc Natl Acad Sci USA 108:2322–2327

Hu G, Hawkins J, Grover C, Wendel J (2010) The history and disposition of transposable elements in polyploid *Gossypium*. Genome 53:599–607

Jiao YN, Wickett NJ, Ayyampalayam S, Chanderbali AS, Landherr L, Ralph PE, Tomsho LP, Hu Y, Liang HY, Soltis PS, Soltis DE, Clifton SW, Schlarbaum SE, Schuster SC, Ma H, Leebens-Mack J, dePamphilis CW (2011) Ancestral polyploidy in seed plants and angiosperms. Nature 473:97–100

Josefsson C, Dilkes B, Comai L (2006) Parent-dependent loss of gene silencing during interspecies hybridization. Curr Biol 16:1322–1328

Kalendar R, Flavell AJ, Ellis THN, Sjakste T, Moisy C, Schulman AH (2011) Analysis of plant diversity with retrotransposon-based molecular markers. Heredity 106:520–530

Kashkush K, Khasdan V (2007) Large-scale survey of cytosine methylation of retrotransposons and the impact of readout transcription from long terminal repeats on expression of adjacent rice genes. Genetics 177:1975–1985

Kashkush K, Feldman M, Levy AA (2002) Gene loss, silencing and activation in a newly synthesized wheat allotetraploid. Genetics 160:1651–1659

Kashkush K, Feldman M, Levy AA (2003) Transcriptional activation of retrotransposons alters the expression of adjacent genes in wheat. Nat Genet 33:102–106

Kenan Eichler M, Leshkowitz D, Tal L, Noor E, Melamed Bessudo C (2011) Wheat hybridization and polyploidization results in deregulation of small RNAs. Genetics 188:263–279

Kraitshtein Z, Yaakov B, Khasdan V, Kashkush K (2010) Genetic and epigenetic dynamics of a retrotransposon after allopolyploidization of wheat. Genetics 186:801–812

Leitch IJ, Bennett MD (2004) Genome downsizing in polyploid plants. Biol J Linn Soc 82:651–663

Leitch AR, Leitch IJ (2008) Genomic plasticity and the diversity of polyploid plants. Science 320:481–483

Levin DA (2002) The role of chromosomal change in plant evolution. Oxford University Press, New York, NY

Levy AA, Feldman M (2002) The impact of polyploidy on grass genome evolution. Plant Physiol 130:1587–1593

Levy AA, Feldman M (2004) Genetic and epigenetic reprogramming of the wheat genome upon allopolyploidization. Biol J Linn Soc 82:607–613

Lim KY, Kovarik A, Matyasek R, Chase M, Clarkson J (2007) Sequence of events leading to near-complete genome turnover in allopolyploid *Nicotiana* within five million years. New Phytol 175:756–763

Liu B, Wendel JF (2003) Epigenetic phenomena and the evolution of plant allopolyploids. Mol Phylogenet Evol 29:365–379

Lu F, Sanyal A, Zhang S, Song R (2009) Comparative sequence analysis of MONOCULM1-orthologous regions in 14 *Oryza* genomes. Proc Natl Acad Sci USA 106:2071–2076

Lynch M (2007) The origins of genome architecture. Sinauer Associates, Sunderland

Mable BK, Alexandrou MA, Taylor MI (2011) Genome duplication in amphibians and fish: an extended synthesis. J Zool 284:151–182

Madlung A, Tyagi AP, Watson B, Jiang HM, Kagochi T (2005) Genomic changes in synthetic *Arabidopsis* polyploids. Plant J 41:221–230

Martienssen RA (2010) Heterochromatin, small RNA and post-fertilization dysgenesis in allopolyploid and interploid hybrids of *Arabidopsis*. New Phytol 186:46–53

Matzke MA, Matzke AJM (1998) Polyploidy and transposons. Trends Ecol Evol 13:241–241

McClintock B (1984) The significance of responses of the genome to challenge. Science 226:792–801

Mestiri I, Chague V, Tanguy A-M, Huneau C, Huteau V (2010) Newly synthesized wheat allohexaploids display progenitor-dependent meiotic stability and aneuploidy but structural genomic additivity. New Phytol 186:86–101

Meudt HM, Clarke AC (2007) Almost forgotten or latest practice? AFLP applications, analyses and advances. Trends Plant Sci 12:106–117

Otto SP (2007) The evolutionary consequences of polyploidy. Cell 131:452–462

Ozkan H, Levy AA, Feldman M (2001) Allopolyploidy-induced rapid genome evolution in the wheat (*Aegilops-Triticum*) group. Plant Cell 13:1735–1747

Parisod C (2012) Polyploids integrate genomic changes and ecological shifts. New Phytol 193:297–300

Parisod C, Salmon A, Zerjal T, Tenaillon M, Grandbastien M-A, Ainouche M (2009) Rapid structural and epigenetic reorganization near transposable elements in hybrid and allopolyploid genomes in *Spartina*. New Phytol 184:1003–1015

Parisod C, Alix K, Just J, Petit M, Sarilar V (2010a) Impact of transposable elements on the organization and function of allopolyploid genomes. New Phytol 186:37–45

Parisod C, Holderegger R, Brochmann C (2010b) Evolutionary consequences of autopolyploidy. New Phytol 186:5–17

Parisod C, Mhiri C, Lim KY, Clarkson JJ, Chase MW, Leitch AR, Grandbastien M-A (2012) Differential dynamics of transposable elements during long-term diploidization of *Nicotiana* section *Repandae* (Solanaceae) allopolyploid genomes. PLoS ONE 7:e50352

Petit M, Lim KY, Julio E, Poncet C, Dorlhac de Borne F, Kovarik A, Leitch AR, Grandbastien M-A, Mhiri C (2007) Differential impact of retrotransposon populations on the genome of allotetraploid tobacco (*Nicotiana tabacum*). Mol Genet Genomics 278:1–15

Petit M, Guidat C, Daniel J, Denis E, Montoriol E, Bui QT, Lim KY, Kovarik A, Leitch AR, Grandbastien M-A, Mhiri C (2010) Mobilization of retrotransposons in synthetic allotetraploid tobacco. New Phytol 186:135–147

Pires JC, Zhao JW, Schranz ME, Leon EJ, Quijada PA (2004) Flowering time divergence and genomic rearrangements in resynthesized *Brassica* polyploids (*Brassicaceae*). Biol J Linn Soc 82:675–688

Renny-Byfield S, Chester M, Kovarik A, LeComber AC, Grandbastien M-A, Deloger M, Nichols RA, Macas J, Novak P, Chase MW, Leitch AR (2011) Next generation sequencing reveals genome downsizing in allotetraploid *Nicotiana tabacum*, predominantly through the elimination of paternally derived repetitive DNAs. Mol Biol Evol 28:2843–2853

Riddle NC, Birchler JA (2003) Effects of reunited diverged regulatory hierarchies in allopolyploids and species hybrids. Trends Genet 19:597–600

Rieseberg LH (2001) Polyploid evolution: Keeping the peace at genomic reunions. Curr Biol 11: R925–R928

Salina EA, Sergeeva EM, Adonina IG, Shcherban AB, Belcram H, Huneau C, Chalhoub B (2011) The impact of Ty3-gypsy group LTR retrotransposons Fatima on B-genome specificity of polyploid wheats. BMC Plant Biol 11:99
Salmon A, Ainouche ML, Wendel JF (2005) Genetic and epigenetic consequences of recent hybridization and polyploidy in *Spartina* (*Poaceae*). Mol Ecol 14:1163–1175
Sarilar V, Marmagne A, Brabant P, Joets J, Alix K (2011) BraSto, a Stowaway MITE from *Brassica*: recently active copies preferentially accumulate in the gene space. Plant Mol Biol 77:59–75
Schnable JC, Springer NM, Freeling M (2011) Differentiation of the maize subgenomes by genome dominance and both ancient and ongoing gene loss. Proc Natl Acad Sci USA 108:4069–4074
Shaked H, Kashkush K, Ozkan H, Feldman M, Levy AA (2001) Sequence elimination and cytosine methylation are rapid and reproducible responses of the genome to wide hybridization and allopolyploidy in wheat. Plant Cell 13:1749–1759
Sharma A, Schneider K, Presting G (2008) Sustained retrotransposition is mediated by nucleotide deletions and interelement recombinations. Proc Natl Acad Sci USA 105:15470–15474
Slotkin RK, Martienssen R (2007) Transposable elements and the epigenetic regulation of the genome. Nat Rev Genet 8:272–285
Soltis DE, Soltis PS (1999) Polyploidy: recurrent formation and genome evolution. Trends Ecol Evol 14:348–352
Song KM, Lu P, Tang KL, Osborn TC (1995) Rapid genome change in synthetic polyploids of *Brassica* and its implications for polyploid evolution. Proc Natl Acad Sci USA 92:7719–7723
Stebbins GL (1971) Chromosomal evolution in higher plants. Edward Arnold, London
Syed NH, Flavell AJ (2006) Sequence-specific amplification polymorphisms (SSAPs): a multi-locus approach for analyzing transposon insertions. Nat Protoc 1:2746–2752
Teixeira FK, Heredia F, Sarazin A, Roudier F, Boccara M, Ciaudo C, Cruaud C, Poulain J, Berdasco M, Fraga MF, Voinnet O, Wincker P, Esteller M, Colot V (2009) A role for RNAi in the selective correction of DNA methylation defects. Science 323:1600–1604
Tenaillon MI, Hollister JD, Gaut BS (2010) A triptych of the evolution of plant transposable elements. Trends Plant Sci 15:471–478
Udall JA, Quijada PA, Osborn TC (2005) Detection of chromosomal rearrangements derived from homeologous recombination in four mapping populations of *Brassica napus* L. Genetics 169:967–979
Vitte C, Panaud O (2005) LTR retrotransposons and flowering plant genome size: emergence of the increase/decrease model. Cytogenet Genome Res 110:91–107
Walsh B (2003) Population-genetic models of the fates of duplicate genes. Genetica 118:279–294
Wang J, Tian L, Lee H-S, Chen ZJ (2006) Nonadditive regulation of FRI and FLC loci mediates flowering-time variation in *Arabidopsis* allopolyploids. Genetics 173:965–974
Waugh R, McLean K, Flavell AJ, Pearce SR, Kumar A, Thomas BBT, Powell W (1997) Genetic distribution of *Bare*-1-like retrotransposable elements in the barley genome revealed by sequence-specific amplification polymorphisms (S-SAP). Mol Gen Genet 253:687–694
Wendel JF (2000) Genome evolution in polyploids. Plant Mol Biol 42:225–249
Wood TE, Takebayashi N, Barker MS, Mayrose I, Greenspoon PB, Rieseberg LH (2009) The frequency of polyploid speciation in vascular plants. Proc Natl Acad Sci USA 106:13875–13879
Yaakov B, Kashkush K (2011) Massive alterations of the methylation patterns around DNA transposons in the first four generations of a newly formed wheat allohexaploid. Genome 54:42–49
Yu Q, Guyot R, de Kochko A, Byers A, Navajas Perez R (2011) Micro-collinearity and genome evolution in the vicinity of an ethylene receptor gene of cultivated diploid and allotetraploid coffee species (*Coffea*). Plant J 67:305–317
Zhao XP, Si Y, Hanson RE, Crane CF, Price HJ (1998) Dispersed repetitive DNA has spread to new genomes since polyploid formation in cotton. Genome Res 8:479–492

Chapter 10
Noise or Symphony: Comparative Evolutionary Analysis of Sugarcane Transposable Elements with Other Grasses

Nathalia de Setta, Cushla J. Metcalfe, Guilherme M.Q. Cruz, Edgar A. Ochoa, and Marie-Anne Van Sluys

Abstract Sugarcane is an important crop worldwide for sugar and biofuel production. Modern sugarcane cultivars have large, highly complex, polyploid genomes, and like other grasses, have a significant transposable element (TE) content. Four sugarcane TE superfamilies, *hAT*, *Mutator*, *Gypsy* and *Copia*, were first described from an EST database and, with the availability of genomic sequence, further characterised and compared with TEs from other grasses. Here we summarise previous work and extend the knowledge of the structure, diversity, evolutionary history, age, transcriptional activity and genomic distribution of sugarcane TEs. We also compare and contrast sugarcane TEs with homologous sequences in rice and sorghum, as well as analyse the age and genomic distribution of sugarcane TEs with related lineages from sorghum and rice. Finally, we discuss the importance of defining sugarcane TE lineages for understanding the contribution of ancestral genomes to modern cultivars, for genome sequencing and annotation and in applied genetics.

Keywords Sugarcane • Transposable elements • Retrotransposon • DNA transposon • Sorghum • Rice • Genome

Abbreviations

BAC	Bacterial Artificial Chromosome
BLASTn	Basic Local Alignment Search Tool nucleotide
EST	Expressed Sequence Tag
FISH	Fluorescent In Situ Hybridisation

Article Note: Both first authors contributed equally to this paper and are joint first co-authors.

N. de Setta (✉) • C.J. Metcalfe (✉) • G.M.Q. Cruz • E.A. Ochoa • M.-A. Van Sluys (✉)
Departamento de Botânica, Instituto de Biociências, Universidade de São Paulo, Rua do Matão, 277, 05508-090 São Paulo, SP, Brazil
e-mail: mavsluys@usp.br

M.-A. Grandbastien and J.M. Casacuberta (eds.), *Plant Transposable Elements*, Topics in Current Genetics 24, DOI 10.1007/978-3-642-31842-9_10,

LTR	Long Terminal Repeats
my	millions of years
mya	millions of years ago
MuLES	*Mutator*-Like ElementS
NCBI	National Center for Biotechnology Information
QTL	Quantitative Trait Loci
SChAT	SugarCane *hAT* sequences
sRNA	small RNA
TE	Transposable Elements
TIRs	Terminal Inverted Repeats
WGD	Whole Genome Duplication

10.1 Introduction

Sugarcane is an important crop worldwide, being a major source of sugar, and is also increasingly being used for the production of renewable energy sources such as ethanol (DCAA/SPAE/MAPA, http://www.agricultura.gov.br/vegetal/estatisticas). Sugarcane has a highly complex genome, which has hindered research, and, unlike many other grasses (International Rice Genome Sequencing Project 2005; Paterson et al. 2009b; Schnable et al. 2009; The International Brachypodium Initiative 2010) the sequencing of sugarcane genome is at the pilot stage. A reduced representation approach, i.e. expressed sequence tags (ESTs) and QTL mapping, has been used to initially characterise the genome (see Souza et al. 2011 for a review) and currently a consortium of laboratories are sequencing, assembling and annotating 300 full-length BACs of the modern sugarcane cultivar R570. In addition, as part of the Brazilian BIOEN project, another 700 BACs will be sequenced from the SP80-3280 cultivar, with the aim to increase our knowledge of sugarcane genome composition and variability. Analysis of EST libraries (Vettore et al. 2003) and BAC sequences (de Setta et al. 2011), combined with experimental approaches, has enabled our group to start characterising the transposable element component of the sugarcane genome.

10.1.1 Sugarcane and the Evolutionary History of Grasses

The grass family (Poaceae) comprises over 600 genera and more than 10,000 species (Clayton and Renvoize 1986; Kellogg 2001). Although other angiosperm families are more speciose, the grasses are important for their ecological dominance and agricultural and economic significance. The grasses include many important crop species, for example *Oryza sativa* (rice), *Hordeum vulgare* (barley), *Triticum aestivum* (wheat), *Sorghum bicolor* (sorghum), *Zea mays* (maize) and *Saccharum officinarum* L. (sugarcane), as well as the economically important turf grass species (Kellogg 2001; Gaut 2002). A large collaborative effort by the Grass Phylogeny Working Group (GPWG) has resulted in a robust phylogeny of the grasses based on

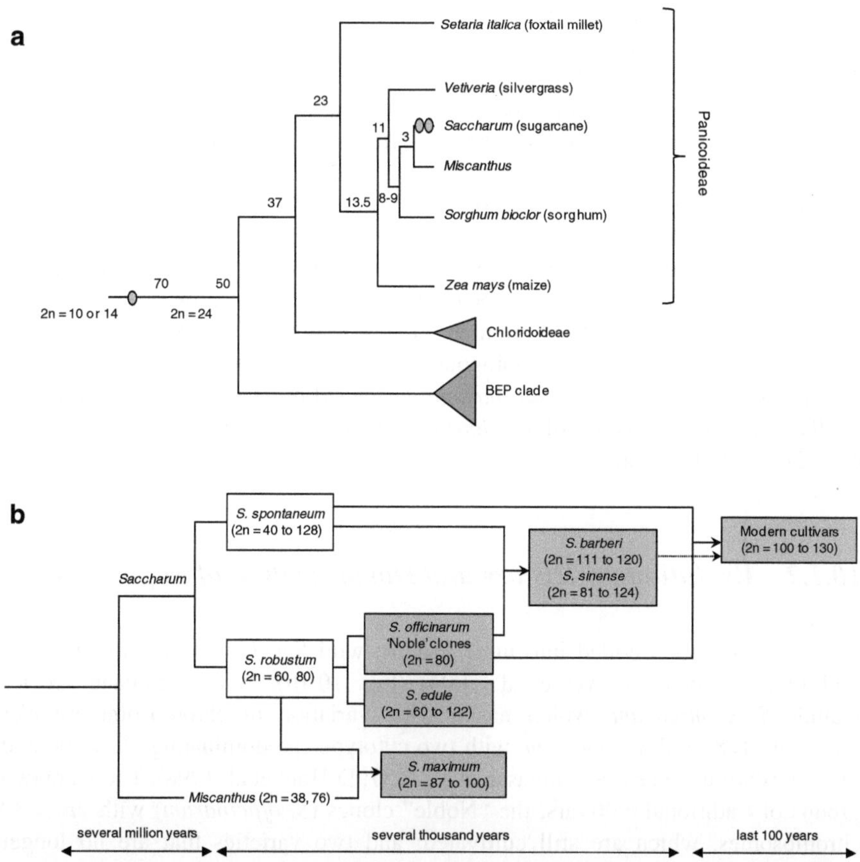

Fig. 10.1 Evolutionary history of grasses and sugarcane. (**a**) Phylogeny of the grasses showing divergence times and whole genome duplications (WGDs) relevant to *Saccharum* spp.. Divergence times and placement of WGDs from Paterson et al. (2010). Numbers above the *line* and *grey oval* indicate time of divergence and a WGD, respectively. BEP clade: Bambusoideae, Ehrhartoideae and Pooideae Subfamilies. (**b**) Evolutionary history of *Saccharum* modified from Grivet et al. (2006). *White box*: wild species, *grey box*: cultivated species. *Line* and *dashed arrows* indicate hybridization events and minor contribution to modern sugarcane cultivars repertoire, respectively. Chromosome numbers from Grivet et al. (2006)

both morphological and molecular data (Grass Phylogeny Working Group 2001). Twelve superfamilies are recognised. *Saccharum* is a member of the large Panicoideae superfamily, which includes the economically important maize, sorghum, pearl millet and foxtail millet. Of these, *Saccharum* is most closely related to sorghum (Fig. 10.1a).

Grasses have wide range in chromosome number and genome size and are considered to have liable genomes (Gaut 2002). The modern sugarcane cultivar genome is particularly complex and large (1C = 5,000 Mb or 5,11 pg) with chromosome numbers ranging from 100 to 130 (D'Hont and Glaszmann 2001). The sugarcane genome is the result of several whole genome duplications (WGD) and recent hybridisation events. The first of these WGDs is thought to occur 20 million years ago (mya) before the diversification of grasses, estimated to be 55–70 mya, based on fossil evidence (Kellogg 2001) (Fig. 10.1a). The ancestor to the grasses is considered to have had a chromosome complement of either $2n = 10$ or 14 (Salse et al. 2008; Devos 2009). After the first WGD this complement underwent various rearrangements resulting in a $2n = 24$ complement, very similar to that seen in rice, which most authors agree is the most likely ancestral karyotype for the grasses (Salse et al. 2008; Devos 2009). The other two WGDs within the sugarcane linage are much more recent. *Saccharum* last shared a common ancestor with *Sorghum* about 8–9 mya (Jannoo et al. 2007). The high chromosome numbers in *Saccharum*, plus the observation that many regions of the *Sorghum* genome corresponds to four or more homologous regions in *Saccharum,* suggest that there was a least two WGDs in the *Saccharum* lineage (Ming et al. 1998; Paterson et al. 2009a), after the divergence of *Saccharum* and *Sorghum* (Jannoo et al. 2007; Wang et al. 2010) (Fig. 10.1a).

10.1.2 Evolutionary History and Domestication of Sugarcane

Saccharum can be divided into three groups: wild *Saccharum*, and modern and traditional cultivars (Grivet et al. 2004) (Fig. 10.1b). Wild sugarcane species include *S. spontaneum*, which has a wide variation in chromosome number ($2n = 40$–128) and *S. robustum*, with two cytotypes predominating, $2n = 60$ and $2n = 80$ chromosomes (Sreenivasan et al. 1987; D'Hont et al. 1998). There are two groups of traditional cultivars, the "Noble" clones (*S. officinarum*) with $2n = 80$ chromosomes, which are still cultivated; and two varieties that are no longer cultivated and exist only in germplasm collections, the North Indian *S. barberi* ($2n = 111$–120 chromosomes) and the Chinese *S. sinense* ($2n = 81$–124 chromosomes) cultivars (D'Hont et al. 1996; Grivet et al. 2004). In addition to these two groups and the modern cultivars, there are two other, less well-described taxa, *S. edule* and *S. maximum*.

10.1.2.1 Evolutionary Path to Sugarcane

Prior to more recent molecular studies, it was thought that other genera, particularly *Miscanthus, Erianthus*, *Schlerostachy* and *Narenga*, contributed in the emergence of sugarcane (reviewed in Daniels and Roach 1987). However, based on several molecular studies, including isozyme, fragment length polymorphism, amplified fragment length polymorphism and repeated species-specific sequence, it is now becoming clearer that the genus *Saccharum* is a well-defined lineage that has diverged over a long period of time from sister genera (reviewed in Grivet et al. 2004, 2006; D'Hont and Glaszmann 2001). The two wild species, *S. spontaneum* and *S. robustum* can also be clearly distinguished both morphologically and at the molecular level (reviewed in Grivet et al. 2004, 2006), they also have different basic chromosome numbers, $x = 8$ in *S. spontaneum* and $x = 10$ in *S. robustum.*

Based on this molecular data, as well as morphological and geographical considerations, Grivet et al. (2004) propose the following scenario. Sugarcane arose in the South-East Asian and Melanesian region. This region has two distinct floras and faunas, first described by Alfred Wallace. The Wallace line or Wallace's line, runs between two continental shelves, the Sunda (mainland Southeast Asia, Java, Sumatra and Kalimantan) and the Sahul (Australia, New Guinea and close islands) shelves. In the scenario proposed by Brandes (1958), the two wild sugarcane species, *S. spontaneum* and *S. robustum*, differentiated on either side of the Wallace line, *S. spontaneum* north of the line on the Sunda shelf, *S. robustum* south of the line on the Sahul shelf. On the island of New Guinea, known for the domestication of several important crops (Lebot 1999), *S. robustum* was domesticated, creating the "Noble clones" (*S. officinarum*).

10.1.2.2 Traditional and Other Cultivars

Several lines of molecular evidence support the origin of the *S. officinarum*, the Noble clones from the wild species *S. robustum* (Fig. 10.1b). Two mitochrondrial haplotypes are found in *S. robustum*, one much more common than the other. This more common mitochondrial is the single haplotype found in a series of *S. officinarum* clones (D'Hont et al. 1993). RFLP analysis of nuclear single copy DNA places *S. officinarum* very close to *S. robustum* (Lu et al. 1994). Finally, while the second wild species, *S. spontaneum*, has a basic chromosome number of $x = 8$, for both *S. officinarum* ($2n = 80$) and *S. robustum* (major cytotypes with $2n = 60$ and 80), the most probable basic chromosome number is $x = 10$ (D'Hont et al. 1998).

There are also several lines of evidence supporting the hypothesis that the traditional cultivars, *S. barberi* and *S. sinense*, are derived from interspecific hybridisation between *S. officinarum* and *S. spontaneum* (reviewed in D'Hont et al. 2002). Nuclear RFLP markers place *S. barberi* and *S. sinense* between *S. officinarum* and *S. spontaneum.* Southern hybridisation with genus-specific sequences from *Erianthus*, *Miscanthus* and *Saccharum* do not support the contribution of *Erianthus* or *Miscanthus* to *S. barberi* and *S. sinense*. Finally, genomic in situ hybridisation using *S. officinarum* and *S. spontaneum* genomic DNA as probes clearly shows homogenous labelling of all chromosomes in *S. barberi* and *S. sinense* (D'Hont et al. 2002). Regarding the less well-described taxa, *S. edule* and *S. maximum*, the little data available suggests that *S. edule* is a series of *S. robustum* mutant clones preserved by humans and that *S. maximum* may be a heterogenous group with different levels of introgression between *Saccharum* and *Miscanthus* (reviewed in Grivet et al. 2006) (Fig. 10.1b).

10.1.2.3 Modern Cultivars

Early in the nineteenth century, breeders in Java and India produced interspecific hybrids between *S. officinarum* as the female parent and *S. spontaneum* and, to a lesser extent *S. barberi*, as the pollen donor. F_1 hybrids were backcrossed with *S. officinarum* in a process known as "nobilisation" (Fig. 10.1b). Hybrids between *S. officinarum* and *S. spontaneum* show a $2n + n$ transmission, where $2n$ is the entire genome of *S. officinarum*. This phenomenon remains true in the first backcross between the $2n + n$ F_1 and the female *S. officinarum* but generally breaks down in subsequent backcrosses (Bremer 1963; Piperidis et al. 2010). Early breeders used this phenomenon to introduce vigour and resistance genes from *S. spontaneum* while quickly recovering the high sugar content of *S. officinarum* (Roach 1972). Further crosses have resulted in the modern sugarcane cultivars, which have highly complex interspecific polyploid genomes, with high chromosome numbers ($2n = 100$–130), 70–80% of which are from *S. officinarum*, 10–23% from *S. spontaneum*, and a small portion being recombinants (D'Hont et al. 1996; Piperidis et al. 2010).

All work presented here is based on sequence data from two modern cultivars, R570 and SP80-3280. The cultivar R570 was chosen by the SUGESI SUgarcane GEnome Sequencing Initiative team as a priority for a reference sequence since it is a typical modern sugarcane cultivar obtained by the Centre d'Essai de Recherche et de Formation (CERF) in La Réunion (http://www.ercane.re/) and for which the genome has been best characterised (see Souza et al. 2011 for a review). The cultivar SP80-3280 is the cultivar that contributed most to the SUCEST SUgarcane EST project and is currently being sequenced by a Brazilian group funded by the FAPESP Bioenergy Research Program BIOEN (http://bioenfapesp.org; Vettore et al. 2003).

10.2 Sugarcane Transposable Elements

10.2.1 Overview

The first analysis of TEs in sugarcane was based on the Sugarcane EST database (SUCEST), which has more than 43,000 putative transcript clusters (Vettore et al. 2003). Rossi et al. (2001) were able to identify 276 cDNAs homologous to TEs, of which 54% were DNA transposons and 46% LTR retrotransposons, classified into 21 TE families. The *Mutator* DNA transposon and *Hopscotch* LTR retrotransposon (*Copia* superfamily; *RLC_scAle*) were the first and the second most expressed families, both in the EST database, and in a validation experiment with callus, leaf roll, apical meristem and flower (Araujo et al. 2005). Araujo et al. (2005) were also able to identify transcripts from *Ac*-like DNA transposons (*hAT* superfamily) and from the *Gypsy* LTR retrotransposon superfamily.

Some of the TEs identified in the EST analyses were further characterised in terms of function, structure and evolutionary history by our research group. This included identification of complete elements in sugarcane BACs, evaluation of transcriptional activity and small RNA (sRNA) targeting, and comparative phylogenetic and synteny analyses with other grass genomes (Rossi et al. 2004; Saccaro-Junior et al. 2007; de Jesus et al. 2012; Domingues et al. 2012; Manetti et al. 2012). The results illustrate that these elements have different diversification and evolutionary histories, involving bursts of transposition and domestication. In this review we summarise the current knowledge of sugarcane TEs. We compare and contrast the four main superfamilies of sugarcane TEs studied by our laboratory, and compare them with homologous sequences in rice and sorghum. Two are Class II superfamilies: *Mutator* (Rossi et al. 2001; Rossi et al. 2004; Saccaro-Junior et al. 2007; Manetti et al. 2012) and *hAT* (Rossi et al. 2001; Araujo et al. 2005; de Jesus et al. 2012); and two are Class I superfamilies, *Copia* and *Gypsy* (Domingues et al. 2012). In order to compare representative families of sugarcane TEs with those from other grasses, we also analysed and compared the chromosomal distribution for all four superfamilies and time of insertion for LTR retrotransposons. We discuss the repetitive content of the sugarcane genome in terms of its impact on the structure of the genome, the importance of describing and cataloguing TEs in terms of genome assembling and annotation, and how a better understanding of the TE content of sugarcane may aid in cultivar development.

10.2.2 Mutator *Transposons*

Mutator has been described as the most mutagenic plant transposon system (Diao and Lisch 2006). The canonical *Mutator* element, *MuDR*, was first described in maize (Robertson 1978) and is composed of two genes, *mudrA* and *mudrB*. The transposase protein necessary for *MuDR* mobilisation is encoded by *mudrA*. On the other hand, no function has been determined for *mudrB*, which is restricted to the *Zea* genus (Diao and Lisch 2006). *Mutator*-like elements (*MuLEs*) are widely distributed in angiosperms (Yu et al. 2000; Lisch et al. 2001; Rossi et al. 2004) and are transcriptionally active (Lisch 2002). In sugarcane, *MuLE*s were first identified in the SUCEST database as the most highly transcribed TEs (Rossi et al. 2001, 2004). Thirty-four poly-A derived clones with homology to the maize *mudrA* were identified. Phylogenetic analysis with *mudrA* sequences of sugarcane, maize, rice and *Arabidopsis thaliana* showed that sugarcane *mudrAs* fall into four well-defined phylogenetically distinct clades, Classes I–IV. Each Class is comprised of either putative domesticated or *bona fide* transposons, and appear to be very old, since they diverged before the monocot–eudicot split (Rossi et al. 2004).

10.2.2.1 Domesticated *Mutators*

MUSTANGs are *mudrA*-derived domesticated genes previously described in *Arabidopsis* and rice (Cowan et al. 2005). Several lines of evidence suggest that Class III and IV are putative domesticated elements. Phylogenetic analysis of sugarcane Class III and IV *Mutator* sequences, *Arabidopsis* and rice MUSTANGs showed a topology according to the host phylogeny, which suggests that they are MUSTANG orthologues, and that a single MUSTANG domestication event occurred prior to the Monocot and Eudicot divergence (Saccaro-Junior et al. 2007). The low copy number of these sequences (Saccaro-Junior et al. 2007) and the relatively high expression level of MUSTANG genes in sugarcane and rice also support the domestication hypothesis (Cowan et al. 2005; Saccaro-Junior et al. 2007). Two TE evolutionary mechanisms were evidenced by the phylogenetic analysis of *Mutator* elements from sugarcane, rice and *Arabidopsis* (Saccaro-Junior et al. 2007). First, MUSTANG genes from sugarcane, *Arabidopsis* and rice evolved and differentiated by a series of duplications. Second, the authors described several sugarcane MUSTANG haplotypes for each rice orthologue, which are probably the result of interspecific hybridisation/polyploidy/aneuploidy of sugarcane. This was the first time that sugarcane TE alleles have been clearly characterised.

10.2.2.2 Bona Fide *Mutator* Transposons

Class I and II *Mutator* elements were identified as *bona fide* transposons, with Class I sequences being the most closely related to the canonical *MuDR* (Saccaro-Junior et al. 2007). Based on hybridisation of a BAC library membrane set using a representative transcript from each class, Class II has the highest number of copies, indicating that these elements underwent a recent or even continuing burst of transcriptional activity in sugarcane, resulting in copy number amplification (Saccaro-Junior et al. 2007) (Table 10.1). In order to better understand class-specific amplification, Saccaro-Junior et al. (2007) did an in silico search in the rice genome (Table 10.1). Here, we extend this information with an in silico search in sorghum, using as queries the sequences of the probes hybridised against the BAC library by Saccaro-Junior et al. (2007). Like sugarcane, rice and sorghum have more Class II elements than Class I (Table 10.1). In sugarcane, rice and sorghum, Class I and II *Mutator* elements are most frequently located along the chromosome arms, with most copies found in euchromatic regions (Saccaro-Junior et al. 2007, Manetti et al. 2012) (Fig. 10.2). Although maize genomes have not been examined specifically for the chromosomal distribution of *Mutator* elements, DNA transposons are mainly found in non-centromeric regions (Schnable et al. 2009). All these results indicate that Class I and II *Mutator* elements have similar chromosome distribution and copy number patterns in grasses.

Mutator transposons have high levels of sequence diversity and are an old component of plant genomes (Yu et al. 2000; Lisch et al. 2001; Rossi et al. 2004;

Table 10.1 Monoploid copy number of *Mutator* and *hAT* superfamily elements in sugarcane, sorghum and rice, according to Saccaro-Junior et al. (2007) and our analyses

TE	Sugarcane	Sorghum	Rice
Mutator			
Class I	28[a]	15[b]	50[c]
Class II	172[a]	76[b]	386[c]
hAT			
191 Lineage	3[a]	2[d]	0[d]
257 Lineage	2[a]	3[d]	0[d]

The putative domesticated *Mutator* and *hAT* elements were not included in this table

[a]Copy number evaluated by Saccaro-Junior et al. (2007) for *Mutator* and in this work for *hAT* using membrane hybridisation of the SHCRBa library membrane set as recommended by manufacture's instruction (http://www.genome.clemson.edu) and PCR-based probes specific for each Class/Lineage. The *hAT* probes were PCR amplified fragments specific for Lineages 191 (207 bp) and 257 (164 bp), respectively (Lineage 191: Schat3F—GGAGAGTATG-GAAGTGTCCC and Schat4R—CCTATCATACTCGCTGTTTCT; Lineage 257: Schat5F—GAGAATGCAGAAGCGGAA and Schat6R—CCACGAAGTCCAGAAGAACT). The total copy number was calculated by dividing the number of positive clones/1.3 (coverage of the library) and the monoploid copy number was calculated by copy number/10 (sugarcane decaploid genome)

[b]Copy number evaluated in this work by an in silico search (BLASTn) in the sorghum genome (http://www.phytozome.net, v1.0) using as queries the specific probes for Class I (TE165) and Class II (TE109), used in the SHCRBa library screening by Saccaro-Junior et al. (2007) (cut-off $e \leq -57$)

[c]Copy number evaluated by Saccaro-Junior et al. (2007) by an in silico search (BLASTn) in the rice using as queries sugarcane *MULE* transcript sequences homologous to maize *mudrA* (cut-off $e \leq -57$, coverage $\geq$60%)

[d]Copy number evaluated in this work by an in silico search (BLASTn) in the rice (http://www.phytozome.net, Build 4.0) and sorghum genomes (http://www.phytozome.net, v1.0) using as queries the TE191and TE257 sequences used as probes in item a (cut-off $e \leq -32$)

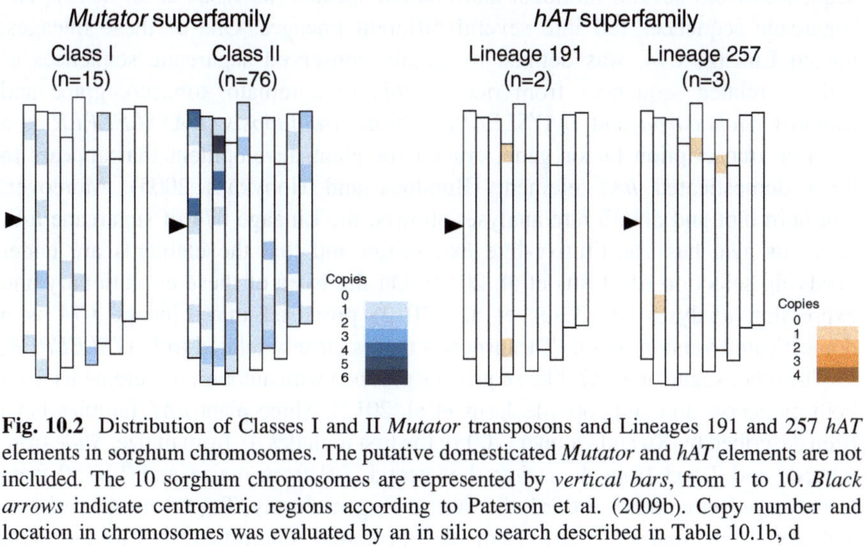

Fig. 10.2 Distribution of Classes I and II *Mutator* transposons and Lineages 191 and 257 *hAT* elements in sorghum chromosomes. The putative domesticated *Mutator* and *hAT* elements are not included. The 10 sorghum chromosomes are represented by *vertical bars*, from 1 to 10. *Black arrows* indicate centromeric regions according to Paterson et al. (2009b). Copy number and location in chromosomes was evaluated by an in silico search described in Table 10.1b, d

Saccaro-Junior et al. 2007), evolving mainly by vertical transmission and family differentiation. This conclusion is based on the high similarity between sequences of the same Class and low similarity between sequences of different Classes. There are families with autonomous and non-autonomous elements, which remain capable of transcription, mobilisation and having an impact on genome structure. However, there are other kinds of *Mutator* insertions, those that have been recruited by the host genome, which have lost mobilisation characteristics and have become domesticated TEs. In sugarcane, both types of families were described (Araujo et al. 2005; Rossi et al. 2004). Thus, sugarcane *MULEs* contribute to the dynamics of the genome, because they can effectively create genetic variability.

10.2.3 **hAT** ***Transposons***

hAT is a DNA transposon superfamily, named after three well-described elements, *Ac* from *Zea mays* (Döring and Starlinger 1984), *hobo* from *Drosophila melanogaster* (McGinnis et al. 1983) and *Tam3* from *Antirrhinum majus* (Hehl et al. 1991). *Ac* is part of the *Ac/Ds* system, first discovered by McClintock as "controlling elements" (McClintock 1951). The *hAT* superfamily is widely distributed in eukaryotes and very ancient, probably originating before the early stages of the plant–animal–fungi divergence (Kempken and Windhofer 2001; Rubin et al. 2001). Currently, there is not a lot of information about their evolution in plants.

hAT-like elements were the second most expressed DNA transposons identified in the SUCEST database (Rossi et al. 2001). Twenty-one cDNA transcripts and five genomic elements were sequenced and compared phylogenetically with *hAT* sequences from several monocot and eudicot species (de Jesus et al. 2012). The sugarcane sequences fell into several different lineages. One of these lineages, named Lineage 074, was defined by highly conserved sugarcane sequences as well as related sequences from rice, *Arabidopsis*, tomato, tobacco, grape and *Populus trichocarpa* and *DAYSLEEPER* from *Arabidopsis*. *DAYSLEEPER* is a putative transcription factor gene, crucial for plant development that appears to be a domesticated *hAT* element (Bundock and Hooykaas 2005). Moreover, Northern blot and dN/dS rate analyses showed the Lineage 074 of sugarcane *hAT* elements also has constitutive-like expression and that the elements are under purifying selection (de Jesus et al. 2012). On the basis of these evolutionary and expression analyses, de Jesus et al. (2012) proposed that Lineage 074 is a domesticated version of a *hAT* transposon that is closely related to *DAYSLEEPER*.

The other sugarcane *hAT*-like sequences grouped with autonomous elements from both monocots and eudicots (de Jesus et al. 2012). Three plant *hAT* families have been described by Xu and Dooner (2005); the first includes *Ac* from maize, *Slide* from tobacco and *Tam3* from *A. majus*; the second, *I-R* from maize and *Tip100* from *Ipomoea purpurea* and the third, the most divergent family, *Tag1* from *Arabidopsis* and *Bg* from maize. Apart from the proposed domesticated Lineage 074, transcriptionally active sugarcane sequences (named SChAT) therefore appear to be most closely related to the classic *Ac* transposon family. We cannot be sure if the *I-R/Tip100* and *Tag1/Bg* families are absent or transcriptionally inactive in sugarcane. Two lineages of these sugarcane *Ac*-like elements, Lineages 191 and 257, were further analysed by de Jesus et al. (2012) and were identified in all examined modern sugarcane cultivars. They had different Southern blot hybridisation patterns in the parental species *S. officinarum* and *S. spontaneum.* Lineage 257 is present in both parental species while Lineage 191 is found only in *S. officinarum.* The identification of the Lineage 257 in both sugarcane parental species suggests that these elements were present in the genome of the *S. officinarum* and *S. spontaneum* ancestor. On the other hand, the absence of Lineage 191 elements in *S. spontaneum* indicates that these elements evolved after the sugarcane parental species diversified, demonstrating that there have been different evolutionary histories of these TEs in closely related species (de Jesus et al. 2012).

To advance our knowledge of the SChAT Lineages 191 and 257, we estimated their copy number in sugarcane by hybridisation to a SHCRBa membrane set, using the approach described by Saccaro-Junior et al. (2007). Results showed that Lineages 191 and 257 have low copy numbers, on average 28.5 and 21.5 copies, respectively, in the polyploid genome, that is 3 and 2 copies in the monoploid genome (Table 10.1). To compare the copy number in sugarcane with sorghum and rice, we performed an in silico search using as queries the same lineage specific probes hybridised against the SHCRBa membrane set. The results indicated that Lineages 191 and 257 also have low copy numbers in sorghum, 2 and 3 copies in the monoploid genome, respectively. These five copies are located on the sorghum chromosome arms (Fig. 10.2). No sequences homologous to Lineages 191 and 257 were identified in the rice genome, either by Southern blot hybridisation experiments (de Jesus et al. 2012) or by in silico searches (Table 10.1).

It is clear from the information above that *hAT* and *Mutator* share some evolutionary features; however, particularly in the grasses discussed here, they also exhibit some differences. Like *Mutator*, *hAT* is an ancient component of plant genomes and is evolving by vertical inheritance concomitant with diversification of families (Xu and Dooner 2005). They have similar chromosomal distributions in both families and there are examples of putative domesticated elements. On the other hand, in sugarcane *hAT* appears to be less active than *Mutator*, as there are lower numbers of both *hAT* transcripts and genomic copies.

10.2.4 LTR Retrotransposons

10.2.4.1 Classification and Structure of LTR Retrotransposons

LTR retrotransposons are ubiquitous in plant genomes. They are the main component of large plant genomes (Kumar and Bennetzen 1999) and in some cases differences in LTR retrotransposon content may account for differences in genome size in closely related species (Hawkins et al. 2006). They are predominately from two large superfamilies, *Copia* and *Gypsy*, which differ both in sequence similarity, in the order of genes in the *pol* domain, and their chromosomal distribution (International Rice Genome Sequencing Project 2005; Paterson et al. 2009b; Schnable et al. 2009; Llorens et al. 2011). Within the two superfamilies, LTR retrotransposons can be further sub-divided into lineages on the basis of reverse-transcriptase sequence identity (Wicker et al. 2007; Du et al. 2010; Llorens et al. 2011).

BACs derived from the R570 sugarcane cultivar (Tomkins et al. 1999) sequenced for the BIOEN Project (de Setta et al. 2011) and also those available at the National Center for Biotechnology Information (NCBI) Web site as at February 2011, were screened for full-length LTR elements by Domingues et al. (2012). Sixty sequences, 32 *Copia* and 28 *Gypsy* elements, were retrieved. These were classified into 35 families based on LTR sequence identity within 7 known (4 *Copia* and 3 *Gypsy*) lineages (Wicker et al. 2007; Du et al. 2010; Llorens et al. 2011). Differences in length of elements within lineages was chiefly due to differences in the size of LTRs and the presence and size of spacer regions between the coding domains and the LTRs, a feature observed in *Copia* elements from wheat, rice and *Arabidopsis* (Wicker and Keller 2007).

10.2.4.2 Transcriptional Activity of LTR Retrotransposons and their Associated sRNAs

Araujo et al. (2005) identified an LTR retrotransposon, *Hopscotch* (*Copia* superfamily), as the second most highly expressed TE family in the SUCEST database. Microarrays with callus, apical meristem, leaf roll and flower tissues confirmed the expression of these elements and identified callus as the tissue with the highest expression levels. In a functional experiment with leaf, callus and root the ability of the U3 region in the LTR to act as an active promoter was demonstrated for four *Hopscotch* ESTs (Araujo et al. 2005). Domingues et al. (2012) re-analysed ESTs from the SUCEST database according to the new classification. They showed that the highest number of ESTs were associated with *Ale1* family elements and that almost all the *Hopscotch* ESTs described by Araujo et al. (2005) were from the same family. Although transcripts from all other LTR retrotransposons lineages were also identified (Domingues et al. 2012), there were more than double the number the ESTs associated with *Ale1* compared with other families, confirming the findings of Araujo et al. (2005).

The activity of LTR retrotransposons is usually controlled by the host genome through the siRNA machinery. Two main classes of siRNAs are generated, the 21-nt class regulates post-transcriptionally related mRNAs, while the 24-nt class suppresses gene expression at the transcriptional level (Baulcombe 2004). If very few sRNAs are mapped to an element, it indicates that this element is not transcriptionally active, or it is very recently activated and has not yet triggered the host small RNA-dependent silencing machinery. If it is not transcriptionally active, silencing may be being maintained by ancient methylation. (Zhang et al. 2006). Previous studies mapping sRNAs to LTR retrotransposons in wheat and maize genomes (Nobuta et al. 2008; Cantu et al. 2010) showed a pattern of concentration of 24-nt sRNA in the LTRs when analysing all LTR retrotransposons together.

Domingues et al. (2012) mapped sRNAs to a single reference sequence (sequence 1) for each family. For almost half of the families (18 out of 33) very few sRNAs (<2,000 counts) were mapped to the reference copy. The "24-nt LTR" pattern observed in wheat and maize LTR retrotransposons (Nobuta et al. 2008; Cantu et al. 2010) was seen in all reference sequences from two families (*Dell* and *Tat3*) and one lineage (*Maximus*).

Two other patterns of sRNA mapping were observed, one in which high numbers of 21-nt sRNAs mapped along the coding region and one in which a very large number of 24-nt sRNAs mapped within the coding region, seen only in the *Ale1* family. *Ale1* is the family for which the highest number of ESTs were identified (Domingues et al. 2012) and for which the U3 region in the LTR has been shown to be capable of acting as a promoter (Araujo et al. 2005). Zhang et al. (2006) reported that highly and constitutively expressed *Arabidopsis* genes were methylated in the coding region, but not in the promoter region, the "body-methylated gene" concept. The sRNA pattern in *Ale1* indicates that the methylation machinery is being guided to the coding region of the element, like a gene that is "body-methylated". It is unclear whether these patterns of sRNA mapping are particular to sugarcane, because previous work mapping sRNAs to LTR retrotransposons in maize and wheat presented overall sRNA patterns for entire superfamilies or a single element (Nobuta et al. 2008; Cantu et al. 2010). However, it is intriguing and worthy of further investigation.

10.2.4.3 Chromosomal Distribution of LTR Retrotransposons

Fluorescence in situ hybridisation (FISH), using the sequence from a representative family of each sugarcane lineage as a probe, in general showed localization patterns expected from prior results with *Gypsy* and *Copia* elements in plants (Heslop-Harrison et al. 1997; Paterson et al. 2009b). *Gypsy* elements tended to be more concentrated in heterochromatic regions; *Copia* elements were more dispersed throughout the genome (Domingues et al. 2012). Two elements, *Tat2* (*Gypsy*) and *Ale1* (*Copia*), had regions with stronger in situ hybridisation signals, that is along particular chromosome arms, with no particular pattern in terms of euchromatin or heterochromatin (Fig. 10.3) (Domingues et al. 2012). *Tat2* showed a similar pattern of uneven distribution in two sugarcane cultivars, R570 and SP80-3280 (Domingues et al. 2012). As far as we can tell, they do not cluster together.

It has been suggested that, when an increasing proportion of the genome consists of TEs, new insertions, even if they occur randomly, will be more likely to occur within another TEs, thus expanding TE clusters (Hua-Van et al. 2011). The clusters of *Tat2* and *Ale1* elements along particular chromosome arms in sugarcane may therefore be a matter of chance, a cluster of similar TEs that are selectively neutral. Modern sugarcane cultivars are recent hybrids between *S. officinarum* and

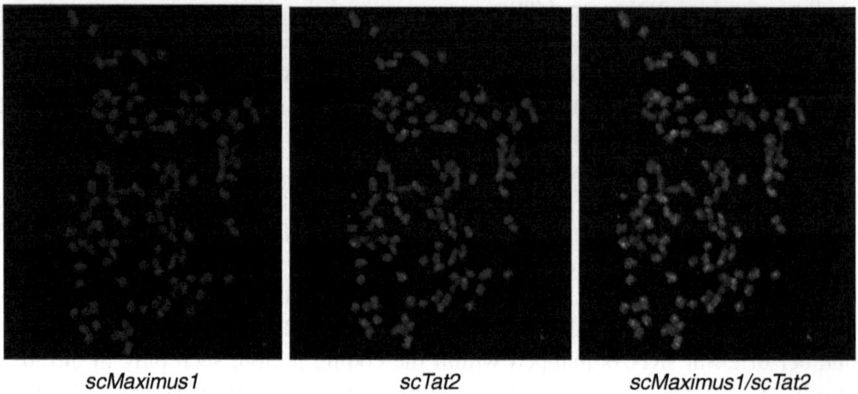

Fig. 10.3 FISH localisation of *Tat2* and *Maximus1* elements to sugarcane chromosomes (R570 cultivar). The LTR retrotransposon probes are 2 and 2.4 kb long, for *Tat2* and *Maximus1*, respectively, and included the reverse transcriptase domain. *Tat2* was labelled with Biotin (*green signal*) and *Maximus1* with Digoxigenin (*red signal*). All FISH procedures were performed according to Domingues et al. (2012)

S. spontaneum, with uneven contributions from the parental genomes. Seventy to eighty percent of the chromosomes of a modern cultivar are estimated to be from *S. officinarum*, only 10–23% from *S. spontaneum*, and a small portion are estimated to be recombinants (D'Hont et al. 1996; Piperidis et al. 2010). The uneven distribution of *Tat2* and *Ale1* elements may therefore be the result of larger copy numbers of these elements in one parental type, similar to that reported in wheat (Salina et al. 2011) or, alternatively, it may also be the preferential loss from one parental genome as seen in tobacco (Renny-Byfield et al. 2011).

To compare the chromosomal distribution of sugarcane LTR retrotransposons with related elements in sorghum and rice, we extracted full-length LTRs from the sorghum and rice genomes (see Fig. 10.4 for method details). Using these sequences, we created in silico heat maps in sorghum and rice for each LTR retrotransposon family. The lower copy number of *Angela1*, *Ivana1* and *Reina1* in sorghum and rice supports Domingues et al. (2012) that the lack of signal for these probes was probably because they are present in lower numbers than other elements. Like sugarcane, *Del1* is found broadly distributed around and within the centromeric region in the sorghum genome, but in rice *Del1* is widely distributed (Fig. 10.4). *Tat2* and *Ale1*, which have clusters of localization in sugarcane, do not show this kind of pattern in sorghum and rice. In these last two species, *Tat2* is widely distributed, with higher copy numbers in heterochromatic regions in sorghum. *Ale1*, on the other hand, is widely distributed in sorghum, but is almost absent in rice. The concentration of signals in various chromosomal regions for *Tat2* and *Ale1* in sugarcane, and the lack of this pattern in rice and sorghum, supports the hypothesis that there has been differential transmission or loss from parental types of some LTR retrotransposons.

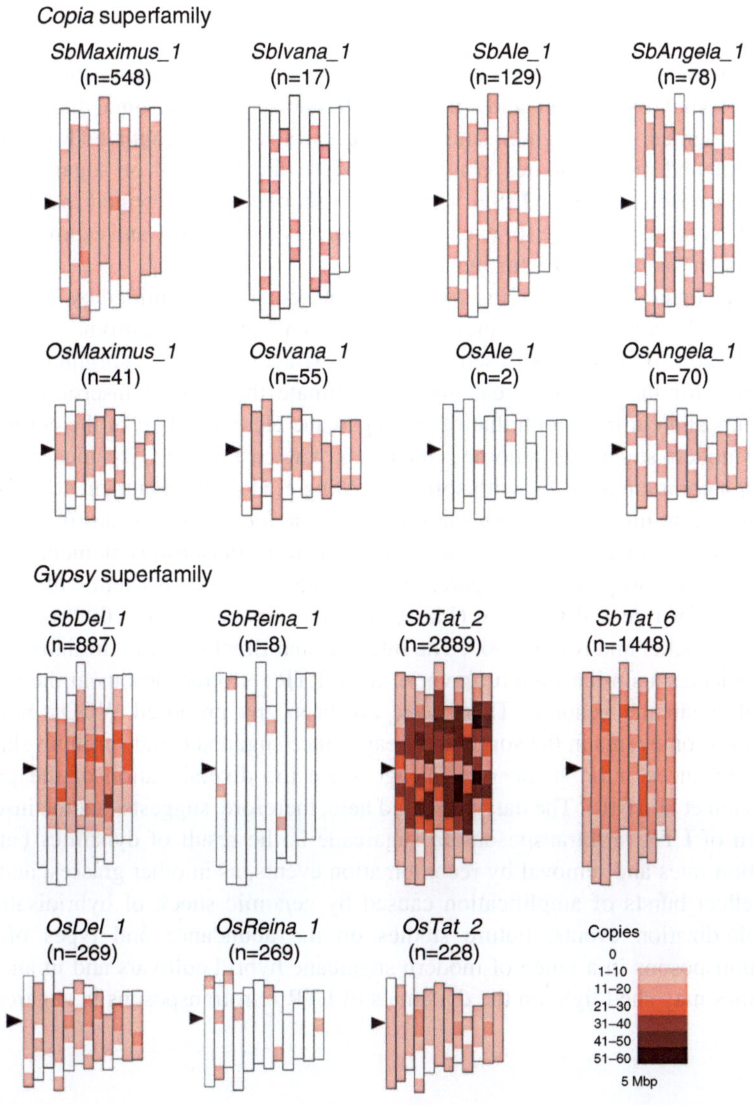

Fig. 10.4 Distribution and copy number of LTR retrotransposons in sorghum (*Sb*) and rice (*Os*) chromosomes. Chromosomes are represented as *vertical bars*, from 1 to 10 for sorghum, and from 1 to 12 for rice. *Black arrows* indicate centromeric regions, according Paterson et al. (2009b) for sorghum, and the International Rice Genome Sequencing Project (2005) for rice. Total copy number identified for each chromosome is shown above each chromosome (*n*). Copy number for each 5 Mbp window is represented in shades of *dark red* to *white*, with 51–60 copies as the most intense *dark red*, to 0 as *white* (*bottom right hand corner*). Copy number and distribution evaluation was done in two steps. First, a single representative sugarcane sequence from each family analysed by FISH by Domingues et al. (2012) was used to identify the closest full element in sorghum or rice (BLASTn, cut off ≥80 % for coverage and identity). Second, the closest full length sorghum or rice sequence was used as query against the sorghum or rice genome to evaluate the copy number and distribution using BLASTn with a cut-off of ≥80 % coverage and identity

10.2.4.4 Time of Insertion of LTR Retrotransposons

To compare insertion time of families of sugarcane LTR retrotransposons with their chromosomal distribution, and with insertion time and chromosomal distribution of related elements in sorghum and rice, we extracted full-length LTRs from the sorghum and rice genomes closely related to the families of sugarcane LTR retrotransposons used for FISH. Using these sequences, we created in silico heat maps in sorghum and rice for each LTR retrotransposon family and estimated times of insertion (see Fig. 10.5 for method details).

We estimated the time of insertion for all 60 sequences identified by Domingues et al. (2012) and one *Tat6* element extracted from a putative centromeric BAC (de Setta et al. 2011). We also used sequences extracted from the sorghum and rice genomes for the in silico heat maps to estimate the time of insertion of these elements and compare with that of the sugarcane elements. Part of the complexity of the modern sugarcane genome is the result of at least two WGDs since sugarcane shared a common ancestor with sorghum (Ming et al. 1998; Paterson et al. 2009a) and the recent interspecific hybridisation between *Saccharum* species (Grivet et al. 2004). Our estimates indicate most of the LTR retrotransposons elements are two my old in rice, sorghum and sugarcane (Fig. 10.5). This is consistent with previous estimates for rice and sorghum (Ma et al. 2004; Paterson et al. 2009b) and other grasses (Wicker and Keller 2007; The International Brachypodium Initiative 2010), and indicates a similar high turnover of most LTR retrotransposons in the modern hybrid sugarcane genome. There have not been any proposed WGDs or recent hybridisation events in the sorghum lineage since sugarcane and sorghum shared a common ancestor or in the rice lineage since the diversification of the grasses (Paterson et al. 2010). The data presented here, therefore, suggests that the insertion pattern of LTR retrotransposons in sugarcane is the result of dynamics between insertion rates and removal by recombination events, as in other grasses, and does not reflect bursts of amplification caused by genomic shock of hybridisation or polyploidization events. Future studies on the abundance and types of LTR retrotransposons in a range of modern sugarcane hybrid cultivars and in ancestral genomes may shed light on the dynamics of LTR retrotransposons in sugarcane.

10.2.4.5 LTR Retrotransposons as Components of the Centromere

Plant pericentromeric regions are known to be enriched in *Gypsy* elements, such as those of the *CRM* and *Tat/Athila* lineages (Theuri et al. 2005; Mizuno et al. 2006; Weber and Schmidt 2009). It is known that pericentromeric LTR retrotransposons persist longer in the genome (Paterson et al. 2009b). For the LTR retrotransposons analysed, the oldest elements in rice and sorghum, that is those with an insertion date >3 my, were those from the *Tat2* and *Del1* lineages in sorghum, which show patterns of concentration in heterochromatic regions (Fig. 10.4). FISH results for sugarcane *Del1* suggests that it is found in and around heterochromatic regions,

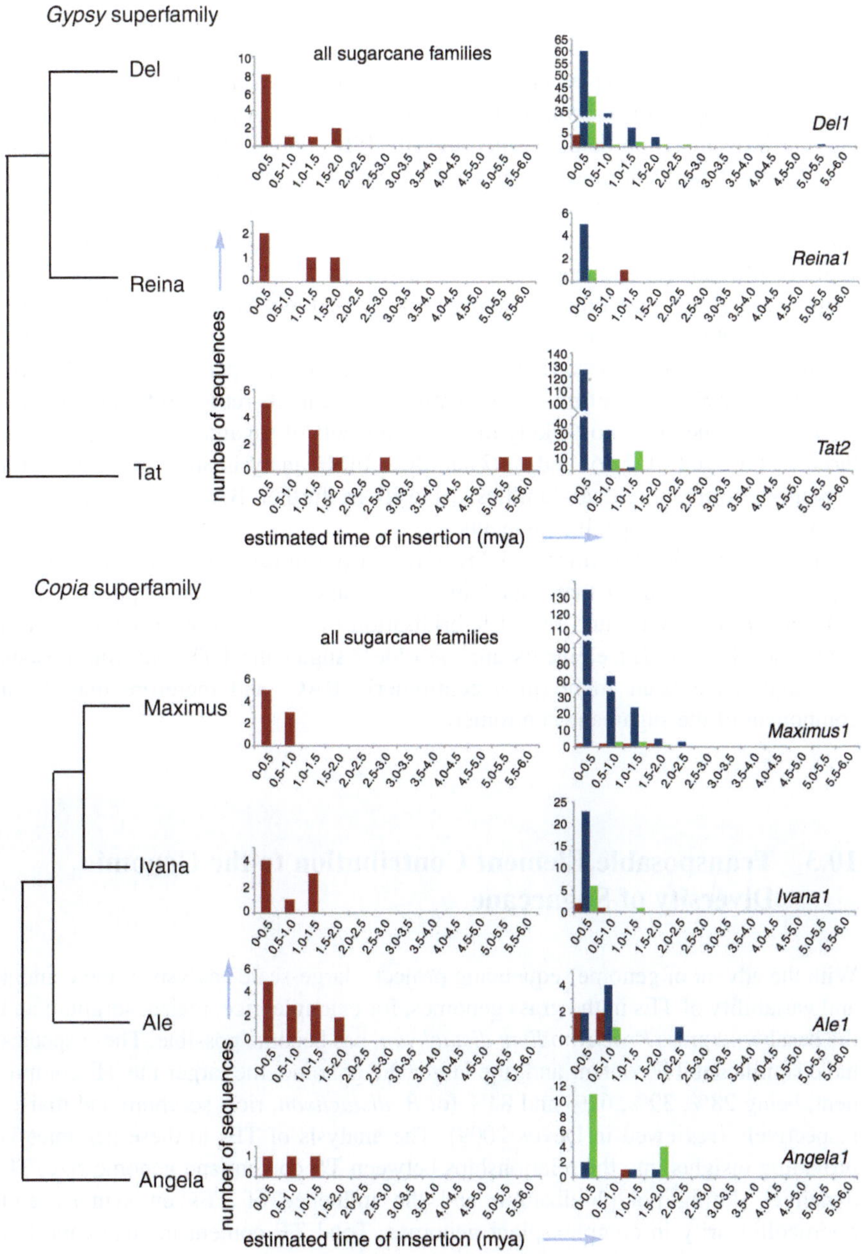

Fig. 10.5 Histogram of the date of insertion of LTR retrotransposons in sugarcane (*red*), sorghum (*blue*) and rice (*green*). The phylogenetic relationships for the *Copia* and *Gypsy* superfamilies are based on Domingues et al. (2012). Complete elements (with both LTRs) retrieved from Fig. 10.4 analysis were evaluated. The date of insertion was calculated using 5′ and 3′ LTR divergence (Kimura 2-parameters method) as implemented by MEGA5 (Tamura et al. 2011), using the molecular clock equation $T = k/2r$, where T is the date of insertion, k is the divergence between LTR sequences, and r is the evolutionary rate, using the rate of 1.3×10^{-8} substitutions per site per year, as described by Vitte et al. (2007)

rather than being strictly centromeric specific. *Tat* elements from sugarcane fall into two clades, one with *Tat2* and *Tat3* sequences, and the other with *Tat1, 4, 5, 6,* and *Tat7* sequences (Domingues et al. 2012). Elements from the 1st clade, *Tat2* and *Tat3*, do not appear to be components of pericentromeric or centromeric regions, as they all have insertion dates <3 my and in FISH analysis *Tat2* does not localise specifically to pericentromeric regions. Elements from the 2nd clade, *Tat1, 4, 5, 6,* and *Tat7,* however, have the oldest insertion times estimated for sugarcane, 2.89 my for a *Tat4* sequence, and 5.1 and 5.5 my for two *Tat6* sequences. We do not have FISH results for any sequences from the 2nd clade, but *Tat6* elements have been found in a BAC identified as centromeric, based on hybridisation to regions consistent with it being centromeric (Domingues et al. 2012) and on the presence of *CRM* LTR retrotransposons and sugarcane centromeric specific repeats (Nagaki and Murata 2005). Therefore, of the families of elements retrieved by Domingues et al. (2012), the ones most likely to be components of sugarcane centromeres are those of the *Tat1, 4, 5, 6,* and *Tat7* families. FISH analysis and fibre-FISH with sequences from individual families, as well as further BAC sequencing, may identify centromeric-specific elements.

In summary, the dynamics of LTR retrotransposon turnover in modern hybrid sugarcane is similar to that found in other grass genomes and appears to be independent of WGDs and recent hybridisation events. Finally, some sequences from one clade of *Tat* elements are the oldest sugarcane LTR retrotransposons identified, have been found in a centromeric BAC, and therefore may be a component of the sugarcane centromere.

10.3 Transposable Element Contribution to the Genomic Diversity of Sugarcane

With the advent of genome sequencing projects, large-scale analysis of the content and variability of TEs in the grass genomes, for example, rice, maize, sorghum and the basal wild grass *Brachypodium distachyon*, has become possible. These species have significant TE content and, the larger the genome, the larger the TE component, being 28%, 39%, 62% and 84% for *B. distachyon*, rice, sorghum and maize, respectively (reviewed in Devos 2009). The analysis of TEs in these genomes is providing insights into the relationships between TE content and genome size, TE dynamics during polyploidization and the influence of TEs on synteny and microcollinearity in complex plant genomes. Total TE content in sugarcane has been evaluated for 100 BACs selected using both TEs and genes as probes (de Setta et al. 2011) and for another 35 BACs sequenced to analyse specific genes and microcollinearity (Jannoo et al. 2007; Wang et al. 2010; Garsmeur et al. 2011; Manetti et al. 2012). All these studies analysed BACs from the only BAC library available, the SHCRBa library (Tomkins et al. 1999), which provides 1.3× coverage of the polyploid genome. de Setta et al. (2011) estimated that the average TE content of the 100 BACs was 55%. On the other hand, for the other 35 BACs

(Jannoo et al. 2007; Wang et al. 2010; Garsmeur et al. 2011; Manetti et al. 2012), the average TE content was estimated to be 22–35%. The differences in estimates could be explained by the bias in BAC selection for the 35 BACs, since most of them belong to gene-rich regions.

Understanding and describing TEs is important for the correct annotation and assembly of a genome. TEs have been mistakenly annotated as hypothetical genes (Bennetzen et al. 2004) and can disrupt collinearity between genomes. It has been proposed that the sorghum genome could be used as a tool for the assembly of the sugarcane genome, because these genomes have a relatively good microcollinearity in genic regions (Wang et al. 2010). As would be expected from the composition of grass genomes in general, TEs, other repetitive sequences and non-coding DNA, occasionally interrupt microcollinearity between sorghum and sugarcane (Jannoo et al. 2007; Wang et al. 2010; Garsmeur et al. 2011). Analysis of BAC sequence microcollinearity in our lab confirms these results and suggests that rice and *B. distachyon* may also be useful tools for the assembly of gene-rich regions (Manetti et al. 2012, unpublished data). On the other hand, the use of maize is not appropriate, since compared to rice, sorghum and *B. distachyon*, it has longer intergenic sequences, and more rearrangements and TEs, in the regions we compared. Although all this data reinforces the fact that TEs can make the assembly of genomes by comparative approaches difficult; they also emphasise the importance of TEs in terms of variability and structuring of the sugarcane genome.

Several lines of evidence suggest that the parental genomes *S. officinarum* and *S. spontaneum* have made different contributions to the TE component of modern sugarcane cultivars. Modern sugarcane cultivars are complex polyploids with varying contributions of the parental genomes *S. officinarum* and *S. spontaneum* (Piperidis et al. 2010). The parental genomes have different distribution of particular *hAT* superfamily lineages; Lineage 257 is found in both *S. officinarum* and *S. spontaneum*, while Lineage 191 is found only in *S. officinarum*. For the *Gypsy* and *Copia* superfamilies, the FISH probe for two families, *Tat2* and *Maximus1* showed clustering of signals on particular chromosomes, not seen in the rice or sorghum in silico map, which may indicate higher copy numbers of that TE from one parental genome, or loss, during recent hybridisation events. Further work with genomic in situ hybridisation (D'Hont et al. 1996) and FISH using TEs as probes should allow us to describe the relative TEs contribution of parental genomes to different sugarcane cultivars. Potentially, it could aid in determining if TEs are segregating within populations, which could be used by breeders as selectively neutral markers.

While research in sugarcane is driven by economic interests, the modern hybrid sugarcane genome is interesting from a purely research point of view because of its origins. Hybridisation is a trigger for genomic restructuring (see Shapiro 2010 for a review), for example, changes in methylation status (Marfil et al. 2006), chromosomal restructuring (Lim et al. 2008) and proliferation or loss of TEs (Ungerer et al. 2009; Renny-Byfield et al. 2011), all of which may be interrelated. Modern sugarcane cultivars are recent hybrids, it has been estimated that they are less than ten meioses (between five and seven) since the first interspecific crosses (Jannoo et al. 1999). While the focus of this review has not been on this aspect of the sugarcane genome, future work on TEs in sugarcane may also help elucidate the fate and effect of TEs in hybridisation events.

10.4 Perspectives

Economic interest has driven grass genome sequencing efforts because they provide most of the world's food. Sugarcane is economically important, producing two-thirds of the world's sugar (Marconi et al. 2011) and interest is increasing in its use as a biofuel (Somerville 2006). It takes at least 250,000 seedlings and 12 years to create a commercially viable cultivar in traditional breeding programmes (Cheavegatti-Gianotto et al. 2011). Currently molecular techniques are not widely used due to the lack of physical or saturated genetic maps and information on the sugarcane genome in general (Souza et al. 2011). The annotation of TEs in sugarcane can assist breeding programmes in several different ways, for example by providing selectively neutral molecular markers, as an aid in the correct annotation of genes and assembling of the genome, and in the development of sugarcane-specific mutagenesis tools.

Acknowledgements We gratefully acknowledge funding from FAPESP-BIOEN (08/52074-0) and CNPq to MAVS. NS, CJM and GMQC are supported by FAPESP fellowships; EAOC is supported by CNPq fellowship.

References

Araujo PG, Rossi M, de Jesus EM, Saccaro NL Jr, Kajihara D, Massa R, de Felix JM, Drummond RD, Falco MC, Chabregas SM, Ulian EC, Menossi M, Van Sluys MA (2005) Transcriptionally active transposable elements in recent hybrid sugarcane. Plant J 44:707–717

Baulcombe D (2004) RNA silencing in plants. Nature 431:356–363

Bennetzen JL, Coleman C, Liu R, Ma J, Ramakrishna W (2004) Consistent over-estimation of gene number in complex plant genomes. Curr Opin Plant Biol 7:732–736

Brandes EW (1958) Origin, classification and characteristics in sugarcane (*Saccharum officinarum* L.). In: Artschwager E, Brandes EW (eds) U. S. Department of agriculture handbook. USDA, Washington DC, pp 1–35

Bremer G (1963) Problems in breeding and cytology of sugarcane. 4. Origin of increase of chromosome number in species hybrids of Saccharum. Euphytica 10:325

Bundock P, Hooykaas P (2005) An *Arabidopsis hAT*-like transposase is essential for plant development. Nature 436:282–284

Cantu D, Vanzetti LS, Sumner A, Dubcovsky M, Matvienko M, Distelfeld A, Michelmore RW, Dubcovsky J (2010) Small RNAs, DNA methylation and transposable elements in wheat. BMC Genomics 11:408

Cheavegatti-Gianotto A, Abreu HMC, Arruda P, Filho JCB, Burnquist WL, Creste S, Ciero L, Ferro JA, Figueira AVO, Filgueiras TS, Grossi-de-Sá MF, Guzzo EC, Hoffmann HP, Landell MGA, Macedo N, Matsuoka S, Reinach FC, Romano E, Silva WJ, Filho MCS, Ulian EC (2011) Sugarcane (*Saccharum X officinarum*): A reference study for the regulation of genetically modified cultivars in Brazil. Trop Plant Biol 4:62–89

Clayton WD, Renvoize SA (1986) Genera graminum. Her Majesty's Stationery Office, London
Cowan RK, Hoen DR, Schoen DJ, Bureau TE (2005) *MUSTANG* is a novel family of domesticated transposase genes found in diverse Angiosperms. Mol Biol Evol 22:2084–2089
D'Hont A and Glaszmann JC (2001) Sugarcane genome analysis with molecular markers: a first decade of research. International Society of Sugar Cane Technologists. Proceedings of the XXIV Congress, Brisbane, Australia, pp 556–559
D'Hont A, Grivet L, Feldmann P, Glaszmann JC, Rao S, Berding N (1996) Characterisation of the double genome structure of modern sugarcane cultivars (*Saccharum spp.*) by molecular cytogenetics. Mol Gen Genet 250:405–413
D'Hont A, Ison D, Alix K, Roux C, Glaszmann JC (1998) Determination of basic chromosome numbers in the genus *Saccharum* by physical mapping of ribosomal RNA genes. Genome 41:221–225
D'Hont A, Paulet F, Glaszmann JC (2002) Oligoclonal interspecific origin of "North Indian" and "Chinese" sugarcanes. Chromosome Res 10:253–262
Daniels J, Roach BT (1987) Taxonomy and evolution. In: Heinz DJ (ed) Sugarcane improvement through breeding. Elsevier, Amsterdam, pp 7–84
de Jesus EM, Cruz EA, Cruz GM, Van Sluys MA (2012) Diversification of *hAT* transposase paralogues in the sugarcane genome. Mol Genet Genomics 287:205–219
de Setta N, Cruz G, Cruz E, Gomes K, Campos R, Hotta C, Vilela M, Vincentz M, Vautrin S, Souza G, Bérgès H, Gaiarsa J, Kitajima J, Van Sluys MA (2011) Sugarcane genome: a snapshot from 100 sequenced BACs. In: Plant and animal genomes XIX conference, San Diego, USA
Devos KM (2009) Grass genome organization and evolution. Curr Opin Plant Biol 13:139–145
D'Hont A, Lu AH, Feldmann P, Glaszmann JC (1993) Cytoplasmic diversity in sugar cane revealed by heterologous probes. Sugar Cane 1:12–15
Diao XM, Lisch D (2006) Mutator transposon in maize and *MULEs* in the plant genome. Yi Chuan Xue Bao 33:477–487
Domingues DS, Cruz GMQ, Metcalfe CJ, Nogueira FTS, Vicentini R, Alves CS, Van Sluys MA (2012) Analysis of plant LTR retrotransposons at the fine-scale family level reveals individual molecular patterns. BMC Genomics 13:137
Döring HP, Starlinger P (1984) Barbara McClintock's controlling elements: now at the DNA level. Cell 39:253–259
Du J, Tian Z, Hans CS, Laten HM, Cannon SB, Jackson S, Shoemaker RC, Ma J (2010) Evolutionary conservation, diversity and specificity of LTR retrotransposons in flowering plants: insights from genome-wide analysis and multi-specific comparison. Plant J 63:584–598
Garsmeur O, Charron C, Bocs S, Jouffe V, Samain S, Couloux A, Droc G, Zini C, Glaszmann JC, Van Sluys MA, D'Hont A (2011) High homologous gene conservation despite extreme autopolyploid redundancy in sugarcane. New Phytol 189:629–642
Gaut BS (2002) Evolutionary dynamics of grass genomes. New Phytol 154:15–28
Grass Phylogeny Working Group (2001) Phylogeny and subfamilial classification of the grasses (Poaceae). Ann Mo Bot Gard 88:373–457
Grivet L, Daniels C, Glaszmann J, D'Hont A (2004) A review of recent molecular genetics evidence for sugarcane evolution and domestication. Ethnobot Res Appl 2:9–17
Grivet L, Glaszmann J, D'Hont A (2006) Molecular evidence of sugarcane evolution and domestication. In: Motley T, Nyree Z, Cross H (eds) Darwin's harvest, new approaches to the origins, evolution and conservation of crops. Columbia University Press, New York, NY, pp 49–66
Hawkins JS, Kim H, Nason JD (2006) Differential lineage-specific amplification of transposable elements is responsible for genome size variation in *Gossypium*. Genome Res 16:1252–1261
Hehl R, Nacken WK, Krause A, Saedler H, Sommer H (1991) Structural analysis of *Tam3*, a transposable element from *Antirrhinum majus*, reveals homologies to the *Ac* element from maize. Plant Mol Biol 16:369–371
Heslop-Harrison JS, Brandes A, Taketa S, Schmidt T, Vershinin A, Alkhimova EG, Kamm A, Doudrick RL, Schwarzacher T, Katsiotis A, Kubis S, Kumar A, Pearce SR, Flavell A, Harrison GE (1997) The chromosomal distributions of *Ty1-copia* group retrotransposable elements in higher plants and their implications for genome evolution. Genetica 100:197–204

Hua-Van A, Le Rouzic A, Boutin TS, Filée J, Capy P (2011) The struggle for life of the genome's selfish architects. Biol Direct 6:19

International Rice Genome Sequencing Project (2005) The map-based sequence of the rice genome. Nature 436:793–800

Jannoo N, Grivet L, Seguin M, Paulet F, Domaingue R, Rao PS, Dookun A, D'Hont A, Glaszmann JC (1999) Molecular investigation of the genetic base of sugarcane cultivars. Theor Appl Genet 99:171–184

Jannoo N, Grivet L, Chantret N, Garsmeur O, Glaszmann JC, Arruda P, D'Hont A (2007) Orthologous comparison in a gene-rich region among grasses reveals stability in the sugarcane polyploid genome. Plant J 50:574–585

Kellogg EA (2001) Evolutionary history of the grasses. Plant Physiol 125:1198

Kempken F, Windhofer F (2001) The *hAT* family: a versatile transposon group common to plants, fungi, animals, and man. Chromosoma 110:1–9

Kumar A, Bennetzen JL (1999) Plant Retrotransposons. Annu Rev Genet 33:479–532

Lebot V (1999) Biomolecular evidence for crop domestication on Sahul. Genet Resour Crop Evol 46:619–628

Lim KY, Soltis DE, Soltis PS, Tate J, Matyasek R, Srubarova H, Kovarik A, Pires JC, Xiong Z, Leitch AR (2008) Rapid chromosome evolution in recently formed polyploids in *Tragopogon* (Asteraceae). PLoS One 3:e3353

Lisch DR (2002) *Mutator* transposons. Trends Plant Sci 7:498–504

Lisch DR, Freeling M, Langham RJ, Choy MY (2001) *Mutator* transposase is widespread in the grasses. Plant Physiol 125:1293–1303

Llorens C, Futami R, Covelli L, Domínguez-Escribá L, Viu JM, Tamarit D, Aguilar-Rodríguez J, Vicente-Ripolles M, Fuster G, Bernet GP, Maumus F, Munoz-Pomer A, Sempere JM, Latorre A, Moya A (2011) The *Gypsy* database (GyDB) of mobile genetic elements: release 2.0. Nucleic Acids Res 39:D70–D74

Lu YH, D'Hont AD, Walker DIT, Rao PS, Felmann P, Glaszmann JC (1994) Relationships among ancestral species of sugarcane revealed using RFLP using single copy maize nuclear probes. Euphytica 78:7–18

Ma J, Devos KM, Bennetzen JL (2004) Analyses of LTR retrotransposon structures reveals recent and rapid genomic DNA loss in rice. Genome Res 14:860–869

Manetti ME, Rossi M, Cruz GM, Saccaro NL Jr, Nakabashi M, Altebarmakian V, Rodier-Goud M, Domingues D, D'Hont A, Van Sluys MA (2012) Mutator system derivatives isolated from sugarcane genome sequence. Trop Plant Biol 5:233–243

Marconi TG, Costa EA, Miranda HR, Mancini MC, Cardoso-Silva CB, Oliveira KM, Pinto LR, Mollinari M, Garcia AA, Souza AP (2011) Functional markers for gene mapping and genetic diversity studies in sugarcane. BMC Res Notes 4:264

Marfil CF, Masuelli RW, Davison J, Comai L (2006) Genomic instability in *Solanum tuberosum* x *Solanum kurtzianum* interspecific hybrids. Genome 49:104–113

McClintock B (1951) Chromosome organization and genic expression. Cold Spring Harb Symp Quant Biol 16:13–47

McGinnis W, Shermoen AW, Beckendorf SK (1983) A transposable element inserted just 5′ to a *Drosophila* glue protein gene alters gene expression and chromatin structure. Cell 34:75–84

Ming R, Liu SC, Lin YR, da Silva J, Wilson W, Braga D, van Deynze A, Wenslaff TF, Wu KK, Moore PH, Burnquist W, Sorrells ME, Irvine JE, Paterson AH (1998) Detailed alignment of saccharum and sorghum chromosomes: comparative organization of closely related diploid and polyploid genomes. Genetics 150:1663–1682

Mizuno H, Ito K, Wu J, Tanaka T, Kanamori H, Katayose Y, Sasaki T, Matsumoto T (2006) Identification and mapping of expressed genes, simple sequence repeats and transposable elements in centromeric regions of rice chromosomes. DNA Res 13:267–274

Nagaki K, Murata M (2005) Characterization of CENH3 and centromere-associated DNA sequences in sugarcane. Chromosome Res 13:195–203

Nobuta K, Lu C, Shrivastava R, Pillay M, De Paoli E, Accerbi M, Arteaga-Vazquez M, Sidorenko L, Jeong DH, Yen Y, Green PJ, Chandler VL, Meyers BC (2008) Distinct size distribution of endogenous siRNAs in maize: evidence from deep sequencing in the *mop1-1* mutant. Proc Natl Acad Sci USA 105:14958–14963

Paterson AH, Bowers JE, Feltus FA, Tang H, Lin L, Wang X (2009a) Comparative genomics of grasses promises a bountiful harvest. Plant Physiol 149:125–131

Paterson AH, Bowers JE, Bruggmann R, Dubchak I, Grimwood J, Gundlach H, Haberer G, Hellsten U, Mitros T, Poliakov A, Schmutz J, Spannagl M, Tang H, Wang X, Wicker T, Bharti AK, Chapman J, Feltus FA, Gowik U, Grigoriev IV, Lyons E, Maher CA, Martis M, Narechania A, Otillar RP, Penning BW, Salamov AA, Wang Y, Zhang L, Carpita NC, Freeling M, Gingle AR, Hash CT, Keller B, Klein P, Kresovich S, McCann MC, Ming R, Peterson DG, Mehboob-ur-Rahman WD, Westhoff P, Mayer KF, Messing J, Rokhsar DS (2009b) The *Sorghum bicolor* genome and the diversification of grasses. Nature 457:551–556

Paterson AH, Freeling M, Tang H, Wang X (2010) Insights from the comparison of plant genome sequences. Annu Rev Plant Biol 61:349–372

Piperidis G, Piperidis N, D'Hont A (2010) Molecular cytogenetic investigation of chromosome composition and transmission in sugarcane. Mol Genet Genomics 284:65–73

Renny-Byfield S, Chester M, Kovarík A, Le Comber SC, Grandbastien M-A, Deloger M, Nichols RA, Macas J, Novák P, Chase MW, Leitch AR (2011) Next generation sequencing reveals genome downsizing in allotetraploid *Nicotiana tabacum*, predominantly through the elimination of paternally derived repetitive DNAs. Mol Biol Evol 28:2843–2854

Roach BT (1972) Nobilisation of sugarcane. Proc Int Soc Sugar Cane Technol 14:206–216

Robertson DS (1978) Characterization of a *Mutator* system in maize. Mutat Res 51:21–28

Rossi M, Araujo PG, Van Sluys MA (2001) Survey of transposable elements in sugarcane expressed sequence tags (ESTs). Genet Mol Biol 24:147–154

Rossi M, Araujo PG, de Jesus EM, Varani AM, Van Sluys MA (2004) Comparative analysis of *Mutator*-like transposases in sugarcane. Mol Genet Genomics 272:194–203

Rubin E, Lithwick G, Levy AA (2001) Structure and evolution of the *hAT* transposon superfamily. Genetics 158:949–957

Saccaro-Junior NL, Van Sluys MA, Varani AM, Rossi M (2007) *MudrA*-like sequences from rice and sugarcane cluster as two *bona fide* transposon clades and two domesticated transposases. Gene 392:117–125

Salina EA, Sergeeva EM, Adonina IG, Shcherban AB, Belcram H, Huneau C, Chalhoub B (2011) The impact of *Ty3-gypsy* group LTR retrotransposons *Fatima* on B-genome specificity of polyploid wheats. BMC Plant Biol 11:99

Salse J, Bolot S, Throude M, Jouffe V, Piegu B, Quraishi UM, Calcagno T, Cooke R, Delseny M, Feuillet C (2008) Identification and characterization of shared duplications between rice and wheat provide new insight into grass genome evolution. Plant Cell 20:11–24

Schnable PS, Ware D, Fulton RS, Stein JC, Wei F, Pasternak S, Liang C, Zhang J, Fulton L, Graves TA, Minx P, Reily AD, Courtney L, Kruchowski SS, Tomlinson C, Strong C, Delehaunty K, Fronick C, Courtney B, Rock SM, Belter E, Du F, Kim K, Abbott RM, Cotton M, Levy A, Marchetto P, Ochoa K, Jackson SM, Gillam B, Chen W, Yan L, Higginbotham J, Cardenas M, Waligorski J, Applebaum E, Phelps L, Falcone J, Kanchi K, Thane T, Scimone A, Thane N, Henke J, Wang T, Ruppert J, Shah N, Rotter K, Hodges J, Ingenthron E, Cordes M, Kohlberg S, Sgro J, Delgado B, Mead K, Chinwalla A, Leonard S, Crouse K, Collura K, Kudrna D, Currie J, He R, Angelova A, Rajasekar S, Mueller T, Lomeli R, Scara G, Ko A, Delaney K, Wissotski M, Lopez G, Campos D, Braidotti M, Ashley E, Golser W, Kim H, Lee S, Lin J, Dujmic Z, Kim W, Talag J, Zuccolo A, Fan C, Sebastian A, Kramer M, Spiegel L, Nascimento L, Zutavern T, Miller B, Ambroise C, Muller S, Spooner W, Narechania A, Ren L, Wei S, Kumari S, Faga B, Levy MJ, McMahan L, Van Buren P, Vaughn MW, Ying K, Yeh CT, Emrich SJ, Jia Y, Kalyanaraman A, Hsia AP, Barbazuk WB, Baucom RS, Brutnell TP, Carpita NC, Chaparro C, Chia JM, Deragon JM, Estill JC, Fu Y, Jeddeloh JA, Han Y, Lee H, Li P, Lisch DR, Liu S, Liu Z, Nagel DH, McCann MC, SanMiguel P, Myers AM, Nettleton D, Nguyen J, Penning BW, Ponnala L, Schneider KL, Schwartz DC, Sharma A, Soderlund C, Springer NM, Sun Q, Wang H, Waterman M, Westerman R, Wolfgruber TK, Yang L, Yu Y, Zhang L, Zhou S, Zhu Q, Bennetzen JL, Dawe RK, Jiang J, Jiang N, Presting GG, Wessler SR, Aluru S, Martienssen RA, Clifton SW, McCombie WR, Wing RA, Wilson RK (2009) The B73 maize genome: complexity, diversity, and dynamics. Science 326:1112–1115

Shapiro JA (2010) Mobile DNA and evolution in the 21st century. Mob DNA 1:4

Somerville C (2006) The billion-ton biofuels vision. Science 312:1277

Souza GM, Berges H, Bocs S, Casu R, D'Hont A, Ferreira JE, Henry R, Ming R, Potier B, Sluys MAV, Vincentz M, Paterson AH (2011) The sugarcane genome challenge: strategies for sequencing a highly complex genome. Trop Plant Biol 4:145–156

Sreenivasan TV, Ahloowalia BS, Heinz DJ (1987) Cytogenetics. In: Heinz DJ (ed) Sugarcane improvement through breeding. Elsevier, Amsterdam, pp 211–253

Tamura K, Peterson D, Peterson N, Stecher G, Nei M, Kumar S (2011) MEGA5: molecular evolutionary genetics analysis using maximum likelihood, evolutionary distance, and maximum parsimony methods. Mol Biol Evol 28:2731–2739

The International Brachypodium Initiative (2010) Genome sequencing and analysis of the model grass *Brachypodium distachyon*. Nature 463:763–810

Theuri J, Phelps-Durr T, Mathews S, Birchler J (2005) A comparative study of retrotransposons in the centromeric regions of A and B chromosomes of maize. Cytogenet Genome Res 110:203–208

Tomkins J, Yu Y, Miller-Smith H, Frisch D, Woo S, Wing R (1999) A bacterial artificial chromosome library for sugarcane. Theor Appl Genet 99:419–424

Ungerer MC, Strakosh SC, Stimpson KM (2009) Proliferation of *Ty3/gypsy*-like retrotransposons in hybrid sunflower taxa inferred from phylogenetic data. BMC Biol 7:40

Vettore AL, da Silva FR, Kemper EL, Souza GM, da Silva AM, Ferro MI, Henrique-Silva F, Giglioti EA, Lemos MV, Coutinho LL, Nobrega MP, Carrer H, Franca SC, Bacci Junior M, Goldman MH, Gomes SL, Nunes LR, Camargo LE, Siqueira WJ, Van Sluys MA, Thiemann OH, Kuramae EE, Santelli RV, Marino CL, Targon ML, Ferro JA, Silveira HC, Marini DC, Lemos EG, Monteiro-Vitorello CB, Tambor JH, Carraro DM, Roberto PG, Martins VG, Goldman GH, de Oliveira RC, Truffi D, Colombo CA, Rossi M, de Araujo PG, Sculaccio SA, Angella A, Lima MM, de Rosa Júnior VE, Siviero F, Coscrato VE, Machado MA, Grivet L, Di Mauro SM, Nobrega FG, Menck CF, Braga MD, Telles GP, Cara FA, Pedrosa G, Meidanis J, Arruda P (2003) Analysis and functional annotation of an expressed sequence tag collection for tropical crop sugarcane. Genome Res 13:2725–2735

Vitte C, Panaud O, Quesneville H (2007) LTR retrotransposons in rice (Oryza sativa, L.): recent burst amplifications followed by rapid DNA loss. BMC Genomics 8:218

Wang J, Roe B, Macmil S, Yu Q, Murray JE, Tang H, Chen C, Najar F, Wiley G, Bowers J, Van Sluys MA, Rokhsar DS, Hudson ME, Moose SP, Paterson AH, Ming R (2010) Microcollinearity between autopolyploid sugarcane and diploid sorghum genomes. BMC Genomics 11:261

Weber B, Schmidt T (2009) Nested *Ty3-gypsy* retrotransposons of a single Beta procumbens centromere contain a putative chromodomain. Chromosome Res 17:379–396

Wicker T, Keller B (2007) Genome-wide comparative analysis of copia retrotransposons in Triticeae, rice, and *Arabidopsis* reveals conserved ancient evolutionary lineages and distinct dynamics of individual *copia* families. Genome Res 17:1072–1081

Wicker T, Sabot F, Hua-Van A, Bennetzen JL, Capy P, Chalhoub B, Flavell A, Leroy P, Morgante M, Panaud O, Paux E, SanMiguel P, Schulman AH (2007) A unified classification system for eukaryotic transposable elements. Nat Rev Genet 8:973–982

Xu Z, Dooner HK (2005) *Mx-rMx*, a family of interacting transposons in the growing *hAT* superfamily of maize. Plant Cell 17:375–388

Yu Z, Wright SI, Bureau T (2000) Mutator elements in *Arabidopsis thaliana*: structure, diversity and evolution. Genetics 156:2019–2031

Zhang X, Yazaki J, Sundaresan A, Cokus S, Chan SW-L, Chen H, Henderson IR, Shinn P, Pellegrini M, Jacobsen SE, Ecker JR (2006) Genome-wide high-resolution mapping and functional analysis of DNA methylation in *Arabidopsis*. Cell 126:1189–1201

Chapter 11
Helitron Proliferation and Gene-Fragment Capture

Yubin Li and Hugo K. Dooner

Abstract *Helitrons* stand out as rare transposons discovered by bioinformatic, rather than genetic, studies. Although they comprise an ancient superfamily of transposons found in plants, animals, and fungi, it is in plants where they have been studied most extensively. Well-annotated plant genomes contain increasingly higher numbers of identified *Helitrons*, including putative autonomous elements and nonautonomous elements with and without gene fragments. The molecular structure of the autonomous *Helitron* and the postulated rolling circle mode of transposition remain hypothetical, and recent evidence suggests that *Helitrons* may transpose by both copy-and-paste and cut-and-paste mechanisms. Two *Helitron* properties, in particular, have caught the imagination of biologists: their ability to undergo sudden bursts of transposition and their ability to capture fragments from different genes to make chimeric transcripts. In this chapter, we provide an overview of what we have learned in the past decade about the biology of these intriguing, newly discovered plant genome residents.

Keywords *Helitrons* • Transposons • Plants

11.1 Introduction

Transposable elements (TEs) are DNA fragments that can move from one site of the genome to another. Though ubiquitous in nature, they were first discovered in maize more than 60 years ago (McClintock 1947). This eventual Nobel-Prize-winning

---

Y. Li
Waksman Institute, Rutgers University, Piscataway NJ, 08854, USA
e-mail: yubin@waksman.rutgers.edu

H.K. Dooner (✉)
Waksman Institute, Rutgers University, Piscataway NJ, 08854, USA

Department of Plant Biology, Rutgers University, New Brunswick, NJ 08901, USA
e-mail: dooner@waksman.rutgers.edu

M.-A. Grandbastien and J.M. Casacuberta (eds.), *Plant Transposable Elements*, Topics in Current Genetics 24, DOI 10.1007/978-3-642-31842-9_11,

discovery began to be acknowledged broadly only three decades later and gained increasingly wider appreciation in the "omics" era (Craig et al. 2002). Today, TEs are considered to have played an intrinsic role in genome structure evolution through the multiple chromosome rearrangements that are brought about by the chromosome cutting properties noted by McClintock (1952). TEs have been proposed as a major driving force in the process of gene creation by providing the raw material needed for the evolution of new gene functions (Dooner and Weil 2012; Feschotte and Pritham 2007) and have turned out to be the major component of most sequenced eukaryotic genomes (Craig et al. 2002).

At the turn of the twenty-first century, the known classes of TEs (Feschotte et al. 2002) were expanded to include the newly hypothesized *Helitron* transposable elements. Unlike Class I elements (retrotransposons) that transpose through RNA, Class II elements (DNA transposons) transpose through DNA. *Helitrons* were postulated to transpose via a hypothetical rolling circle (RC) replication mechanism (Kapitonov and Jurka 2001) and, therefore, fall into the latter class. A more recent classification of eukaryotic transposons places them under a special Subclass 2 among DNA transposons (Wicker et al. 2007). In the past decade, a considerable effort has been made to better understand these elusive TEs from all different angles. Our goal in this chapter is to summarize our current knowledge about these DNA transposons in the plant kingdom and to provide a personal view of further explorations in this emerging field.

11.2 Discovery of *Helitrons*

Shortly before their discovery as unique eukaryotic transposons, *Helitrons* had been described as repetitive sequences in *Arabidopsis thaliana*, one of the three genomes analyzed by Kapitonov and Jurka (2001) in their seminal paper. The first such repeat detected was *Aie* (Arabidopsis insertion element), a 527-bp element insertion present downstream of the polyadenylation site of *AtRAD51* in the Columbia ecotype but absent in its Landsberg *erecta* counterpart (Doutriaux et al. 1998). *Aie* is AT-rich, contains no ORFs, has a stem-and-loop sequence on the 3′ side (5 unpaired bases in a 21-bp stem, with a 4-bp loop), and shows some short duplications around the insertion site. Because it lacked terminal inverted repeats (TIRs), *Aie* was taken to be a remnant of an imperfect transposition event, an interpretation supported by its multicopy presence in the two ecotypes.

Due to their abundance in the genome, elements closely related to *Aie* were readily uncovered in subsequent computational analyses of *Arabidopsis* repetitive sequences. *AthE1* was the most abundant class of repetitive elements in the *A. thaliana* 1998 sequence database (Surzycki and Belknap 1999). Although they could be as long as 2 kb, these elements lacked any detectable coding capacity for known transposases. While the 5′ and 3′ ends of *AthE1* family members were highly conserved, they did not represent either inverted or direct repeats. Direct repeats flanking transposons, also known as target site duplications (TSD), are a common feature of retrotransposons and DNA transposons. Their absence in *AthE1* elements suggested that these elements differed from most other known transposons in being unable to recombine into the genome by introducing staggered cuts in the target DNA.

In a comprehensive analysis of potential transposon sequences in chromosome 2 of *Arabidopsis*, sequences resembling *AthE* were found to make up 1.1 % of the chromosome. No detectable TSDs or TIRs flanked these unusual repeats, which were named ATREP1-10 and classified as ten families of nonautonomous DNA transposons (Kapitonov and Jurka 1999). Another analysis of transposon diversity in a much larger *Arabidopsis* dataset (≈17.2 Mb) grouped 179 *AthE*-like or *ATREP*-like elements into seven families based on common structural features and identified them as members of a novel superfamily of transposons, named *Basho*, that moved by an unknown transposition mechanism (Le et al. 2000). A *Basho*-like group was also identified in maize, supporting the concept of a new plant transposon superfamily. Completion of the whole genome sequence of *Arabidopsis* (Arabidopsis Genome Initiative 2000) revealed the existence of 1,265 *Basho* elements. In contrast with the class I elements that primarily occupy the centromere, but consistent with other class II transposons, *Basho* elements predominate on the periphery of pericentromeric domains. Novel elements resembling the structurally unusual *Basho* elements were also found in rice, suggesting a wide distribution of these elements in plants (Turcotte et al. 2001). Similar to *Basho* elements in *Arabidopsis*, the rice elements are small (<2 kb), lack coding capacities, TSDs or TIR, and are highly conserved at both termini. The big outstanding question after these studies was: by what mechanism does this new superfamily of transposons multiply and transpose in the host genome?

In 2001, this question was answered hypothetically when Kapitonov and Jurka (2001) carried out an in silico reconstruction of putative autonomous transposons from inactive copies accumulated in the three genomes analyzed, *Arabidopsis thaliana*, *Caenorhabditis elegans*, and *Oryza sativa*. Deletions, insertions, and premature stop codons were removed from the consensus sequences of the transposons by computational approaches, in a reconstruction process reminiscent of that of *Sleeping Beauty* (Ivics et al. 1997). Finally, rolling circle (RC) replication, a transposition mechanism until then restricted to prokaryotes, was proposed to explain movement of this previously unknown category of eukaryotic DNA transposons. The new elements were designated *Helitrons* because the protein encoded by the putative autonomous elements had a conserved DNA helicase domain.

11.3 Genomics of *Helitrons*

11.3.1 Molecular Structure of Putatively Autonomous and Nonautonomous Helitrons

Helitrons have been found in every plant genome where they have been carefully looked for (Table 11.1). As a consequence of their in silico detection, the majority of *Helitrons* identified in a given species share distinct structural features with other

Table 11.1 Dynamic distribution of *Helitron* transposons in sequenced plant genomes

Organism	Genome size (Mb)	Family no.	Putative autonomous	Occurrence no.	Genome fraction (%)	References
Arabidopsis thaliana	115	NA		1,265	NA	Arabidopsis Genome Initiative (2000)
	115	4	+	910	~2.0	Kapitonov and Jurka (2001)
	115	10		1,242	1.30	Yang and Bennetzen (2009b)
	115	NA		3,437	1.85	Hollister et al. (2011)
	119	34		12,947	6.72	Ahmed et al. (2011)
Arabidopsis lyrata	206.7	NA		10,452	2.64	Hollister et al. (2011)
Brachypodium distachyon	272	48		120	0.18	International Brachypodium Initiative (2010)
Brassica rapa	284	NA		6,214	0.60	Wang et al. (2011)
Glycine max	975	NA		7,128	0.53	Schmutz et al. (2010)
				82	NA	Du et al. (2010)
Medicago truncatula	243	10	+	1,386	1.29	Yang and Bennetzen (2009b)
Oryza sativa var. japonica	389	NA		552	NA	Sweredoski et al. (2008)
		NA		3,037	0.33	Paterson et al. (2009)
		23	+	6,947	2.09	Yang and Bennetzen (2009b)
Oryza sativa var. indica		NA		604	NA	Sweredoski et al. (2008)
Physcomitrella patens	480	1	+	19	0.12	Rensing et al. (2008)
Populus trichocarpa	485	2		NA	0.06	Tuskan et al. (2006)
Selaginella moellendorffii	213	4	+	5,394	1.57	Banks et al. (2011)
Sorghum bicolor	748	1		1,017	0.81	Paterson et al. (2009)
		11	+	4875	3.00	Yang and Bennetzen (2009b)
Vitis vinifera	487	NA		109	0.01	Jaillon et al. (2007)
Zea mays	2,400	29[a]		2,791	2.00	Du et al. (2009)
	2,050	8		1,930	2.20	Yang and Bennetzen (2009a)

NA not available
[a]Family number of genic *Helitrons*

elements in the same species and in closely related species. The putative autonomous *Helitrons* reconstructed from nonautonomous ones in *Arabidopsis thaliana* (*Helitron1* and *Helitron2*) and *Caenorhabditis elegans* (*Helitron1_CE*) encode a large protein denominated RepHel that contains a Rep domain homologous to RC replication initiators and a Hel domain homologous to DNA helicases (Kapitonov and Jurka 2001). Because the predicted RepHel proteins share motifs with the transposases of bacterial RC transposons, *Helitrons* were postulated to transpose by RC replication. The enzymatic core of the ~100-aa Rep domain contains three motifs that are conserved in a wide diversity of eukaryotes (Feschotte and Pritham 2007; Kapitonov and Jurka 2007). The larger, ~400-aa Hel domain contains eight universally conserved motifs in all putative autonomous *Helitrons* (Fig. 11.1a). Examples of these conserved motifs are shown in Fig. 11.1d. Conservation of the RepHel protein has been used as the criterion to identify hypothetical autonomous *Helitrons* in all plant host genomes (Table 11.1).

Shorter nonautonomous *Helitrons* are far more abundant and correspond to the non-TIR-, non-TSD-containing highly repetitive sequences that were noted earlier in *Arabidopsis* and rice. They have been grouped into multiple families based on the degree of sequence conservation at both 5′ and 3′ termini (Fig. 11.1b). Most of these elements are smaller than 2 kb and encode no detectable proteins. Longer elements with extra protein-coding capacity (Fig. 11.1c) occur in some species. For example, in *Arabidopsis* and rice, the putative autonomous *Helitrons* also encode subunits of RPA70, a single-stranded-DNA-binding protein. These are absent in *C. elegans*, making it unlikely that they are part of the transposition machinery (Kapitonov and Jurka 2001). Though RPA-like proteins have also been identified in some animal *Helitrons* (Feschotte and Pritham 2007; Kapitonov and Jurka 2007), their exact function remains unknown.

11.3.2 Biological and Computational Identification of Helitrons

Among the dozens of known eukaryotic DNA transposons (Feschotte and Pritham 2007; Kapitonov and Jurka 2008; Wicker et al. 2007), *Helitrons* stand out as a rare example of TEs discovered purely by computational, rather than genetic, studies. Though only recently identified, *Helitrons* are an ancient superfamily of eukaryotic DNA transposons, as evidenced by their cross-kingdom presence in plants (Table 11.1), fungi (Galagan et al. 2005), and animals (Cocca et al. 2011; Kapitonov and Jurka 2001; Pritham and Feschotte 2007). *Helitrons* are the only eukaryotic transposons that lack TIRs, do not generate TSDs upon integration in the host genome, and do not encode any known transposases. Furthermore, until their computational discovery, none had been found to be the causative agent of a mutation. These unusual features delayed their discovery, although *Helitrons* resemble other eukaryotic DNA transposons in terms of their impact on the host genome. Following their discovery, *Helitrons* have been identified by both biological and computational approaches.

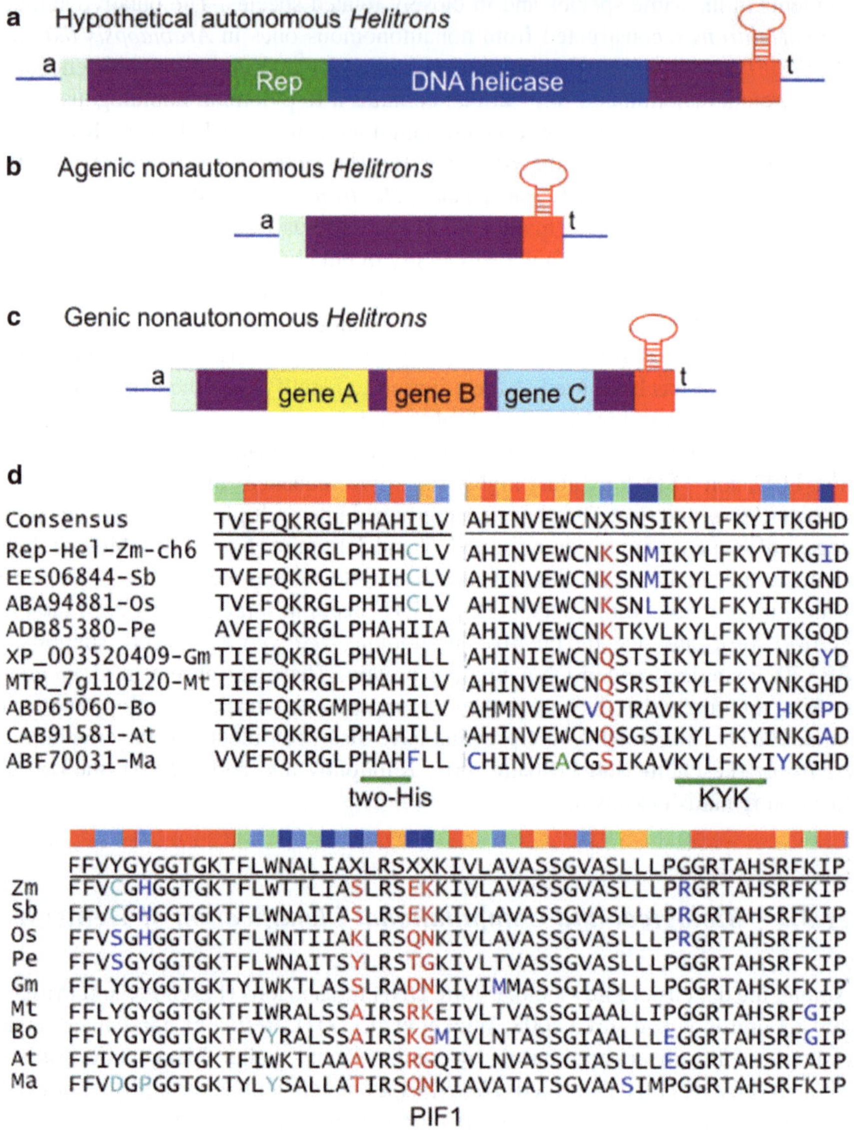

Fig. 11.1 Generic structure of identified *Helitrons* in different eukaryotes. (**a**) Hypothetical autonomous *Helitron* with coding capacity for a RepHel protein. Rep (Replication motifs are in *green*, and Hel (Helicase motifs are in *blue*. The conserved 5′ TC terminus is shown in *light green*. The conserved 3′ CTRR terminus is shown in *red*, with a stem-loop structure formed from a palindromic sequence in the 3′ subterminal region. The insertion is targeted to an AT dinucleotide shown in lowercase above a *blue line* representing the flanking sequence. The vast majority of *Helitrons* are nonautonomous elements with similar terminal structures as the autonomous copies. (**b**) Agenic nonautonomous *Helitrons* lack any known coding capacity. (**c**) Genic nonautonomous *Helitrons* carry fragments from a variable number of genes in the host genome (*yellow*, *orange*, and *light blue boxes*). (**d**) Multiple alignments of the conserved motifs of Rep domain (two-His and KYK and PIF1 helicase domain) in plant *Helitrons*. At, *Arabidopsis thaliana*; Bo, *Brassica oleracea*; Gm, *Glycine max*; Ma, *Musa acuminata*; Mt, *Medicago truncatula*; Os, *Oryza sativa*; Pe, *Phyllostachys edulis*; Sb, *Sorghum bicolor*; Zm, *Zea mays*

11.3.2.1 Biological Identification of *Helitrons*

Helitrons have been detected biologically in only a handful of cases, either as insertional mutagens causing spontaneous mutations (Table 11.2) or as colinearity disruptors contributing to haplotypic diversity within a species.

Molecular characterization of the spontaneous *sh2-7057* mutant allele in maize (Lal et al. 2003) revealed that the mutation carried a large *Helitron* insertion in the 11th intron of the *sh2* gene. This was the first case to demonstrate the mutagenicity of *Helitron* transposons. Though the insertion in this mutant was larger than 12 kb, it lacked coding capacity for known transposases and, instead, carried several gene fragments, including four exons with similarity to a plant DEAD box RNA helicase.

The strong terminal sequence similarity of the insertion in the spontaneous mutation *ba1-ref* (*barren stalk-1*) with the *Helitron* transposon in *sh2-7057* led to the realization that this classical mutation, identified more than three quarters of a century ago, had been caused by a *Helitron* insertion. In contrast to the insertion in *sh2-7057*, the 6.5 kb *Helitron* element in *ba1-ref* inserted in the proximal promoter region of the *ba1* gene (Gupta et al. 2005). Though the 6.5-kb insertion also carried multiple pseudogene fragments, these differed from those in the *Helitron* transposon of *sh2-7057*. The conserved 5′ and 3′ termini of these *Helitrons* were found to be repetitive in the maize genome, suggesting that they play an important role in *Helitron* amplification.

More strikingly, three independent *ts4* mutations, which develop carpels in the florets of the tassel, were found to carry *Helitron* insertions in the promoter of the *zma-MIR172e* gene (Chuck et al. 2007). These mutations arose at different times in different genetic backgrounds. Since only the ends of the insertions were sequenced, it is not possible to speculate on the relationships among these elements. However, the similarity in size between the insertions in *ts4-TP* and *ba1-ref* (~6 kb) suggests that the former may also carry gene fragments.

Mutations caused by *Helitron* insertions have been identified in other plant genomes, as well (Table 11.2). *Hel-It1*, the first mutagenic *Helitron* described in dicots, interrupts the anthocyanin pigmentation gene *DFR-B* in the *pearly-s* mutant of *Ipomoea tricolor* (Choi et al. 2007). This 11.5-kb *Helitron* shows the structure predicted for a plant autonomous element, with conserved 5′ and 3′ termini and genes for Rep/Hel and RPA proteins. A frameshift mutation in the former and a nonsense mutation in the latter would render this element nonautonomous, but several related elements are found in the *Ipomoea* genome. In fact, *RPA* transcripts not containing the nonsense mutation of *Hel-It1* were detected in the *pearly-s* mutant and were proposed to originate from a hypothetical autonomous element present in that line.

The 3′-UTR of genes appeared to be an underrepresented target for *Helitron* insertion until a recent study on the S-RNase-based gametophytic self-incompatibility system in the tetraploid sour cherry (*Prunus cerasus*). A 306-bp nonautonomous *Helitron* element was identified 38 bp downstream of the stop codon of the *SFB* gene in four nonfunctional (self-compatible) S_{36} variants (Tsukamoto et al. 2010). The vast majority of *SFB* transcripts in S_{36} do not have

Table 11.2 Characterized variants resulting from *Helitron* insertions

Variant name	Targeted gene	Species	Gene function	Insertion site	References
sh2-7527	*shrunken-2*	*Zea mays*	Large subunit of the tetrameric maize endersperm ADP-glucose pyrophosphorylase	Eleventh intron	Lal et al. (2003)
ba1-ref	*barren stalk1*	*Zea mays*	An atypical bHLH transcription factor that affects every axillary meristem	Proximal promoter	Gupta et al. (2005)
Hel-ts4-TP	*tassel seed4*	*Zea mays*	microRNA 172e involved in sex determination and meristem cell fate	TATA box	Chuck et al. (2007)
Hel-ts4-A	*tassel seed4*	*Zea mays*	microRNA 172e involved in sex determination and meristem cell fate	TATA box	Chuck et al. (2007)
Hel-ts4-ref	*tassel seed4*	*Zea mays*	microRNA 172e involved in sex determination and meristem cell fate	883 bp upstream	Chuck et al. (2007)
AtREP2	MEDEA	*Arabidopsis thaliana*	SET domain protein of polycomb group	3,809 bp upstream	Spillane et al. (2004)
Hel-It1	DFR-B	*Ipomoea tricolor*	Dihydroflavonol 4-reductase for anthocyanin biosynthesis	Fifth intron	Choi et al. (2007)
Helitron-Os	TnpA	*Oryza sativa cv. japonica*	CACTA element transposase	Seventh intron	Greco et al. (2005)
Helitron-Pc	SFB	*Prunus cerasus*	Pollen self-incompatibility locus	3-′UTR	Tsukamoto et al. (2010)

a poly (A) tail, suggesting that the presence of the *Helitron* element interferes with the polyadenylation process. *Helitron* elements have also been found associated with certain *S* haplotypes in the self-compatible species *Arabidopsis thaliana* (Liu et al. 2007; Sherman-Broyles et al. 2007), raising the intriguing prospect that they may have played a widespread role in the evolution of self-compatibility. However, further studies are needed to establish conclusively that the *Helitron* insertion was the real cause of the loss of function of the S_{36} variants in sour cherry.

Genome components other than genes, such as DNA transposons, can also be targeted by *Helitrons*. In *OsES1*, a rice homolog of the maize *En*/*Spm* transposon, a 1,280-bp nonautonomous *Helitron* transposon, is located in the seventh intron of the gene encoding the TnpA transposase (Greco et al. 2005). The *Helitron* insertion seems to induce alternative splicing, as do many other transposon insertions in transcribed regions (Dooner and Weil 2012). Thus, *Helitrons* may play a role in the regulation of the transpositional activity of *CACTA* elements, the most abundant superfamily of DNA transposons in rice (Paterson et al. 2009).

Because many maize *Helitrons* carry segments of multiple genes, they have been identified much more frequently as disruptors of genetic colinearity among different maize inbred lines (Brunner et al. 2005a, b; Fu and Dooner 2002; Lai et al. 2005; Morgante et al. 2005; Song and Messing 2003; Wang and Dooner 2006). The so-called "intraspecific violation of genetic colinearity" (Fu and Dooner 2002) or "plus–minus variation" (Lai et al. 2005) resulting from *Helitron* insertions in maize led to community efforts to achieve a more detailed and precise identification and annotation of *Helitrons* (Du et al. 2008, 2009; Yang and Bennetzen 2009a). This effort was essential to a proper annotation of the actual gene content in the maize genome (Schnable et al. 2009) because of the gene-fragment-rich property of the widely prevalent nonautonomous elements (Lal et al. 2009a).

Recently, a maize-type of *Helitron* transposon was discovered in the Pooideae grass *Lolium perenne* (perennial ryegrass). Large (~7.5 kb) *Helitron* elements were identified that had trapped fragments, including exons and introns, from three genes: *GIGANTEA* (*GI*), succinate dehydrogenase, and ribosomal protein S7 (Langdon et al. 2009). All three fragmented genes shared the same transcription orientation as the *Helitron* elements. Highly similar *Helitrons* were detected in the closely related grass species *Festuca pratensis* (meadow fescue), indicating a likely common ancestral origin of these elements.

11.3.2.2 Computational Identification of *Helitrons* in Sequenced Organisms

The vast majority of *Helitrons* were identified from in silico studies of sequenced genomes either manually or via investigator-designed ad hoc mining programs, such as *DomainOrganizer* (Tempel et al. 2006), *HelitronFinder* (Du et al. 2008, 2009), *HelSearch* (Yang and Bennetzen 2009b), and *Helitron_scan* (Feschotte et al. 2009). The contribution of *Helitrons* to plant genomes varies widely, from none to as high as ~7 %. However, determining an exact figure for the *Helitron* content of any given host genome is chancy. Due to the extremely limited sequence conservation among *Helitrons*, it is not surprising to find quite different figures in updated versions of the same genome sequence (e.g., Du et al. 2010; Schmutz et al. 2010).

The published programs for automated computational identification and classification of *Helitrons* utilize either a homology-based or a structure-based approach. The latter approach (Du et al. 2008; Yang and Bennetzen 2009b) has been applied only recently in the analysis of whole genomes (Du et al. 2009, 2010; Yang and Bennetzen 2009a).

Initially, the homology-based approach was used to compare sequences at both the nucleotide and amino acid levels, as demonstrated by Kapitonov and Jurka (2001) in their original paper. *Helitron*-like transposons in rice were classified as *Helitrons* based on their capacity to code for proteins homologous to Rep/helicase and RPA (Kapitonov and Jurka 2001) and their shared structure hallmarks with *Arabidopsis Helitrons* (AT insertion site, 5′-TC, and 3′-CTRR and the 15- to 20-nucleotide palindrome close to the 3′-end). In an analogous approach, 21 *Helitron* elements were identified in the model legume *Lotus japonicus* by using as queries the RC motif and domain-5 of the RepHelicase from Arabidopsis *Helitrons*. Altogether, *Helitron* elements made up 0.4 % of the 32.4 Mb examined sequences (Holligan et al. 2006).

Novel *Helitrons* were also identified by nucleotide similarity to whole *Helitron* elements or to just the termini (Du et al. 2008, 2009; Kapitonov and Jurka 2001; Sweredoski et al. 2008; Tempel et al. 2007; Yang and Bennetzen 2009a, b). Other prevalent criteria implemented in genome-wide annotations of *Helitron* transposons include nonallelic locations in a given host genome and presence/absence of polymorphisms revealed from vertical comparison of colinear regions in closely related genomes (Wicker et al. 2010).

In addition to the two model plant genomes where *Helitrons* were originally identified, *Helitrons* have been detected in many other flowering and nonflowering plants. Paralleling the 20-fold variation in genome size, *Helitron* content varies from 0.01 % in grape to 6.72 % in the latest annotation of the *Arabidopsis thaliana* genome (Table 11.1). The estimated contribution of *Helitron* elements to a particular host genome also varies in different databases analyzed by different researchers, as seen *Arabidopsis thaliana*, rice, sorghum, and soybean.

Helitrons are poorly conserved among species, even of the same genus; this has made it hard to determine their presence systematically. Nevertheless, comparisons of the *Helitron* content of closely related species have been carried out in *Arabidopsis* and rice. The former involved the whole genomes of *A. thaliana* and *A. lyrata* (Hollister et al. 2011) and the latter, the partial genomes of 13 *Oryza* species (Gill et al. 2010).

As shown in a recent study on TE evolutionary dynamics in *Arabidopsis* employing the powerful transposon display method, *Basho Helitrons* were amplifiable in *A. thaliana* but were apparently absent from *A. lyrata*. This led to the suggestion of a recent burst of *Basho* insertions specifically within *A. thaliana* (Lockton and Gaut 2010). However, a subsequent sequence annotation effort revealed that *Helitrons* are actually the most abundant TEs in the fully sequenced *A. lyrata* genome (Hollister et al. 2011).

In an attempt to examine the relative abundance and distribution of TE classes across the genus *Oryza*, DNA transposons were identified by homology-based searches of BAC-end sequences from 13 species representing 8–17 % of each of the ten *Oryza* genome types. The *Helitron* content in the genus was found to vary greatly, from 0.29 % in *O. australiensis* to 3.15 % in *O. glaberrima* (Gill et al. 2010).

The identification of *Helitrons* from newly sequenced genomes remains a challenging endeavor despite the availability of several refined programs for detecting them. As shown in Table 11.1, *Helitron*-related sequences make up as much as 1.6 % of the *Selaginella* genome (Banks et al. 2011), but less than 0.2 % of the *Brachypodium* (International Brachypodium Initiative 2010) and *Physcomitrella* (Rensing et al. 2008) genomes. The lesson learned from other genomes, such as sorghum, suggests that the *Helitron* content of the latter two genomes will increase upon future careful annotation.

Glimpses of ongoing sequencing projects reveal that *Helitrons* are major components of some other plant genomes, as they are in sequenced model genomes. For example, *Helitron* transposons constitute ~1 % of 1.2 Mb of sequences from the tetraploid moso bamboo (*Phyllostachys pubescens* E. Mazel ex H. de Leh.) (Gui et al. 2010). In wheat (*Triticum aestivum*), 3,222 TEs have been annotated in 18.2 Mb of sequence from chromosome 3B. Only five families of agenic nonautonomous *Helitrons* were identified, representing just 0.07 % of the genomic sample sequences, in contrast to the 81.4 % contribution from all other TEs (Choulet et al. 2010). The only *Helitron* found so far in barley (Scherrer et al. 2005) is present in about 20–30 copies in the genome, based on 574 Mb of high-throughput sequences representing about 10 % of a genome equivalent (Wicker et al. 2008). Very recently, a putative *Helitron* sequence was first reported in sunflower and its insertion was dated to 1.14 million years ago (Buti et al. 2011).

In spite of the ever-growing numbers of identified *Helitrons* in newly sequenced genomes, a much more careful characterization of *Helitron* composition is necessary for sequenced plant genomes where *Helitrons* have not been yet identified, such as *Carica papaya* (Ming et al. 2008), *Cucumis sativus* (Huang et al. 2009), and *Solanum tuberosum* (The Potato Genome Sequencing Consortium 2011). Given the ubiquitous presence of these elements in all carefully annotated plant genomes, *Helitron*-free plant genomes are unlikely to exist.

11.3.3 Coding Capacity

The structure of the hypothetical autonomous *Helitron* proposed by Kapitonov and Jurka (2001) is fairly sound since elements with a similar structure continue to be found in an increasing number of genomes (Choi et al. 2007; Morgante et al. 2005). However, all of the *Helitrons* identified so far are nonautonomous and, oftentimes, bear gene fragments coding for proteins other than the REP-HEL transposase proposed for the RC transposition of *Helitrons* (Brunner et al. 2005a, b; Gupta et al. 2005; Lai et al. 2005; Lal et al. 2003; Morgante et al. 2005; Wang and Dooner 2006; Xu and Messing 2006).

In maize, two research groups have scanned the nearly complete genome sequence using similar computational approaches (Du et al. 2009; Yang and Bennetzen 2009a) and concluded that the majority of the ~2,000 genic *Helitrons* identified carried fragments from genes located in different chromosomes, with a few exceptions coming from neighboring genes. The tendency of *Helitrons* to gene-fragment capture seen in maize may be not a general property of plant *Helitrons*. For instance, in *A. thaliana*, very few *Helitron* families were found to have acquired gene fragments (Hollister and Gaut 2007; Yang and Bennetzen 2009b). A similar low propensity to capture genes was found among *Helitrons* from rice, sorghum, and *Medicago* (Yang and Bennetzen 2009b).

As is the case with most other transposon superfamilies (Levin and Moran 2011), small RNAs generated from endogenous *Helitron* sequences have the potential to inhibit TE mobility through the posttranscriptional degradation of transposon mRNA. As recently reported in *Physcomitrella patens*, 6 % of the nucleotides within 48 23-nucleotide RNA loci overlapped with regions similar to *Helitron* elements, which make up just 0.12 % of the genome (Cho et al. 2008).

11.3.4 Target Preference

The insertion site preference of *Helitron* transposons has been analyzed at the nucleotide level (target site sequence specificity), gene level (coding capacity of target sequence), and genome level (chromosomal distribution).

Plant *Helitrons* insert almost invariably in a 5′-AT-3′ dinucleotide (Brunner et al. 2005a, b; Choi et al. 2007; Gupta et al. 2005; Kapitonov and Jurka 2001; Lai et al. 2005; Lal et al. 2003; Morgante et al. 2005; Wang and Dooner 2006; Xu and Messing 2006) and, exceptionally, in a 5′-NT-3′ dinucleotide (Du et al. 2008, 2009; Morgante et al. 2005; Yang and Bennetzen 2009a). In addition, plant *Helitron* insertion sites are notably AT-enriched on either side of the insertion (Du et al. 2009; Yang and Bennetzen 2009a).

The discovery over the last decade that *Helitron* insertions have been the cause of spontaneous mutations in several plant species would suggest that *Helitrons* target genic regions (see Table 11.2), at least in these host genomes. Supporting this inference, maize *Helitrons* were found to be most abundant in gene-rich regions across the genome (Du et al. 2009; Yang and Bennetzen 2009a). However, this may not be a general pattern in plants.

In *Arabidopsis*, for example, *Helitrons* are enriched in gene-poor pericentromeric regions (Yang and Bennetzen 2009b), thus showing a pattern opposite to that of other DNA transposons, which are frequently associated with gene-rich regions. However, in a different study that compared the proximity of transposons of different ages to genes in *A. thaliana*, *Helitrons*, and other recently active TE families, such as MITEs, tended to be closer to genes than ancient families, such as CACTA-like elements (Hollister and Gaut 2009). Moreover, nonautonomous *Helitrons*, many as small as MITEs, were unmethylated in higher proportions than most other TE families.

These observations were explained by a model in which host silencing of TEs near genes has deleterious effects on neighboring gene expression, resulting in the preferential loss of methylated TEs from gene-rich chromosomal regions.

In rice, *Helitron* elements are more scattered along the chromosomes and not enriched in all pericentromeric regions (Yang and Bennetzen 2009b). As with other TEs, the distribution of *Helitrons* in present-day genomes probably reflects a combination of factors, such as continued mobility, insertion specificity, purifying selection against insertion in genes, and rates of DNA removal in gene-poor heterochromatic regions.

11.3.5 Differential Amplification and Contribution to Host Genome

The variable patterns of *Helitron* accumulation in sequenced plant genomes suggest different dynamics of *Helitron* proliferation across species and differential contributions to the present structure of their host genomes.

Helitrons make up a wide fraction of the plant genomes sequenced so far, from barely detectable to as much as 1/16 (Table 11.1). As has been well documented, TE proliferation and polyploidization are the two major processes that increase plant genome size (Bennetzen 2005). *Cornucopious*, the most abundant *Helitron* transposon subfamily in maize, consists of thousands of copies of ~1-kb agenic elements with variable sequence identity to the consensus (Du et al. 2009). These relatively small maize *Helitrons* may be actively transposing after a recent escape from transposition suppression, like the *mPing* MITEs suddenly amplified during rice domestication (Naito et al. 2006), whereas the amplification of the vast majority of *Helitron* families in maize, rice, and *Sorghum* peaked about 0.25 million years ago (Yang and Bennetzen 2009a).

In the recent annotation of the *A. thaliana* genome (Ahmed et al. 2011), *Helitron*-related sequences made up 6.7 % of the genome, more than the sum of all other DNA transposons (Table 11.1). In agreement with earlier results (Hollister and Gaut 2009), elements from the *Helitron* and *Tc1*/mariner superfamilies had the highest proportion of unmethylated sequences, whereas those from the *Gypsy* and *CACTA* superfamilies had the lowest.

As with *Helitron* content, different numbers of *Helitron* families have been identified the same organism (Table 11.1). In general, *Helitrons* with a smaller size tend to be amplified to a high degree (Ahmed et al. 2011; Du et al. 2009; Hollister and Gaut 2007). And, as noted in *Arabidopsis* and maize, longer *Helitrons* are less likely to persist in the genome (Hollister and Gaut 2007; Yang and Bennetzen 2009a), presumably because they are selected against in order to avoid the deleterious effects of inter *Helitron* ectopic recombination. However, other explanations may be possible because no recombination was detected within the heavily methylated gene fragments borne on maize *Helitrons* in a large-scale experiment specifically designed for that purpose (He and Dooner 2009).

In addition to their effect on genome size through massive amplification of agenic families, *Helitrons* contribute to haplotype variability through transposition and chromosome rearrangements (Ahmed et al. 2011; Brunner et al. 2005a; Lai et al. 2005; Morgante et al. 2005; Wang and Dooner 2006). The mechanism of gene movement that results in the erosion of colinearity between closely related species was recently investigated in a three-way comparison of the *Brachypodium*, rice, and sorghum genomes (Wicker et al. 2010). Gene capture by TEs, including *Helitrons*, was not found to have contributed significantly to gene movements within the grass family. On the other hand, TEs of many superfamilies, including *Helitrons*, were found at the borders of the noncolinear (i.e., mobilized) regions, suggesting that repair of TE-induced double strand breaks through synthesis-dependent strand annealing (SDSA) may have been involved in the change of position of genes in related genomes.

11.4 The Genetics of *Helitrons*

Being a member of the rare group of transposons that have been discovered computationally (Feschotte and Pritham 2007), it is not surprising that *Helitron* genetics trails its genomics. Yet, a genetic approach will be needed to identify a functional autonomous *Helitron* transposon, discern the actual mode(s) of transposition, assess the regulation of and by captured gene fragments, and elucidate other aspects of basic *Helitron* biology.

11.4.1 Transposition Mechanism: Rolling Circle and/or Cut-and-Paste?

A rolling circle replication mechanism has been proposed for the amplification of this novel class of transposons (Kapitonov and Jurka 2001). The putative autonomous *Helitrons* from the three genomes originally examined shared two conserved domains: the cross-kingdom DNA helicase domain and the replicator initiator proteins of RC plasmids and certain ssDNA viruses (Fig. 11.1a). Though still a hypothetical mechanism, RC replication is supported by the conserved structure of putative autonomous copies from several sequenced model plant genomes (Table 11.1).

The genome-wide distribution of *Helitron* elements favors a dispersive transposition model, although occasional *Helitron* clusters have been reported in some plant genomes (Lai et al. 2005; Yang and Bennetzen 2009a). Some peculiar head-to-head, head-to-tail, and tail-to-tail *Helitron* configurations have been identified in the maize genome (Du et al. 2008; Yang and Bennetzen 2009a), but they are composed of dissimilar *Helitrons* with similar terminal sequences, which differ from the perfect head-to-tail *Helitron* configurations expected from a RC replication mechanism and, so far, found only in the *Myotis lucifugus* genome (Pritham and Feschotte 2007).

As discussed in Sect. 11.3.5, *Helitrons* have contributed to the frequent loss of genetic colinearity in related plant genomes. Many recently duplicated fragments in the grasses are bordered by transposable elements (TEs), including *Helitrons* (Wicker et al. 2010). Other chromosomal rearrangements, such as inversions, are also oftentimes associated with *Helitron* transposons. Of the 154 inversions identified between *Arabidopsis thaliana* and *Arabidopsis lyrata*, one-third are flanked by inverted repeats from *Helitron* elements (Hu et al. 2011).

In addition to RC replication, a *Helitron* cut-and-paste transposition mechanism, like the one used by most known DNA transposons, was recently proposed. Li and Dooner (2009) found that, unexpectedly, some maize *Helitrons* could excise somatically. The somatic excision products or footprints left by removal of a 6-kb *Helitron* consisted of a variable number of TA repeats at the prior insertion site, an unlikely consequence of a RC replication mechanism. Somatic excision products were also detected from other genic and agenic *Helitron* elements (Du et al. 2008; Li and Dooner 2009). This finding suggests that, like *Tn7* (Craig 2002) and *Mutator* (Walbot and Rudenko 2002), *Helitrons* may exhibit both replicative and excisive modes of transposition.

11.4.2 Gene Capture

Transduplication or the capture of host gene sequences, first reported for *Mutator* elements (Jiang et al. 2004; Talbert and Chandler 1988), is a common feature of several families of plant transposons (Dooner and Weil 2007). However, *Helitrons* may contribute the largest portion of transduplicated sequences in some plant genomes, like maize (Brunner et al. 2005b; Du et al. 2009; Lai et al. 2005; Morgante et al. 2005; Wicker et al. 2010; Yang and Bennetzen 2009a, b).

In contrast to the broad-spectrum of captured genes in maize, only a few genes have been captured by *Helitrons* in *A. thaliana* (Hollister and Gaut 2007; Yang and Bennetzen 2009b). Gene-capture by *Helitrons* is also a rare event in *Medicago*, *Brachypodium*, *sorghum*, and rice (Fan et al. 2008; Wicker et al. 2010; Yang and Bennetzen 2009b). No correlation has been found between the transcriptional orientation of the captured gene fragments and the orientation of the TE in which they are lodged. In fact, some *Helitrons* contain multiple genes with opposite transcriptional orientations (Lai et al. 2005; Lal et al. 2003; Wang and Dooner 2006; Wicker et al. 2010).

In spite of the well-documented transcriptional activities of genes captured by *Helitrons* from different plant species (Brunner et al. 2005b; Lai et al. 2005; Lal et al. 2003; Morgante et al. 2005 and see Sect. 11.4.3), no cases of functional full-length gene capture by *Helitron* elements have been reported. Although an almost intact cytidine deaminase gene missing only the first six amino acids was found embedded in a maize *Helitron*, no transcripts corresponding to it were detected in any tissue examined (Xu and Messing 2006).

The capture of gene fragments from various genomic locations by the same *Helitron* may give rise to complex networks regulating the donor genes (Brunner et al. 2005b; Lai et al. 2005). The extent to which the host genome could benefit from these potentially deleterious effects (Du et al. 2009) is unclear.

11.4.3 Coevolution with the Host Genome

The potential role of *Helitrons* and other TEs in gene creation in plants has been recently reviewed by Dooner and Weil (2012).

Gene fragments captured by *Helitrons* originate from nonadjacent loci in the genome, yet they tend to be in the same transcriptional orientation relative to each other and to the *Helitron*'s *RepHel* gene. A large collection of gene-fragment-bearing *Helitrons* in maize show a notable bias in the orientation of gene fragments that is compatible with *Helitron* promoter-driven expression (Du et al. 2009; Yang and Bennetzen 2009a). Several chimeric transcripts containing exons from different genes ("exon shuffling") have been detected for maize *Helitrons* (Brunner et al. 2005b; Lai et al. 2005; Morgante et al. 2005). Though many of these transcripts contain premature stop codons in all reading frames and are unlikely to encode functional proteins immediately, *Helitrons* could have contributed to gene creation over evolutionary time (Brunner et al. 2005b). Expression of chimeric transcripts can also be driven by the promoter of the disrupted gene, rather than by a *Helitron* promoter. In maize, chimeric transcripts derived from genes captured by the inserted *Helitron* in the *sh2-7057* mutant are produced from the *sh2* promoter (Lal et al. 2003), rather than from a *Helitron* promoter.

The idea that TEs have been co-opted by the host as regulatory sequences has received considerable experimental support. Many *cis*-regulatory elements involved in transcriptional regulation have characteristics of TEs and some of them are *Helitrons*. For example, the *CArG* motif essential for the transcriptional activation of *LEAFY COTYLEDON2* (*LEC2*), a master regulator of seed development in *A. thaliana*, is located at the beginning of a *Helitron* element (*Helitron3*). This and other TE insertions located in the promoter region of *LEC2* were speculated to control the gene's specific expression pattern (Berger et al. 2011).

TE sequences are also found in transcripts, where they may play an unsuspected regulatory role. In *Arabidopsis thaliana*, more than 2,000 putative TE-gene chimeras, where a TE is found in at least one expressed exon, have been identified and compared to all TEs in a TE database (Lockton and Gaut 2009). *Helitron*-like sequences were strikingly underrepresented (2.4 %) in exons, contrasting with the high abundance (~20 %) of all other TEs. A similar pattern was found for the specific targets of the *MOM1* (*MORPHEUS' MOLECULE1*) regulator of transcriptional gene silencing in *Arabidopsis* (Numa et al. 2010). The majority of

MOM1 targets carry sequences related to TEs of both classes and are clustered at pericentromeric regions, suggesting that *MOM1* acts on regions of heterochromatin in the genome. *Helitron* remnants, on the other hand, were significantly underrepresented among *MOM1*-regulated transcripts. The authors suggested that, because *Helitrons* target active genes undergoing transcription, their low frequency among *MOM1*-target sequences may reflect exclusion of MOM1 from active chromatin environments. As major contributors to the evolution of plant genomes, more in-depth analyses are required to decipher the contributions of TEs to annotated protein-coding regions, an essentially unexplored field (Lal et al. 2009b).

11.4.4 Epigenetic Regulation

There is growing evidence that the proliferation of TEs in plants is under epigenetic regulation and that their biological properties are strongly affected by cycles of methylation and demethylation (Lisch 2009).

The past couple of years have seen a considerable increase in experimental data, mainly from *Arabidopsis*, on the methylation status of TEs. As shown in two earlier bisulfite sequencing studies (Gehring et al. 2006; He and Dooner 2009), *Helitrons* are heavily methylated at CG sites. In the first study, a *Helitron* inserted 4 kb upstream of the start site of the *Arabidopsis MEDEA* gene was heavily methylated, yet did not contribute to the allele-specific DNA hypomethylation in the endosperm (Gehring et al. 2006). In the second study, two maize *Helitrons* shown to be nonrecombinogenic despite the presence of multiple gene fragments were much more methylated than the adjacent recombinogenic gene-rich region (He and Dooner 2009).

Transcriptional reactivation of TEs in the mature pollen of *Arabidopsis* has been detected in microarray assays of TE expression profiles during development (Slotkin et al. 2009). In most tissues and stages, the ORFs of *Helitron2* and six other full-length TEs (including retrotransposons and DNA transposons) were either not expressed or expressed at a very low level, indicating that they are generally silenced. However, all seven full-length TEs examined were coordinately expressed in mature pollen. TE expression coincides with loss of DNA methylation and downregulation of the chromatin remodeler *DDM1*.

A recent study analyzed the contribution of TEs and small RNAs to gene expression variation in *A. thaliana* and *A. lyrata*, a closely related congener with a two to threefold higher copy number for every TE family examined, including *Helitrons* (Hollister et al. 2011). Reassessment of the TE content in the two species revealed that, unexpectedly, *Helitron*s were the highest copy number DNA transposons in both (Table 11.1). The 24-nt siRNA complements from the two species were compared in order to address the possible role of siRNA-guided transcriptional gene silencing in differential TE proliferation. *Helitrons* were found to be less often targeted by unique 24-nt siRNAs in *A. lyrata* than in *A. thaliana*, possibly explaining their higher copy number in the former. An almost concurrent reanalysis of DNA methylation, siRNA, and TE datasets from *Arabidopsis thaliana* concluded that *Helitrons* actually contribute ~7 % of the annotated genome (Table 11.1) and, along with the *Tc1/mariner* superfamily, have the largest fraction (40–50 %) of unmethylated TE sequences (Ahmed et al. 2011).

Around a dozen *Arabidopsis* genes are imprinted, i.e., expressed in a parent-of-origin-dependent manner in the endosperm during seed development (Kermicle 1970). In a couple of cases, *Helitron* insertions have been implicated in imprinting. In a study on the association of TE methylation with gene imprinting during seed development in *A. thaliana*, TE fragments were found to be extensively demethylated in the endosperm (Gehring et al. 2009). Two imprinted members of the class IV homeodomain transcription factors contain remnants of *Helitron* elements at the 5′end. Although these genes showed reciprocal imprinting, i.e., predominant expression of the maternal allele in one and of the paternal allele in the other, methylation of the *Helitron* remnants was lost from the maternal alleles in both cases. Other imprinted genes are also neighbored by TEs. *AGL36*, a maternally expressed gene, contains remnants of *Helitrons* and other TE sequences within a 1.7-kb promoter fragment that is sufficient to confer parent-of-origin-specific expression of a reporter (Shirzadi et al. 2011). Paternally expressed genes, as well, are enriched for cis-proximal transposons, particularly for *Helitrons* (Wolff et al. 2011). It has been proposed that imprinting may have evolved from targeted methylation of TE insertions near genes followed by positive selection when the resulting expression change was advantageous (Gehring et al. 2009).

Whether a TE can exert a regulatory effect on a nearby gene obviously depends on the distance between the transposon and the gene. A methylated *AtREP2 Helitron* inserted 3.8 kb upstream of the imprinted *MEA* gene in the Col-0 and Ler-0 ecotypes of *Arabidopsis thaliana* was considered a candidate for imprinting control elements until ecotypes were found where *MEA* was still imprinted, though they lacked the upstream *Helitron* (Spillane et al. 2004). In a recent study relating gene expression to distance from the nearest TE in *A. thaliana*, average gene expression increased with distance up to about 2.5 kb (Hollister et al. 2011).

11.5 Perspective

The huge number of annotated *Helitron* transposons in plant genomes, including both putative autonomous elements and nonautonomous elements with and without gene fragments (Table 11.1), represents only the tip of the iceberg.

The molecular structure of the autonomous *Helitron* and the RC mechanism of transposition (Kapitonov and Jurka 2001) remain hypothetical, but are supported, respectively, by the conservation of structure of the putative autonomous element across evolutionarily widely divergent species and the identification of occasional head-to-tail configurations that make RC replication a credible transposition mechanism. Whether the RepHel protein is necessary and/or sufficient for RC transposition needs to be confirmed experimentally. The discovery of *Helitron* somatic excision products in maize (Li and Dooner 2009) suggests that *Helitrons* may transpose by both copy-and-paste and cut-and-paste mechanisms.

As is evident from successive sequence annotations of the same genome, determination of the overall *Helitron* contents in a given genome is a challenging and uncertain exercise (Feschotte and Pritham 2009). The conserved sequence and structure of the 3′ end of known *Helitrons* has served as the basis for the development of a number of ad hoc programs for specific genome-wide surveys of this highly divergent family of transposons. However, their cross-species applications are still not efficient in identifying *Helitrons* in new species and novel programs, possibly based on the recognition of conserved nucleotide patterns, are desirable for the efficient de novo identification of *Helitrons* from all genome sequencing projects.

Only a few cases of gene-fragment-bearing *Helitrons* have been identified in plants other than maize. The high frequency of gene fragment capture by maize *Helitrons* is enigmatic, but it has been suggested to result from a RepHel enzyme with a different replication/repair fidelity (Yang and Bennetzen 2009b). The identification and characterization of an autonomous *Helitron* in maize would be highly desirable because maize is an excellent experimental genetic system and has currently active elements, as is evident from several recently arisen mutations (Table 11.2).

The dynamic evolution of *Helitron* is best exemplified by the discovery in maize of a new group of *Helitron*-like sequences, designated *Heltir*, which end in perfect 37-bp TIRs (Du et al. 2009). The sequence variability of *Helitrons* and the presence in the genome of other forms, like *Heltirs*, complicate the accurate estimation of the contribution of this transposon superfamily to plant genomes.

References

Ahmed I, Sarazin A, Bowler C, Colot V, Quesneville H (2011) Genome-wide evidence for local DNA methylation spreading from small RNA-targeted sequences in Arabidopsis. Nucleic Acids Res 39:6919–6931

Arabidopsis Genome Initiative (2000) Analysis of the genome sequence of the flowering plant *Arabidopsis thaliana*. Nature 408:796–815

Banks JA, Nishiyama T, Hasebe M, Bowman JL, Gribskov M et al (2011) The Selaginella genome identifies genetic changes associated with the evolution of vascular plants. Science 332:960–963

Bennetzen JL (2005) Transposable elements, gene creation and genome rearrangement in flowering plants. Curr Opin Genet Dev 15:621–627

Berger N, Dubreucq B, Roudier F, Dubos C, Lepiniec L (2011) Transcriptional regulation of Arabidopsis *LEAFY COTYLEDON2* involves RLE, a *cis*-element that regulates trimethylation of Histone H3 at Lysine-27. Plant Cell 23:4065–4078

Brunner S, Fengler K, Morgante M, Tingey S, Rafalski A (2005a) Evolution of DNA sequence nonhomologies among maize inbreds. Plant Cell 17:343–360

Brunner S, Pea G, Rafalski A (2005b) Origins, genetic organization and transcription of a family of non-autonomous helitron elements in maize. Plant J 43:799–810

Buti M, Giordani T, Cattonaro F, Cossu RM, Pistelli L et al (2011) Temporal dynamics in the evolution of the sunflower genome as revealed by sequencing and annotation of three large genomic regions. Theor Appl Genet 123:779–791

Cho SH, Addo-Quaye C, Coruh C, Arif MA, Ma Z et al (2008) *Physcomitrella patens DCL3* is required for 22–24 nt siRNA accumulation, suppression of retrotransposon-derived transcripts, and normal development. PLoS Genet 4:e1000314

Choi JD, Hoshino A, Park KI, Park IS, Iida S (2007) Spontaneous mutations caused by a *Helitron* transposon, *Hel-It1*, in morning glory, *Ipomoea tricolor*. Plant J 49:924–934

Choulet F, Wicker T, Rustenholz C, Paux E, Salse J et al (2010) Megabase level sequencing reveals contrasted organization and evolution patterns of the wheat gene and transposable element spaces. Plant Cell 22:1686–1701

Chuck G, Meeley R, Irish E, Sakai H, Hake S (2007) The maize *tasselseed4* microRNA controls sex determination and meristem cell fate by targeting *Tasselseed6/indeterminate spikelet1*. Nat Genet 39:1517–1521

Cocca E, De Iorio S, Capriglione T (2011) Identification of a novel helitron transposon in the genome of Antarctic fish. Mol Phylogenet Evol 58:439–446

Craig NL (2002) *Tn7*. In: Craig NL, Craigie R, Gellert M, Lambowitz AM (eds) Mobile DNA II. ASM Press, Washington, D.C., pp 422–456

Craig NL, Craigie R, Gellert M, Lambowitz AM (2002) Mobile DNA II. ASM Press, Washington, D.C

Dooner HK, Weil CF (2007) Give-and-take: interactions between DNA transposons and their host plant genomes. Curr Opin Genet Dev 17:486–492

Dooner HK, Weil CF (2012) Transposons and gene creation. In: Fedoroff N (ed) Molecular genetics and epigenetics of plant transposons: sculpting genes and genomes. Wiley, Hoboken, NJ

Doutriaux MP, Couteau F, Bergounioux C, White C (1998) Isolation and characterisation of the RAD51 and DMC1 homologs from *Arabidopsis thaliana*. Mol Gen Genet 257:283–291

Du C, Caronna J, He L, Dooner HK (2008) Computational prediction and molecular confirmation of *Helitron* transposons in the maize genome. BMC Genomics 9:51

Du C, Fefelova N, Caronna J, He L, Dooner HK (2009) The polychromatic *Helitron* landscape of the maize genome. Proc Natl Acad Sci USA 106:19916–19920

Du J, Grant D, Tian Z, Nelson RT, Zhu L et al (2010) SoyTEdb: a comprehensive database of transposable elements in the soybean genome. BMC Genomics 11:113

Fan C, Zhang Y, Yu Y, Rounsley S, Long M et al (2008) The subtelomere of *Oryza sativa* chromosome 3 short arm as a hot bed of new gene origination in rice. Mol Plant 1:839–850

Feschotte C, Pritham EJ (2007) DNA transposons and the evolution of eukaryotic genomes. Annu Rev Genet 41:331–368

Feschotte C, Pritham EJ (2009) A cornucopia of Helitrons shapes the maize genome. Proc Natl Acad Sci USA 106:19747–19748

Feschotte C, Jiang N, Wessler SR (2002) Plant transposable elements: where genetics meets genomics. Nat Rev Genet 3:329–341

Feschotte C, Keswani U, Ranganathan N, Guibotsy ML, Levine D (2009) Exploring repetitive DNA landscapes using REPCLASS, a tool that automates the classification of transposable elements in eukaryotic genomes. Genome Biol Evol 1:205–220

Fu H, Dooner HK (2002) Intraspecific violation of genetic colinearity and its implications in maize. Proc Natl Acad Sci USA 99:9573–9578

Galagan JE, Calvo SE, Cuomo C, Ma LJ, Wortman JR, Batzoglou S, Lee SI, Baştürkmen M, Spevak CC, Clutterbuck J, Kapitonov V, Jurka J, Scazzocchio C, Farman M, Butler J, Purcell S, Harris S, Braus GH, Draht O, Busch S, D'Enfert C, Bouchier C, Goldman GH, Bell-Pedersen D, Griffiths-Jones S, Doonan JH, Yu J, Vienken K, Pain A, Freitag M, Selker EU, Archer DB, Peñalva MA, Oakley BR, Momany M, Tanaka T, Kumagai T, Asai K, Machida M, Nierman WC, Denning DW, Caddick M, Hynes M, Paoletti M, Fischer R, Miller B, Dyer P, Sachs MS, Osmani SA, Birren BW (2005) Sequencing of *Aspergillus nidulans* and comparative analysis with *A. fumigatus* and *A. oryzae*. Nature 438:1105–1115

Gehring M, Huh JH, Hsieh TF, Penterman J, Choi Y, Harada JJ, Goldberg RB, Fischer RL (2006) DEMETER DNA glycosylase establishes MEDEA polycomb gene self-imprinting by allele-specific demethylation. Cell 124:495–506

Gehring M, Bubb KL, Henikoff S (2009) Extensive demethylation of repetitive elements during seed development underlies gene imprinting. Science 324:1447–1451

Gill N, SanMiguel P, Dhillon BDS, Abernathy B, Kim H et al (2010) Dynamic *Oryza* genomes: repetitive DNA sequences as genome modeling agents. Rice 3:251–269

Greco R, Ouwerkerk PB, Pereira A (2005) Suppression of an atypically spliced rice CACTA transposon transcript in transgenic plants. Genetics 169:2383–2387

Gui YJ, Zhou Y, Wang Y, Wang S, Wang SY, Hu Y, Bo SP, Chen H, Zhou CP, Ma NX, Zhang TZ, Fan LJ (2010) Insights into the bamboo genome: syntenic relationships to rice and sorghum. J Integr Plant Biol 52:1008–1015

Gupta S, Gallavotti A, Stryker GA, Schmidt RJ, Lal SK (2005) A novel class of *Helitron*-related transposable elements in maize contain portions of multiple pseudogenes. Plant Mol Biol 57:115–127

He L, Dooner HK (2009) Haplotype structure strongly affects recombination in a maize genetic interval polymorphic for *Helitron* and retrotransposon insertions. Proc Natl Acad Sci USA 106:8410–8416

Holligan D, Zhang X, Jiang N, Pritham EJ, Wessler SR (2006) The transposable element landscape of the model legume *Lotus japonicus*. Genetics 174:2215–2228

Hollister JD, Gaut BS (2007) Population and evolutionary dynamics of *Helitron* transposable elements in *Arabidopsis thaliana*. Mol Biol Evol 24:2515–2524

Hollister JD, Gaut BS (2009) Epigenetic silencing of transposable elements: a trade-off between reduced transposition and deleterious effects on neighboring gene expression. Genome Res 19:1419–1428

Hollister JD, Smith LM, Guo YL, Ott F, Weigel D, Gaut BS (2011) Transposable elements and small RNAs contribute to gene expression divergence between *Arabidopsis thaliana* and *Arabidopsis lyrata*. Proc Natl Acad Sci USA 108:2322–2327

Hu TT, Pattyn P, Bakker EG, Cao J, Cheng JF, Clark RM, Fahlgren N, Fawcett JA, Grimwood J, Gundlach H, Haberer G, Hollister JD, Ossowski S, Ottilar RP, Salamov AA, Schneeberger K, Spannagl M, Wang X, Yang L, Nasrallah ME, Bergelson J, Carrington JC, Gaut BS, Schmutz J, Mayer KF, Van de Peer Y, Grigoriev IV, Nordborg M, Weigel D, Guo YL (2011) The *Arabidopsis lyrata* genome sequence and the basis of rapid genome size change. Nat Genet 43:476–481

Huang S, Li R, Zhang Z, Li L, Gu X, Fan W, Lucas WJ, Wang X, Xie B, Ni P, Ren Y, Zhu H, Li J, Lin K, Jin W, Fei Z, Li G, Staub J, Kilian A, van der Vossen EA, Wu Y, Guo J, He J, Jia Z, Ren Y, Tian G, Lu Y, Ruan J, Qian W, Wang M, Huang Q, Li B, Xuan Z, Cao J, Asan WZ, Zhang J, Cai Q, Bai Y, Zhao B, Han Y, Li Y, Li X, Wang S, Shi Q, Liu S, Cho WK, Kim JY, Xu Y, Heller-Uszynska K, Miao H, Cheng Z, Zhang S, Wu J, Yang Y, Kang H, Li M, Liang H, Ren X, Shi Z, Wen M, Jian M, Yang H, Zhang G, Yang Z, Chen R, Liu S, Li J, Ma L, Liu H, Zhou Y, Zhao J, Fang X, Li G, Fang L, Li Y, Liu D, Zheng H, Zhang Y, Qin N, Li Z, Yang G, Yang S, Bolund L, Kristiansen K, Zheng H, Li S, Zhang X, Yang H, Wang J, Sun R, Zhang B, Jiang S, Wang J, Du Y, Li S (2009) The genome of the cucumber, *Cucumis sativus* L. Nat Genet 41:1275–1281

International Brachypodium Initiative (2010) Genome sequencing and analysis of the model grass *Brachypodium distachyon*. Nature 463:763–768

Ivics Z, Hackett PB, Plasterk RH, Izsvak Z (1997) Molecular reconstruction of *Sleeping Beauty*, a *Tc1*-like transposon from fish, and its transposition in human cells. Cell 91:501–510

Jaillon O, Aury JM, Noel B, Policriti A, Clepet C, Casagrande A, Choisne N, Aubourg S, Vitulo N, Jubin C, Vezzi A, Legeai F, Hugueney P, Dasilva C, Horner D, Mica E, Jublot D, Poulain J, Bruyère C, Billault A, Segurens B, Gouyvenoux M, Ugarte E, Cattonaro F, Anthouard V, Vico V, Del Fabbro C, Alaux M, Di Gaspero G, Dumas V, Felice N, Paillard S, Juman I, Moroldo M, Scalabrin S, Canaguier A, Le Clainche I, Malacrida G, Durand E, Pesole G, Laucou V, Chatelet P, Merdinoglu D, Delledonne M, Pezzotti M, Lecharny A, Scarpelli C, Artiguenave F, Pè ME, Valle G, Morgante M, Caboche M, Adam-Blondon AF, Weissenbach J, Quétier F, Wincker P (2007) The grapevine genome sequence suggests ancestral hexaploidization in major angiosperm phyla. Nature 449:463–467

Jiang N, Bao Z, Zhang X, Eddy SR, Wessler SR (2004) Pack-MULE transposable elements mediate gene evolution in plants. Nature 431:569–573

Kapitonov VV, Jurka J (1999) Molecular paleontology of transposable elements from *Arabidopsis thaliana*. Genetica 107:27–37

Kapitonov VV, Jurka J (2001) Rolling-circle transposons in eukaryotes. Proc Natl Acad Sci USA 98:8714–8719

Kapitonov VV, Jurka J (2007) Helitrons on a roll: eukaryotic rolling-circle transposons. Trends Genet 23:521–529

Kapitonov VV, Jurka J (2008) A universal classification of eukaryotic transposable elements implemented in Repbase. Nat Rev Genet 9:411–412, author reply 414

Kermicle JL (1970) Dependence of the R-mottled aleurone phenotype in maize on mode of sexual transmission. Genetics 66:69–85

Lai J, Li Y, Messing J, Dooner HK (2005) Gene movement by *Helitron* transposons contributes to the haplotype variability of maize. Proc Natl Acad Sci USA 102:9068–9073

Lal SK, Giroux MJ, Brendel V, Vallejos CE, Hannah LC (2003) The maize genome contains a *Helitron* insertion. Plant Cell 15:381–391

Lal SK, Georgelis N, Hannah LC (2009a) *Helitrons*: their impact on maize genome evolution and diversity. In: Bennetzen JL, Hake SC (eds) Handbook of maize: genetics and genome, vol 2. Springer, New York, pp 329–339

Lal SK, Oetjens M, Hannah LC (2009b) *Helitrons*: enigmatic abductors and mobilizers of host genome sequences. Plant Sci 176:181–186

Langdon T, Thomas A, Huang L, Farrar K, King J, Armstead I (2009) Fragments of the key flowering gene *GIGANTEA* are associated with helitron-type sequences in the Pooideae grass *Lolium perenne*. BMC Plant Biol 9:70

Le QH, Wright S, Yu Z, Bureau T (2000) Transposon diversity in *Arabidopsis thaliana*. Proc Natl Acad Sci USA 97:7376–7381

Levin HL, Moran JV (2011) Dynamic interactions between transposable elements and their hosts. Nat Rev Genet 12:615–627

Li Y, Dooner HK (2009) Excision of *Helitron* transposons in maize. Genetics 182:399–402

Lisch D (2009) Epigenetic regulation of transposable elements in plants. Annu Rev Plant Biol 60:43–66

Liu P, Sherman-Broyles S, Nasrallah ME, Nasrallah JB (2007) A cryptic modifier causing transient self-incompatibility in *Arabidopsis thaliana*. Curr Biol 17:734–740

Lockton S, Gaut BS (2009) The contribution of transposable elements to expressed coding sequence in *Arabidopsis thaliana*. J Mol Evol 68:80–89

Lockton S, Gaut BS (2010) The evolution of transposable elements in natural populations of self-fertilizing *Arabidopsis thaliana* and its outcrossing relative *Arabidopsis lyrata*. BMC Evol Biol 10:10

McClintock B (1947) Cytogenetic studies of maize and Neurospora. Carnegie Inst Wash Yearbook 46:146–152

McClintock B (1952) Chromosome organization and gene expression. Cold Spring Harb Symp Quant Biol 16:13–47

Ming R, Hou S, Feng Y, Yu Q, Dionne-Laporte A, Saw JH, Senin P, Wang W, Ly BV, Lewis KL, Salzberg SL, Feng L, Jones MR, Skelton RL, Murray JE, Chen C, Qian W, Shen J, Du P, Eustice M, Tong E, Tang H, Lyons E, Paull RE, Michael TP, Wall K, Rice DW, Albert H, Wang ML, Zhu YJ, Schatz M, Nagarajan N, Acob RA, Guan P, Blas A, Wai CM, Ackerman CM, Ren Y, Liu C, Wang J, Wang J, Na JK, Shakirov EV, Haas B, Thimmapuram J, Nelson D, Wang X, Bowers JE, Gschwend AR, Delcher AL, Singh R, Suzuki JY, Tripathi S, Neupane K, Wei H, Irikura B, Paidi M, Jiang N, Zhang W, Presting G, Windsor A, Navajas-Pérez R, Torres MJ, Feltus FA, Porter B, Li Y, Burroughs AM, Luo MC, Liu L, Christopher DA, Mount SM, Moore PH, Sugimura T, Jiang J, Schuler MA, Friedman V, Mitchell-Olds T, Shippen DE, dePamphilis CW, Palmer JD, Freeling M, Paterson AH, Gonsalves D, Wang L, Alam M (2008) The draft genome of the transgenic tropical fruit tree papaya (*Carica papaya* Linnaeus). Nature 452:991–996

Morgante M, Brunner S, Pea G, Fengler K, Zuccolo A et al (2005) Gene duplication and exon shuffling by helitron-like transposons generate intraspecies diversity in maize. Nat Genet 37:997–1002

Naito K, Cho E, Yang G, Campbell MA, Yano K, Okumoto Y, Tanisaka T, Wessler SR (2006) Dramatic amplification of a rice transposable element during recent domestication. Proc Natl Acad Sci USA 103:17620–17625

Numa H, Kim JM, Matsui A, Kurihara Y, Morosawa T, Ishida J, Mochizuki Y, Kimura H, Shinozaki K, Toyoda T, Seki M, Yoshikawa M, Habu Y (2010) Transduction of RNA-directed DNA methylation signals to repressive histone marks in *Arabidopsis thaliana*. EMBO J 29:352–362

Paterson AH, Bowers JE, Bruggmann R, Dubchak I, Grimwood J, Gundlach H, Haberer G, Hellsten U, Mitros T, Poliakov A, Schmutz J, Spannagl M, Tang H, Wang X, Wicker T, Bharti AK, Chapman J, Feltus FA, Gowik U, Grigoriev IV, Lyons E, Maher CA, Martis M, Narechania A, Otillar RP, Penning BW, Salamov AA, Wang Y, Zhang L, Carpita NC, Freeling M, Gingle AR, Hash CT, Keller B, Klein P, Kresovich S, McCann MC, Ming R, Peterson DG, Mehboob-ur-Rahman WD, Westhoff P, Mayer KF, Messing J, Rokhsar DS (2009) The *Sorghum bicolor* genome and the diversification of grasses. Nature 457:551–556

Pritham EJ, Feschotte C (2007) Massive amplification of rolling-circle transposons in the lineage of the bat *Myotis lucifugus*. Proc Natl Acad Sci USA 104:1895–1900

Rensing SA, Lang D, Zimmer AD, Terry A, Salamov A, Shapiro H, Nishiyama T, Perroud PF, Lindquist EA, Kamisugi Y, Tanahashi T, Sakakibara K, Fujita T, Oishi K, Shin-I T, Kuroki Y, Toyoda A, Suzuki Y, Hashimoto S, Yamaguchi K, Sugano S, Kohara Y, Fujiyama A, Anterola A, Aoki S, Ashton N, Barbazuk WB, Barker E, Bennetzen JL, Blankenship R, Cho SH, Dutcher SK, Estelle M, Fawcett JA, Gundlach H, Hanada K, Heyl A, Hicks KA, Hughes J, Lohr M, Mayer K, Melkozernov A, Murata T, Nelson DR, Pils B, Prigge M, Reiss B, Renner T, Rombauts S, Rushton PJ, Sanderfoot A, Schween G, Shiu SH, Stueber K, Theodoulou FL, Tu H, Van de Peer Y, Verrier PJ, Waters E, Wood A, Yang L, Cove D, Cuming AC, Hasebe M, Lucas S, Mishler BD, Reski R, Grigoriev IV, Quatrano RS, Boore JL (2008) The Physcomitrella genome reveals evolutionary insights into the conquest of land by plants. Science 319:64–69

Scherrer B, Isidore E, Klein P, Kim JS, Bellec A, Chalhoub B, Keller B, Feuillet C (2005) Large intraspecific haplotype variability at the *Rph7* locus results from rapid and recent divergence in the barley genome. Plant Cell 17:361–374

Schmutz J, Cannon SB, Schlueter J, Ma J, Mitros T, Nelson W, Hyten DL, Song Q, Thelen JJ, Cheng J, Xu D, Hellsten U, May GD, Yu Y, Sakurai T, Umezawa T, Bhattacharyya MK, Sandhu D, Valliyodan B, Lindquist E, Peto M, Grant D, Shu S, Goodstein D, Barry K, Futrell-Griggs M, Abernathy B, Du J, Tian Z, Zhu L, Gill N, Joshi T, Libault M, Sethuraman A, Zhang XC, Shinozaki K, Nguyen HT, Wing RA, Cregan P, Specht J, Grimwood J, Rokhsar D, Stacey G, Shoemaker RC, Jackson SA (2010) Genome sequence of the palaeopolyploid soybean. Nature 463:178–183

Schnable PS, Ware D, Fulton RS, Stein JC, Wei F, Pasternak S, Liang C, Zhang J, Fulton L, Graves TA, Minx P, Reily AD, Courtney L, Kruchowski SS, Tomlinson C, Strong C, Delehaunty K, Fronick C, Courtney B, Rock SM, Belter E, Du F, Kim K, Abbott RM, Cotton M, Levy A, Marchetto P, Ochoa K, Jackson SM, Gillam B, Chen W, Yan L, Higginbotham J, Cardenas M, Waligorski J, Applebaum E, Phelps L, Falcone J, Kanchi K, Thane T, Scimone A, Thane N, Henke J, Wang T, Ruppert J, Shah N, Rotter K, Hodges J, Ingenthron E, Cordes M, Kohlberg S, Sgro J, Delgado B, Mead K, Chinwalla A, Leonard S, Crouse K, Collura K, Kudrna D, Currie J, He R, Angelova A, Rajasekar S, Mueller T, Lomeli R, Scara G, Ko A, Delaney K, Wissotski M, Lopez G, Campos D, Braidotti M, Ashley E, Golser W, Kim H, Lee S, Lin J, Dujmic Z, Kim W, Talag J, Zuccolo A, Fan C, Sebastian A, Kramer M, Spiegel L, Nascimento L, Zutavern T, Miller B, Ambroise C, Muller S, Spooner W, Narechania A, Ren L, Wei S, Kumari S, Faga B, Levy MJ, McMahan L, Van Buren P, Vaughn MW, Ying K, Yeh CT, Emrich SJ, Jia Y, Kalyanaraman A, Hsia AP, Barbazuk WB, Baucom RS, Brutnell TP, Carpita NC, Chaparro C, Chia JM, Deragon JM, Estill JC, Fu Y, Jeddeloh JA, Han Y, Lee H, Li P, Lisch DR, Liu S, Liu Z, Nagel DH, McCann MC, SanMiguel P, Myers AM, Nettleton D, Nguyen J, Penning BW, Ponnala L, Schneider KL, Schwartz DC, Sharma A, Soderlund C,

Springer NM, Sun Q, Wang H, Waterman M, Westerman R, Wolfgruber TK, Yang L, Yu Y, Zhang L, Zhou S, Zhu Q, Bennetzen JL, Dawe RK, Jiang J, Jiang N, Presting GG, Wessler SR, Aluru S, Martienssen RA, Clifton SW, McCombie WR, Wing RA, Wilson RK (2009) The B73 maize genome: complexity, diversity, and dynamics. Science 326:1112–1115

Sherman-Broyles S, Boggs N, Farkas A, Liu P, Vrebalov J, Nasrallah ME, Nasrallah JB (2007) *S* locus genes and the evolution of self-fertility in *Arabidopsis thaliana*. Plant Cell 19:94–106

Shirzadi R, Andersen ED, Bjerkan KN, Gloeckle BM, Heese M, Ungru A, Winge P, Koncz C, Aalen RB, Schnittger A, Grini PE (2011) Genome-wide transcript profiling of endosperm without paternal contribution identifies parent-of-origin-dependent regulation of *AGAMOUS-LIKE36*. PLoS Genet 7:e1001303

Slotkin RK, Vaughn M, Borges F, Tanurdzić M, Becker JD, Feijó JA, Martienssen RA (2009) Epigenetic reprogramming and small RNA silencing of transposable elements in pollen. Cell 136:461–472

Song R, Messing J (2003) Gene expression of a gene family in maize based on noncollinear haplotypes. Proc Natl Acad Sci USA 100:9055–9060

Spillane C, Baroux C, Escobar-Restrepo JM, Page DR, Laoueille S, Grossniklaus U (2004) Transposons and tandem repeats are not involved in the control of genomic imprinting at the MEDEA locus in Arabidopsis. Cold Spring Harb Symp Quant Biol 69:465–475

Surzycki SA, Belknap WR (1999) Characterization of repetitive DNA elements in Arabidopsis. J Mol Evol 48:684–691

Sweredoski M, DeRose-Wilson L, Gaut BS (2008) A comparative computational analysis of nonautonomous helitron elements between maize and rice. BMC Genomics 9:467

Talbert LE, Chandler VL (1988) Characterization of a highly conserved sequence related to mutator transposable elements in maize. Mol Biol Evol 5:519–529

Tempel S, Giraud M, Lavenier D, Lerman IC, Valin AS, Couée I, Amrani AE, Nicolas J (2006) Domain organization within repeated DNA sequences: application to the study of a family of transposable elements. Bioinformatics 22:1948–1954

Tempel S, Nicolas J, El Amrani A, Couee I (2007) Model-based identification of Helitrons results in a new classification of their families in *Arabidopsis thaliana*. Gene 403:18–28

The Potato Genome Sequencing Consortium (2011) Genome sequence and analysis of the tuber crop potato. Nature 475:189–195

Tsukamoto T, Hauck NR, Tao R, Jiang N, Iezzoni AF (2010) Molecular and genetic analyses of four nonfunctional S haplotype variants derived from a common ancestral S haplotype identified in sour cherry (*Prunus cerasus* L.). Genetics 184:411–427

Turcotte K, Srinivasan S, Bureau T (2001) Survey of transposable elements from rice genomic sequences. Plant J 25:169–179

Tuskan GA, Difazio S, Jansson S, Bohlmann J, Grigoriev I, Hellsten U, Putnam N, Ralph S, Rombauts S, Salamov A, Schein J, Sterck L, Aerts A, Bhalerao RR, Bhalerao RP, Blaudez D, Boerjan W, Brun A, Brunner A, Busov V, Campbell M, Carlson J, Chalot M, Chapman J, Chen GL, Cooper D, Coutinho PM, Couturier J, Covert S, Cronk Q, Cunningham R, Davis J, Degroeve S, Déjardin A, Depamphilis C, Detter J, Dirks B, Dubchak I, Duplessis S, Ehlting J, Ellis B, Gendler K, Goodstein D, Gribskov M, Grimwood J, Groover A, Gunter L, Hamberger B, Heinze B, Helariutta Y, Henrissat B, Holligan D, Holt R, Huang W, Islam-Faridi N, Jones S, Jones-Rhoades M, Jorgensen R, Joshi C, Kangasjärvi J, Karlsson J, Kelleher C, Kirkpatrick R, Kirst M, Kohler A, Kalluri U, Larimer F, Leebens-Mack J, Leplé JC, Locascio P, Lou Y, Lucas S, Martin F, Montanini B, Napoli C, Nelson DR, Nelson C, Nieminen K, Nilsson O, Pereda V, Peter G, Philippe R, Pilate G, Poliakov A, Razumovskaya J, Richardson P, Rinaldi C, Ritland K, Rouzé P, Ryaboy D, Schmutz J, Schrader J, Segerman B, Shin H, Siddiqui A, Sterky F, Terry A, Tsai CJ, Uberbacher E, Unneberg P, Vahala J, Wall K, Wessler S, Yang G, Yin T, Douglas C, Marra M, Sandberg G, Van de Peer Y, Rokhsar D (2006) The genome of black cottonwood, *Populus trichocarpa* (Torr. & Gray). Science 313:1596–1604

Walbot V, Rudenko GN (2002) *MuDR/Mu* transposable elements of maize. In: Craig NL, Craigie R, Gellert M, Lambowitz AM (eds) Mobile DNA II. ASM Press, Washington, D.C., pp 533–564

Wang Q, Dooner HK (2006) Remarkable variation in maize genome structure inferred from haplotype diversity at the *bz* locus. Proc Natl Acad Sci USA 103:17644–17649

Wang X, Wang H, Wang J, Sun R, Wu J, Liu S, Bai Y, Mun JH, Bancroft I, Cheng F, Huang S, Li X, Hua W, Wang J, Wang X, Freeling M, Pires JC, Paterson AH, Chalhoub B, Wang B, Hayward A, Sharpe AG, Park BS, Weisshaar B, Liu B, Li B, Liu B, Tong C, Song C, Duran C, Peng C, Geng C, Koh C, Lin C, Edwards D, Mu D, Shen D, Soumpourou E, Li F, Fraser F, Conant G, Lassalle G, King GJ, Bonnema G, Tang H, Wang H, Belcram H, Zhou H, Hirakawa H, Abe H, Guo H, Wang H, Jin H, Parkin IA, Batley J, Kim JS, Just J, Li J, Xu J, Deng J, Kim JA, Li J, Yu J, Meng J, Wang J, Min J, Poulain J, Wang J, Hatakeyama K, Wu K, Wang L, Fang L, Trick M, Links MG, Zhao M, Jin M, Ramchiary N, Drou N, Berkman PJ, Cai Q, Huang Q, Li R, Tabata S, Cheng S, Zhang S, Zhang S, Huang S, Sato S, Sun S, Kwon SJ, Choi SR, Lee TH, Fan W, Zhao X, Tan X, Xu X, Wang Y, Qiu Y, Yin Y, Li Y, Du Y, Liao Y, Lim Y, Narusaka Y, Wang Y, Wang Z, Li Z, Wang Z, Xiong Z, Zhang Z (2011) The genome of the mesopolyploid crop species *Brassica rapa*. Nat Genet 43:1035–1039

Wicker T, Narechania A, Sabot F, Stein J, Vu GT, Graner A, Ware D, Stein N (2007) A unified classification system for eukaryotic transposable elements. Nat Rev Genet 8:973–982

Wicker T, Narechania A, Sabot F, Stein J, Vu GT et al (2008) Low-pass shotgun sequencing of the barley genome facilitates rapid identification of genes, conserved non-coding sequences and novel repeats. BMC Genomics 9:518

Wicker T, Buchmann JP, Keller B (2010) Patching gaps in plant genomes results in gene movement and erosion of colinearity. Genome Res 20:1229–1237

Wolff P, Weinhofer I, Seguin J, Roszak P, Beisel C, Donoghue MT, Spillane C, Nordborg M, Rehmsmeier M, Köhler C (2011) High-resolution analysis of parent-of-origin allelic expression in the Arabidopsis endosperm. PLoS Genet 7:e1002126

Xu JH, Messing J (2006) Maize haplotype with a helitron-amplified cytidine deaminase gene copy. BMC Genet 7:52

Yang L, Bennetzen JL (2009a) Distribution, diversity, evolution, and survival of Helitrons in the maize genome. Proc Natl Acad Sci USA 106:19922–19927

Yang L, Bennetzen JL (2009b) Structure-based discovery and description of plant and animal *Helitrons*. Proc Natl Acad Sci USA 106:12832–12837

Chapter 12
Transposable Element Exaptation in Plants

Douglas R. Hoen and Thomas E. Bureau

Abstract While evolution is often understood exclusively in terms of adaptation, innovation often begins when a feature adapted for one function is co-opted for a different purpose, such as when feathers originally adapted for insulation became used for flight. Co-opted features are called exaptations. Transposable elements are often viewed as molecular parasites, yet they are frequently the source of evolutionary innovation. One way in which transposable elements contribute to evolution is that their sequences can be co-opted to perform phenotypically beneficial functions. Transposable element gene exaptations have contributed to major innovations such as the vertebrate adaptive immune system and the mammalian placenta. They also often become transcription factors, and transposable element-derived transcription factor binding sites can form new regulatory networks. In this chapter, we review transposable element coding sequence exaptations in plants.

Keywords Transposable elements • Mobile elements • Transposons • Transposases • Retrotransposons • Molecular domestication • Exaptation • Evolutionary innovation • Gene expression • Transcription factors • Regulatory networks

Abbreviations

CCA1	Circadian clock-associated 1
CENP-B	Centromere-associated protein B
dsRNA	Double-stranded RNA
DTE	Domesticated transposable element

D.R. Hoen (✉) • T.E. Bureau
Department of Biology, McGill University, 1205 du Docteur-Penfield Avenue, Montreal, QC, Canada H3A 1B1
e-mail: douglas.hoen@mcgill.ca; thomas.bureau@mcgill.ca

M.-A. Grandbastien and J.M. Casacuberta (eds.), *Plant Transposable Elements*, Topics in Current Genetics 24, DOI 10.1007/978-3-642-31842-9_12,

ELF4	Early flowering 4
En/Spm	*Enhancer/Suppressor mutator*
FBS	FHY3/FAR1 binding site
FHL	FHY1-like
FHY1	Far-red elongated hypocotyl 1
FHY3	Far-red elongated hypocotyl 3
FRS	FAR1-related sequence
hAT	*hobo/Ac/Tam*
HY5	Long hypocotyl 5
miRNA	microRNA
MIR	microRNA gene
MITE	Miniature inverted-repeat transposable element
MUG	MUSTANG
Muk	Mu killer locus
MULE	*Mutator*-like element
PB1	Phox and Bem1
PHY	Phytochrome
PIL5	Phytochrome interacting factor 3-like 5
sRNA	small RNA
TE	Transposable element
TF	Transcription factor

12.1 Introduction

When Barbara McClintock first discovered transposable elements (TEs), she immediately recognized their potential importance in gene regulation and genome evolution (McClintock 1950). But the discovery that TEs replicate within the genome led to their characterization as selfish (Doolittle and Sapienza 1980) parasitic (Orgel and Crick 1980; Hickey 1982) junk (Ohno 1972). We have since learned that TEs are diverse, abundant, and ubiquitous (Feschotte and Pritham 2007; Wicker et al. 2007; Pritham 2009; Aziz et al. 2010; Levin and Moran 2011) and that they contribute to numerous aspects of eukaryotic genome structure, function, and evolution (Table 12.1). Despite this, it remains common to view TEs as molecular parasites (Kidwell and Lisch 2001; Rose and Oakley 2007; Malone and Hannon 2009; Obbard et al. 2009). This stems from an explanatory framework in which evolution is understood exclusively through adaptationist arguments (Gould and Lewontin 1979). Since TEs may persist in the short term without contributing beneficial phenotypes, the argument goes, that is the "reason" they exist, and any beneficial effects that do happen to come about are therefore incidental (Werren 2011). An alternative view is that self-replication may not fully account for TE ubiquity and abundance. If TEs are able to produce mutations that are important to adaptation, then this ability itself may increase the long-term evolutionary success of TEs. That is, not only may the specific mutations produced by TEs be selected, but at least in certain situations, such as during rapid environmental change, mutability itself may be selected (McClintock 1984; King and Kashi 2007; Le Rouzic et al. 2007; Zeh et al. 2009; Biémont 2010).

Table 12.1 Contribution of TEs to eukaryotic genome structure, function, and evolution

Type of contribution	Examples	References
Genome structure		Parisod et al. (2010)
	Genome size	Agren and Wright (2011)
	Chromatin organization	Lunyak and Atallah (2011)
Genome maintenance	DNA repair	Cordaux and Batzer (2009)
	Centromere maintenance	Lisch (2009)
	Telomere maintenance	Pardue and DeBaryshe (2011)
Generation of variation		Kidwell and Lisch (2001)
	Chromosomal rearrangement	Feschotte and Pritham (2007)
	Copy number variation	Conrad et al. (2010)
	Structural variation	Xing et al. (2009)
	Somatic variation	Levin and Moran (2011)
	Allelic recombination	Gaut et al. (2007)
	Ectopic recombination	Ponting et al. (2011)
Evolutionary innovation	Gene duplication	Flagel and Wendel (2009)
	Novel regulatory networks	Feschotte (2008)
	Epigenetic regulation	Weil and Martienssen (2008), Lisch (2009)
	Origin of sex	Hickey (1982), Rose and Oakley (2007)
	Origin of dedicated germ line	Johnson (2008)
	Response to stress	McClintock (1984), Lisch (2009)
	Speciation	Rebollo et al. (2010)

A narrow focus exclusively on immediate adaptation fails to satisfactorily explain the origin of many important traits. Darwin (1876) himself emphasized that gradual adaptation alone does not account for the incipient stages of many useful structures, but rather that changes in function, such as swim bladders becoming lungs when fish colonized land, are often at the root of innovation. Gould and Vrba (1982) proposed that adaptive features originally built by natural selection for one role, or even nonadaptive features, that have since been co-opted for a new role, be called *exaptations*. Instead of parasites, TEs might better be viewed as exaptation engines. Broadly speaking, many of the aforementioned contributions of TEs to organismal evolution may be considered exaptations, albeit nonspecific or indirect. Indeed, any feature evolved at one level of selection (e.g., TE self-replicative selection; see below) that produces a beneficial effect at another level of selection (e.g., organismal phenotypic selection) is an exaptation (Gould and Lloyd 1999). More narrowly, the genetic constituents (i.e., sequences) of TEs, such as genes, binding sites, and terminal repeats, can be directly exapted for specific phenotypic functions in the organism.

These specific, direct exaptations of TE sequences, which we refer to simply as TE exaptations, and especially protein-coding TE exaptations in plants, are the topic of this chapter. We begin by describing the first exapted TE genes to be discovered, the *Drosophila P* neogenes, to provide not only an historical perspective but also more importantly an illustration of the evolutionary forces underlying TE exaptation. We explore these forces by developing a model of the process. We then turn to plants, describing in detail the first known and best characterized exapted plant TEs, the *FHY3* family. We then review the other confirmed exapted plant TE families, briefly examine other types of TE exaptation, such as chimerization, exonization, and noncoding exaptation, and lastly discuss the significance of TE exaptation for regulatory evolution. For additional perspectives, we highly recommend, in addition to others cited throughout the chapter, the following excellent reviews: Volff (2006), Dooner and Weil (2007), Feschotte and Pritham (2007), Feschotte (2008), Sinzelle et al. (2009), and Hua-Van et al. (2011).

12.2 Discovery of Molecular Domestication

Drosophila P elements were among the first TEs to be extensively characterized (Biémont 2010) and produced the first TE exaptations to be discovered. Like other *Drosophila* TEs (Sánchez-Gracia et al. 2005), *P* elements frequently move between species (Daniels et al. 1990), an evolutionary strategy called horizontal transfer that is common among DNA transposons (Diao et al. 2005; Schaack et al. 2010). *P* elements recently horizontally transferred from *D. willistoni* to *D. melanogaster* (Daniels et al. 1990; Pinsker et al. 2001). Crosses between *D. melanogaster* males carrying autonomous *P* elements and females lacking *P* elements lead to TE activation and decreased fitness. The resultant syndrome, hybrid dysgenesis, sexually isolates strains carrying *P* elements from naive strains lacking *P* elements, a genetic barrier that may contribute to speciation (Kidwell et al. 1977).

Drosophila P element families consist of a few autonomous elements and many nonautonomous elements. Autonomous *P* elements produce a complete 88-kDa protein that catalyzes transposition in the germ line. In somatic cells, differential splicing produces a truncated 66-kDa isoform that may repress transposition (Rio 1990). These truncated "*P* repressors" retain the N-terminal DNA-binding domain but lack the C-terminal domain required for TE excision and may induce repression by binding to a specific subterminal *P* element motif, blocking the promoter and preventing transcription and normal *P* transposase activity (Kaufman et al. 1989). Some nonautonomous *P* elements encode similar truncated products that may act as repressors (Miller et al. 1997). However, although *P* repressor proteins were initially proposed to explain *P* element somatic silencing, it has more recently been shown that silencing is predominantly mediated by small interfering RNAs (Brennecke et al. 2007; Khurana et al. 2011).

Unlike *D. melanogaster*, which has dozens of autonomous *P* elements (Daniels et al. 1990), species in the *D. obscura* group have only a few *P*-like genes. Although these genes are similar to truncated repressors, they are not flanked by *P* terminal sequences, are not mobile, and are arranged in a complex locus that is orthologous in the genomes of *D. guanche*, *D. subobscura*, and *D. madeirensis*. These are not decayed TE fossils, but instead have three undisrupted exons and conserved intronic splice signals, are highly similar between species, and are transcribed (Paricio et al. 1991; Miller et al. 1992). Thus, these *P* homologs are neither functional transposases, nor TE fossils, but are neogenes that must have been immobilized in a common ancestor of the three *obscura* species, exapted to perform a non-transposition function, and thereafter conserved by phenotypic selection (Miller et al. 1999). Miller et al. (1992) termed this process of TE gene exaptation *molecular domestication*. We refer to exapted TE genes as domesticated transposable elements (DTEs).

P elements were also domesticated independently in the *D. montium* subgroup. A single-copy DTE lacking *P* termini, but again with three uninterrupted exons similar to the *obscura* DTEs, is present at orthologous locations in at least nine *montium* species. Both the *obscura* and *montium* DTEs have, subsequent to their domestication, undergone lineage-specific gene rearrangements, including gene duplication, secondary *P* element insertion, and exon shuffling. Additional TE insertions provided *cis*-regulatory elements. In total, there were four independent *P* element immobilization events, and four different products are encoded by the *obscura* and *montium* DTEs (Nouaud and Anxolabehere 1997; Quesneville et al. 2005).

Despite their similarity to *P* repressors, the DTEs likely have functions other than repression (Miller et al. 1992). In the *obscura* group, there are no autonomous *P* elements, so the DTEs cannot act as repressors. While the *montium* group does contain at least one active *P* element subfamily, that family is highly diverged from the DTE, suggesting that repression may not occur. The *montium* DTE is incapable of repressing transcription or the transposition of canonical *P* elements. Instead, it is expressed in the brains and gonads of transgenic fly larvae and adults and binds to multiple sites on the chromosome that are not similar to extant *P* elements, suggesting that domesticated *P* transposases likely serve some function other than *P* repression yet involving DNA binding (Reiss et al. 2005).

P elements were originally thought to be found only in flies (*Diptera*) but have now also been identified in zebrafish (Hammer 2005). DNA-binding domains homologous to those of *P* elements in flies and zebrafish, i.e., THAP domains (Roussigne et al. 2003; Sabogal et al. 2010), have been identified in approximately 100 genes, many of them transcription factors, with various functions including apoptosis, angiogenesis, cell cycle regulation, neurological function, stem cell pluripotency, and epigenetic gene silencing (Clouaire et al. 2005). Thus, *P* elements may have been domesticated multiple times and played an important role in the evolution of animals (Quesneville et al. 2005).

12.3 Frequent Birth Model

How does molecular domestication work? The answer is not as straightforward as it may seem. Consider the two different levels of selection (Gould and Lloyd 1999) acting on TEs and ordinary genes. Ordinary genes (*a.k.a.* host genes or cellular genes) do not self-replicate and are maintained by what Doolittle and Sapienza (1980) termed *phenotypic selection*; that is, by selection on the beneficial phenotypes that they produce at the organism level. TEs, on the other hand, are maintained by what we term *self-replicative selection* (and which Doolittle and Sapienza termed *nonphenotypic selection*); that is, by selection on their replication at the genome level in germ cells. In addition, TEs are subject to phenotypic selection, because TE-induced mutations may cause phenotypes that impair or occasionally benefit the organism, thereby impairing or benefitting their own proliferation. Similarly, TEs produce molecules such as proteins or small RNAs that may affect cell function and thus their own proliferation, either deleteriously or beneficially.

One type of interaction between the phenotypic and self-replicative levels of selection is genetic (or intragenomic) conflict, which can result in the evolution of self-regulatory mechanisms to repress transposition (Werren 2011). For instance, some evidence suggests that alternative splicing of *P* transposase genes can produce different isoforms that in germ cells catalyze transposition and in somatic cells may repress it (Rio 1990). Similarly, maize *Spm* elements have a single gene, which encodes several isoforms, one of which, TnpA, may act as a repressor of unmethylated *Spm* promoters (Fedoroff 1999). Domesticated retroelement genes, such as the mouse *Fv1* gene, may also serve as repressors (Volff 2006). Furthermore, DTEs derived from one type of TE may regulate other types of TE. For instance, CENP-B homologs in fission yeast are derived from DNA transposons, but, in addition to roles in centromere function, they repress retrotransposition (Cam et al. 2008). On a noncoding level, some TEs contain tissue-specific regulatory elements, such as pollen-specific enhancers, that limit somatic mutation without restricting TE proliferation, a form of auto-repression (Raizada et al. 2001; Lisch and Jiang 2009). Finally, other types of noncoding repression also occur, such as the *Mu killer* locus in maize (Lisch and Jiang 2009).

Like TE repression, molecular domestication results from an interplay between phenotypic and self-replicative levels of selection. Molecular domestication is a process in which a TE gene, maintained by self-replicative selection, becomes an ordinary gene, maintained by phenotypic selection. The balance of these competing levels of selection may tip to either side: towards self-replicative selection to resulting in bursts of transposition, or towards phenotypic selection to result in molecular domestication. Various models of TE copy number dynamics and population genetics have been proposed, some of which take into account molecular domestication (Brookfield 1982; Hickey 1982; Charlesworth et al. 1994; Le Rouzic and Deceliere 2005; Le Rouzic et al. 2007). In this section, we develop a conceptual model, which we name the *Frequent Birth* model, with the aim of better understanding the evolutionary forces underlying the process of molecular domestication. We suggest that, rather than being a peculiar side effect of TE activity, molecular domestication is a natural consequence of the interplay of phenotypic and self-replicative selection and may occur far more frequently than is currently known, especially in the short term.

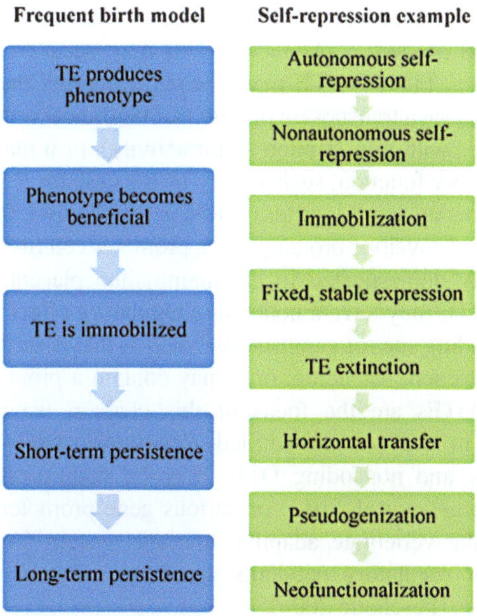

Fig. 12.1 *Frequent Birth* model of molecular domestication and an example based on self-repression of transposition as the initial (short-term) beneficial phenotype, illustrating the evolutionary forces underlying the process. First, the TE produces a phenotype; for example, in somatic tissue it may encode an auto-repressive isoform that reduces mutability, and may subsequently lose its ability to encode the transpositionally competent isoform, becoming nonautonomous. Second, the phenotype becomes beneficial; for example, somatic auto-repression is inherently beneficial, whereas other phenotypes may be beneficial only under certain environmental conditions. Third, the TE loses its ability to mobilize, which eliminates self-replicative selection, leaving phenotypic selection to act on it alone and placing it in a stable genomic environment. If it provides sufficient phenotypic benefit, the TE becomes a nascent DTE, fixed in the population, and can persist in the short term. We propose that short-term persistence may occur relatively frequently, because nascent DTEs may provide only marginal phenotypic benefit and thus be frequently lost due to genetic drift, or because their benefit may be transient (or both). In our auto-repression example, the nascent DTE may lead to the extinction of the cognate TE family in the population, which would remove purifying selection on the DTE. While horizontal transfer of a similar TE family may restore selection, such an occurrence may be unlikely and would in any case provide only a temporary resolution as the new TE family may also go extinct. Finally, to achieve long-term persistence, the phenotypic selection acting on a DTE may need to be increased in strength or stabilized. In our auto-repression example, relaxed selection and pseudogenization following TE extinction may permit a mutation to enable an entirely new function. Alternatively, increased adaptive pressure may result from environmental changes requiring the organism to adapt more quickly, which may increase the benefit of, for example, nascent networks of regulatory DTEs

The first step in molecular domestication is for a TE sequence to produce a selectable phenotype (Fig. 12.1). TEs can produce phenotypes through a variety of mechanisms, not all of which are capable of domestication. For instance, a phenotype produced when a TE inserts into and knocks out a gene is not a potential source of molecular domestication, because it is the disruption of the open reading frame that is the cause, not the TE sequence itself. Instead, to become domesticated, the TE sequence itself must produce a phenotype. This may occur in two different ways: coding or noncoding. First, the TE may encode a product that acts in trans. For example, it may produce a transposase protein that binds near ordinary genes, altering their expression (i.e., with transcription factor activity); or it may produce a product with an auto-regulatory function, such as may be the case for *P* repressor proteins or for TE loci that generate small interfering RNAs (see below), or it may produce an endogenous retroviral envelope protein, which promotes cell fusion, a function which has been repeatedly domesticated in the mammalian placenta (Rawn and Cross 2008). Second, the TE may have a noncoding regulatory effect *in cis*. For example, it may contain a binding site for a transposase such as that described above; or it may be the target of epigenetic silencing; or it may contain a promoter or an enhancer. Although coding DTEs are the focus of this chapter, we also briefly discuss noncoding DTEs in the section entitled *Regulatory Exaptation: Coding and Noncoding*. Coding and noncoding DTEs may be domesticated separately, e.g., syncytin genes (Cornelis et al. 2012) or various gene promoters (Lee et al. 2008), or together, e.g., the vertebrate adaptive immune system (Agrawal et al. 1998; Fugmann 2010) or regulatory networks consisting of transcription factors and cognate binding sites (Feschotte 2008).

The second step in our model, the phenotype produced by a TE must become beneficial to the organism. This may occur in a variety of ways. For instance, auto-regulation is (usually) inherently beneficial to the organism, because it limits somatic mutation. Other phenotypes may be beneficial only rarely. For instance, a transposase that produces a regulatory effect on ordinary genes by binding to interspersed cognate TEs to form a nascent regulatory network may be beneficial only for a small fraction of binding site distributions, a fraction which may be relatively small under stable environmental conditions with stabilizing selection, but which may increase under changing or variable environmental conditions with disruptive or directional selection.

Third, the TE is immobilized. TEs frequently sustain mutations that prevent transposition, such as truncations of one or both termini. Immobilization may contribute to molecular domestication in several ways. It may elevate and stabilize TE expression, for instance by integrating the TE gene with existing *cis*-regulatory elements in the surrounding genome. It may also restrict epigenetic and posttranscriptional silencing, which is mediated by double-stranded RNAs typically generated for DNA transposons from intact terminal inverted repeats, and thus may be abrogated if the TE is truncated. Most importantly, immobilization eliminates self-replicative selection and along with it selective constraints that may limit the phenotypic benefit of the sequence. For example, a mobile TE encoding a strong meiotic auto-repressor would be highly unlikely to evolve, because it would restrict its own ability to persist; however, an immobilized meiotic auto-repressor might increase in strength if conserved exclusively by phenotypic selection, provided sufficient adaptive benefit.

We note that immobilization may occur before the TE begins to produce a beneficial phenotype (i.e., before step 2), rather than after. Furthermore, it is sometimes asked whether molecular domestication even requires TE immobilization, or might a gene encoded by an active TE nevertheless perform a beneficial function for the organism that contributes to the conservation of both the TE and the organism (Hoen et al. 2006; Smith et al. 2011)? The answer seems to hinge largely on the strength of phenotypic vs. self-selective selection on active TEs. Consider germ cell-specific expression, e.g., due to somatic self-repression in *P* elements or tissue-specific enhancers. Germ cell-specific repression may be due to direct phenotypic selection against genomes with TEs that are active in soma, or to self-replicative selection to counteract cellular TE silencing mechanisms, or both. If phenotypic selection does play a major role in producing germ cell-specific transposition, it may be plausible that it could also maintain other cellular functions, unrelated to transposition, within active TEs, and provided they do not interfere too strongly with self-replication. Such a function would need to be simultaneously selected at both the self-replicative and phenotypic levels of selection; otherwise, it could simply become domesticated. While no such cases have yet been reported, it is not clear whether this is because it is impossible, improbable, difficult to detect, or simply unexamined.

Fourth, an immobilized TE may become conserved, at least in the short term, if it provides a phenotypic benefit to the organism. Genomes contain many immobilized TEs, most of which are TE "fossils." Like ordinary genes, immobilized TE-derived sequences are not subject to self-replicative selection, so to persist they must be selected phenotypically. The phenotypic effects of immobilized TEs range from deleterious to beneficial, but most have only weak effects, since those with strong deleterious effect are rapidly eliminated from the population by negative phenotypic selection, and those with strong beneficial effects may originate only rarely. However, nascent DTEs with even relatively weak or transient benefits may persist in a population for at least a short time, depending on the strength of selection and population size, before succumbing to elimination by random drift (Le Rouzic et al. 2007). We hypothesize that DTEs may be born frequently, and there may therefore be a far higher number of weak-effect DTEs with short lifespans than there are long-lived DTEs. Short-lived, weak-effect DTEs would be extremely difficult to detect, since even long-lived DTEs are difficult to discriminate from TEs, as suggested by the dearth of successful systematic searches for them, and as key lines of evidence used to discriminate long-lived DTEs from TEs, such as sequence conservation and inter-species co-linearity, would not be useful in detecting short-lived TEs. Nonetheless, if our hypothesis is correct, weak-effect DTEs may supply a preexisting pool of adaptive potential that may play an important role in evolution, enabling rapid adaptation under certain conditions, such as accelerated environmental change.

Finally, to persist in the long term, a nascent DTE must produce a sufficiently stable and strong benefit. It seems likely that in most cases this would require either a series of mutations to the nascent DTE, to increase its benefit, or specific environmental conditions favoring the innovation of new functions over than the maintenance of stable phenotypes, such as environmental change and directional selection (or both). Some DTEs may even need to adopt entirely new functions by undergoing a second exaptation to persist long term, such as would be the case for *P* neogenes if they were originally domesticated as repressors (Reiss et al. 2005), since such a function would be inherently transient (Fig. 12.1). Thus, only a small fraction of nascent DTEs are likely to reach the stage of long-term persistence.

12.4 Domesticated Transposable Elements in Plants

12.4.1 The FHY3 *Family*

Plants process sensory inputs by modulating gene expression networks affecting development and growth. The perception of light is of particular importance. To determine their position, the time of day, and the season, plants measure the intensity, direction, duration, periodicity, and spectrum of incident light using light-sensitive proteins called photoreceptors (Jiao et al. 2007). Of three major classes of photoreceptors in plants, the phytochromes are the best characterized, consisting in *Arabidopsis thaliana* of a monophyletic family of five genes, *PHYA–PHYE* (*phytochrome A–E*) that has undergone diversification and subfunctionalization throughout angiosperm evolution (Mathews 2006). Phytochromes predominantly absorb red and far-red light to effect various responses including germination, development, dormancy, shade avoidance, and flowering. *PHYA* is highly conserved among angiosperms and some *PHYA* mutations are conditionally lethal. *PHYA* is the primary photoreceptor for both very low fluence irradiance and far-red high irradiance responses and triggers development in a wide range of light conditions (Mathews 2006). In darkness, phyA protein is concentrated in the cytosol. Absorption of red light by phyA triggers structural changes, exposing interaction surfaces to which FHY1 (far-red elongated hypocotyl 1) and FHL (FHY1-LIKE) bind. FHY1/FHL possess nuclear localization signals, causing PHYA, which does not have a nuclear localization signal, to be translocated into the nucleus, where it mediates light responses by triggering transcriptional cascades (Bae and Choi 2008; Chen and Chory 2011) (Table 12.2).

Screens for *A. thaliana* mutants that undergo normal de-etoliation under white light, but have impaired inhibition of hypocotyl growth under far-red light, identified *fhy3* (*far-red elongated hypocotyl 3*) (Whitelam et al. 1993), which specifically disrupts only the high irradiance phyA response (Yanovsky et al. 2000) and *far1* (*far-red-impaired response 1*) (Hudson et al. 1999). The *FAR1* gene sequence is similar to *Mutator* transposase genes, yet lacks TE termini, making it the first DTE to be recognized in plants (Lisch et al. 2001). *FAR1* and *FHY3* are paralogs that probably arose early in eudicot evolution. Each is located at orthologous chromosomal positions in *A. thaliana*, *B. rapa*, and *P. trichocarpa* (Lin et al. 2007). Although both genes affect hypocotyl elongation, *fhy3* has a more pleiotropic phenotype, for example, also affecting cotyledon opening. Double mutants have longer hypocotyls and greater reductions in cotyledon expansion than single mutants, suggesting that FHY3 and FAR1 act in an additive manner. Overexpression of the

Table 12.2 Known domesticated transposable element genes in plants

DTE[a]	Family	TE	Species (taxa)[b]	Method of discovery	Function
FAR1	*FHY3*	MULE	*A. thaliana* (Dicots)	Phenotype screen (impaired de-etoliation under far-red light)	TF[c]; phyA signaling; may regulate hundreds of genes
FHY3 *FRS1-12*			(Angiosperms)	Similarity to FHY3/FAR1	Unknown (probable TFs)
DAYSLEEPER	–	*hAT*	*A. thaliana* (Eudicots)	Yeast one-hybrid screen (Kubox1 motif)	Development (essential)
MUG1-8	*MUG*	MULE	*A. thaliana* (Angiosperms)	*In silico* search	Plant & flower development; pleiotropic
GARY	(~2 paralogs)	*hAT*	Barley, wheat (Poaceae)	*In silico* search	Unknown

[a]In species where discovered
[b]Species where discovered (taxonomic distribution)
[c]Transcription factor (TF)

FHY3 C-terminal fragment, which contains a SWIM zinc finger domain, completely blocks *phyA* signaling. The FHY3 protein can substitute for FAR1, as can the promoters, to completely suppress the *far1* mutant phenotype; conversely, *FAR1* can only partially restore *fhy3* (Wang and Deng 2002; Lin et al. 2008b).

Thus, *FHY3* and *FAR1*, after duplicating early in eudicot diversification, have undergone subfunctionalization to maintain partially overlapping functions, with FHY3 acting more widely. FHY3 and FAR1 directly activate *FHY1* and *FHL* expression, which accumulate in dark and dissipate in light, and are required for the high irradiance phyA response. FHY3 and FAR1 activate expression by homo- and heterodimerizing and binding to a specific *cis*-regulatory sequence, called the FHY3/FAR1-binding site (FBS), found in the promoters of both *FHY1* and *FHL*. An N-terminal C2H2 zinc finger domain mediates DNA binding. Two other domains, the central MULE domain and N-terminal SWIM domain, are required for dimerization and to activate expression (Lin et al. 2007, 2008b).

In addition to de-etoliation, FHY3 and FAR1 have recently been found to play additional roles. During dark to red-light transitions, phyA signaling is rapidly desensitized by multiple transcriptional and posttranslational feedback mechanisms, three of which directly involve FHY3/FAR1. First, under far-red light, FHY3 binds to underphosphorylated phyA, protecting against COP1/SPA-mediated proteolysis (Saijo et al. 2008). Second, *FHY3*/*FAR1* expression is repressed by phyA signaling in light, which in turn reduces *FHY1*/*FHL* expression (Lin et al. 2007). Third, phyA signaling in light activates the transcription of another photomorphogenic transcription factor, HY5 (long hypocotyl 5), which blocks FHY3/FAR1 activity by binding adjacent to the *FHY1*/*FHL* FBS, sterically hindering FHY3/FAR1 binding. HY5 also interacts directly with the DNA-binding domain of FHY3/FAR1, but it is not clear whether this interaction is important in repressing FHY1/FHL expression, or if it plays a role in other FHY3/FAR1 and HY5 co-regulatory activities (Li et al. 2010).

FHY3 also plays multiple roles in circadian clock regulation. The FBS is present in the promoters of more than 200 genes that exhibit diurnal or circadian cycling, including *PHYB*, *CCA1* (*circadian clock-associated 1*), and *ELF4* (*EARLY FLOWERING 4*) (Lin et al. 2007). FHY3 specifically gates red light signaling for circadian clock resetting, playing a crucial role in the maintenance of clock rhythmicity, especially at dawn (Allen et al. 2006). By binding to the *ELF4* promoter and directly interacting with at least three additional transcription factors, including HY5, FHY3 regulates both the rhythmicity and amplitude of the circadian clock central oscillator (Li et al. 2011a). Finally, FHY3 also stimulates chloroplast division (Ouyang et al. 2011).

More functions of FHY3/FAR1 likely remain to be uncovered. Microarray studies revealed that in *fhy3* mutants especially, but also *far1*, the majority of known light-regulated genes, including transcription factor genes and genes involved in cell elongation, have reduced responsiveness to continuous far-red light (Hudson et al. 2003). Furthermore, the FBS is found in hundreds of promoters (Lin et al. 2007). A recent chromatin immunoprecipitation-based sequencing (ChIP-seq) study showed that FHY3 binds to thousands of binding sites associated with over 1,700 genes (within 1,000 base pairs upstream to the 3′ untranslated region) (Ouyang et al. 2011). Nearly 800 genes are bound only in darkness, while over 200 are bound only in light, and nearly 800 in both conditions. The majority of genic-binding sites are in promoters, with density peaking at the transcription start site. About half of the genic-binding sites are FBS, but it remains unknown where exactly FHY3 binds at the remaining sites. Several other types of transcription factor binding site are significantly enriched near the FBS, including 283 genes that have HY5-binding sites in close proximity, and 136 genes that are co-regulated with PIL5 (PHYTOCHROME INTERACTING FACTOR 3-LIKE5) (Ouyang et al. 2011).

In addition, about 40% of FHY3-binding sites are intergenic. Unexpectedly, the majority of intergenic-binding sites are located in a centromeric motif, a pattern unlike any other known plant transcription factor (Ouyang et al. 2011). The functional significance of centromeric binding is not yet known, but we note that it is reminiscent of mammalian CENP-B (CENTROMERE-ASSOCIATED PROTEIN B), a DTE and component of the centromere/kinetochore, which binds to centromeric satellites and regulates centromere formation (Casola et al. 2008; Zaratiegui et al. 2011).

Microarray analyses showed that about one-eighth of the genes with FHY3-binding sites are differentially regulated in dark (197 genes) or far-red light (86 genes) conditions, which is roughly half of all genes differentially regulated in these conditions. Most of these "directly regulated" genes are activated rather than repressed by FHY3, especially in light, where all but a single gene is activated. In dark, FHY3 directly represses the expression of 43 genes, 42 of which are released from repression on light exposure. Among the directly regulated genes, several functional categories are highly enriched, including transcriptional regulation, signal transduction, intracellular signaling, environmental response, hormone response, and development (Ouyang et al. 2011). These results suggest that FHY3 may have roles in diverse regulatory networks, or networks of other types, as yet mostly uncharacterized.

Furthermore, *FHY3* and *FAR1* are but two close paralogs in a gene family that includes 12 additional *FRSs* (*FAR1-related sequences*) in *A. thaliana* (Hudson et al. 2003; Lin and Wang 2004). The *FHY3*/*FAR1*/*FRS* family (hereafter, the *FHY3* family) consists of five widely diverged phylogenetic clades, each containing at least two *A. thaliana* genes (Lin et al. 2007), and each except one with members in both eudicots and monocots. The sole eudicot-specific clade contains both *FHY3* and *FAR1*. Phylogenetic analysis of the *FHY3*/*FAR1* clade suggests that *FHY3* and *FAR1* arose by duplication early in eudicot evolution, well before the divergence of the asterids. Furthermore, *FAR1* and *FHY3* loci are arranged in tandem in the *Populus* genome, possibly the ancestral configuration, suggesting they arose by tandem duplication. Other duplications are also evident. In the *FRS3* clade, *FRS5* and *FRS9* are arranged in tandem in *A. thaliana* suggesting another tandem duplication. In the *FRS7* clade, the *FRS7* and *FRS12* sequences are very similar, suggesting a recent duplication.

All 12 *A. thaliana FRSs* are ubiquitously expressed in all major organs, except *FRS10*, which appears to have a highly unstable transcript. FRS proteins all localize to the nucleus, consistent with transcription factor activity, although FRS1 maintains residual cytosolic distribution. So far, only two clades in addition to the *FHY3*/*FAR1* clade have been characterized. Unlike *fhy3*/*far1*, *frs6* and *frs8* (*FRS6* clade) do not affect hypocotyl elongation, but mutant plants flower early, especially under short-day conditions, suggesting they are positive regulators of phyB-mediated inhibition of floral initiation (Lin and Wang 2004). Conversely, *fhy3*/*far1* mutants also flower early but have greater effect under long-day conditions. *FRS9* (*FRS3* clade) RNAi knockdowns exhibit a hypersensitive response specifically to continuous red light, suggesting FRS9 is a negative regulator of phyB-mediated de-etoliation (Lin and Wang 2004). Thus, in addition to the well-characterized functions of FHY3/FAR1, the other FHY3 family members are likely to play even more diverse roles, at least some of which may involve phytochrome-mediated light responses.

The phylogenetic pattern, chromosomal arrangement, and functions of the *FHY3* gene family show that it underwent successive duplication and subfunctionalization throughout angiosperm evolution (Lin et al. 2007). Intriguingly, it may also have been initially founded in not one but multiple molecular domestication events. The *FHY3* family phylogram published by Lin et al. (2007) includes two internal branches belonging to extant, active TEs (LOM-1 and Jittery). All the five major clades in the phylogram, including the TE branches, have greater than 90% bootstrap support. If this phylogram is correct, then the *PHY3* family must have originated by at least three independent molecular domestication events. If so, and if the descendants of these multiple domestication events are involved in similar functions, as they appear to be (i.e., phytochrome-mediated responses), it raises important questions about the nature of molecular domestication. What might each independent domestication event have had in common, linking them to phytochrome responses? One possibility is pleiotropy. While members of at least three *FHY3* family clades are known to be involved in light responses, FHY3/FAR1 may have additional uncharacterized functions, so the different domestication events may not all be specifically linked to light responses. Another possibility is receptiveness. Over the period in which the domestications occurred, the phytochrome system may have been evolving rapidly, making it especially receptive to the addition of novel transcription factors. Yet it seems implausible that other systems would not also have been evolving quickly during the same period.

A third possibility is that the ancestral Mutator elements themselves, while still active, may have somehow been tied to phytochrome responses. This might not seem surprising, as various TEs are known to respond to specific environmental cues, e.g., stress responses (McClintock 1984; Zeh et al. 2009). Yet if true it might mean that the active TEs would themselves have been able to affect light response phenotypes. Or perhaps, after the first domestication event established a DTE with a phytochrome-related function, the nascent DTE might have continued to interact with cognate TE transposases and binding sites, and vice versa. Could a situation have existed in which DTEs and TEs were coevolving, gradually adding new DTE-binding sites as well as new DTEs? The continued ability of a DTE to bind to sites in TEs would permit it to expand and modulate its functions rather than relying solely on a preexisting distribution of binding sites present at the time of domestication. If such a mode of interaction between phenotypic and self-replicative selection could indeed be maintained, it might accelerate the evolution of the new function, benefitting the organism. Indeed, this may be a plausible model of transcription factor network domestication even in cases involving only a single transposase domestication event.

12.4.2 DAYSLEEPER

DAYSLEEPER was isolated in a yeast one-hybrid screen for proteins binding upstream of a DNA repair gene (*Ku70*), where it binds to multimers of a motif (Kubox1) also found upstream of other genes (Bundock and Hooykaas 2005). DAYSLEEPER has the same conserved domain architecture as *hAT* (*hobo*/*Ac*/*Tam*) transposase, including the *hAT* dimerization domain, but lacks residues required for transposition, and the *DAYSLEEPER* locus is not flanked by TE termini. Unlike most DTEs, *DAYSLEEPER* is located in a pericentromeric region dominated by TEs that, unlike *DAYSLEEPER*, are heavily targeted by small RNAs and DNA methylation and are not expressed (http://www.arabidopsis.org).

Homozygous *daysleeper* knockout mutants have severe developmental defects, which can be rescued by molecular complementation. Overexpressing *DAYSLEEPER* for prolonged periods causes slow growth, delayed flowering, altered cauline leaves, fasciation, partial or total sterility, and altered flower morphology. Overexpression induced for 24 h in wild-type seedlings leads to strong changes in the transcript abundance of dozens of genes, many of which are upregulated by more than an order of magnitude. However, none of the genes with significantly altered expression have a Kubox1 DNA motif, nor do any of the genes identified to have a Kubox1 motif were found to have significantly altered expression, including *Ku70*. Thus, while DAYSLEEPER does bind DNA, it is not clear whether the observed regulatory effects are direct or indirect (Bundock and Hooykaas 2005).

Although these results suggest that DAYSLEEPER may be a transcription factor with multiple roles, unfortunately no additional reports have been published since the original publication by Bundock and Hooykaas (2005). The genes with significantly altered expression when *DAYSLEEPER* is overexpressed are involved in a range of processes, especially response to stimulus, pathogen defense, signaling, metabolism, and development. EST library searches show that *DAYSLEEPER* homologs are found in widely diverged eudicots, both rosids and asterids, but not in monocots (D. Hoen, unpublished results).

12.4.3 The MUSTANG *Family*

To date, most DTEs have been identified through forward genetics. For instance, *FAR1* was identified in a phenotypic screen for far-red light mutants (Lin et al. 2007) and *DAYSLEEPER* in a yeast one-hybrid screen for proteins binding upstream of a DNA repair gene (Bundock and Hooykaas 2005). However, phenotypic screens may be ill suited to the identification of gene families in which close homologs are can compensate for a single knockout, as may be the case for families of DTEs undergoing lineage-specific expansion and subfunctionalization. The problem may be exacerbated in plants, which tend to have large gene families. Furthermore, a general lack of awareness of the existence of molecular domestication may cause some researchers who do observe novel DTEs in forward genetic screens to dismiss them under the false assumption that they are TEs. Because of these limitations, we do not know what ascertainment biases may exist in the set of known DTEs, nor how many DTEs remain to be identified, a potentially large number considering that transposases are the single most abundant and ubiquitous genes in nature (Aziz et al. 2010).

To address this problem, we need systematic screens of genomic data to identify DTE candidates, followed by reverse genetic characterization to determine whether they are *bona fide* DTEs. One of the few *in silico* searches conducted thus far was designed to identify plant *Mutator*-like DTEs that originated prior to the divergence of monocots and eudicots (Cowan et al. 2005). *Mutator*-like genes that lack TE termini were identified through comprehensive searches of rice and Arabidopsis genome sequences. Phylogenetic analysis revealed that while most *Mutator*-like genes cluster into clades found only in rice or Arabidopsis, indicative of lineage-specific transposases, two families are different, having close orthologs in both rice and Arabidopsis that are not associated with TE termini and that are expressed. One of these was the previously identified *FHY3* family (Hudson et al. 2003), a validation that the method could succeed in finding DTEs. The second was a novel family of DTEs, which was named *MUSTANG* (*MUG*). *MUG1* has syntenic orthologs in rice, Arabidopsis, Medicago, and poplar. Synonymous substitution rate analysis suggests that MUG1 has been subject to strong purifying selection at the protein level. Like the majority of characterized DTEs derived from DNA transposons (Sinzelle et al. 2009), including the *FHY3* family, *MUG* has maintained its DNA-binding domain, suggesting it may be a transcription factor. In fact, all of the ancestral *Mutator* conserved domains are present in *MUG*. In addition some MUG genes contain a protein-interaction domain, PB1 (Phox and Bem1), not normally found in transposases. Expressed *MUG* homologs have also been identified in sugarcane (Saccaro et al. 2007).

Investigations subsequent to Cowan et al. (2005) have revealed additional details. The *MUG* family consists of two major clades, each with members in all examined angiosperms, including basal angiosperms, but not in gymnosperms or other plants. The clades have undergone different patterns of diversification in monocots and eudicots, and it is not yet clear whether they originated in a single or multiple molecular domestication events. Microarray experiments show that in *mug* mutants, the expression levels of hundreds of additional genes are significantly altered, similar to the pattern found in *fhy3* (Ouyang et al. 2011), and consistent with the hypothesis that MUGs may function as transcription factors. Different *Arabidopsis mug* mutants have different pleiotropic phenotypes, including increased freezing tolerance, delayed development, delayed flowering time, aberrant flower morphology, and reduced seed set. This phylogenetic distribution of *MUG*, along with its flower-related phenotypes, suggests that it was domesticated early in angiosperm evolution and may have coincided with the origin of flowers (Joly-Lopez et al. 2012). These results were corroborated by a genome-wide survey of the transcription of TE-like sequences in rice, which found only three putative *Mutator*-like families that are highly transcribed, two of which are the *FHY3* and *MUG* families (Jiao and Deng 2007). (The third does not in fact appear to be related to *Mutator*, as it does not contain MuDR or MULE conserved domains; thus, it may be spurious (D. Hoen, unpublished data).)

The successful *in silico* identification and reverse-genetic characterization of *MUG* highlights the importance of performing systematic screens to identify DTEs. Although Cowan et al. (2005) identified only two *Mutator*-like DTE families with orthologs in both monocots and eudicots, more may remain to be discovered. For example, Benjak et al. (2008) identified single-copy grapevine genes that are derived from DNA transposons but lack TE termini, finding 2 *DAYSLEEPER* homologs, 8 *MUG* homologs, and 5 *FHY3* homologs, as well as one *hAT*-like and one *Mutator*-like gene that may be novel DTEs. We ourselves have also undertaken subsequent searches, the results of which suggest that plant genomes may contain many additional unreported DTEs (D. Hoen, unpublished data). Once a sufficiently large and unbiased sample of DTEs has been identified, we may more confidently characterize the nature of molecular domestication and its effect on evolution.

12.4.4 GARY

In addition to *MUG*, one other putative plant DTE was detected *in silico*. Muehlbauer et al. (2006) identified a *hAT*-like EST in barley corresponding to a gene, *GARY*, with only one or two copies in barley, two syntenic copies in the rice, and close EST homologs in several other cereal grasses. *GARY* is not flanked by TE termini, key residues that would be required for mobility are missing, and no transposition was observed in experimental conditions conducive to it (Muehlbauer et al. 2006). The function of *GARY* is not known, but given its phylogenetic distribution and its expression in wheat and barley spikes, it may have a grass-specific reproductive function. Searches of sequenced plant genomes confirm that *GARY* is found only in grasses, not eudicots (D. Hoen, unpublished data). The narrow phylogenetic distributions of *GARY* and *DAYSLEEPER* suggest that molecular domestication is an ongoing process in both monocot and eudicot plants.

12.4.5 Additional Examples

In total, roughly 100 families of eukaryotic DTEs have been identified so far (Volff 2006; Feschotte and Pritham 2007; Sinzelle et al. 2009). Even though retrotransposons greatly outnumber DNA transposons, the majority of known DTEs are derived from DNA transposons, on which we have focused, being the only well-characterized cases in plants. The disproportionate number of known DTEs derived from DNA transposons may be the result of ascertainment biases, as few systematic searches for DTEs have yet been conducted, or it may be due to intrinsic differences between DNA transposons and retrotransposons. For instance, the molecular functions of DNA transposons may be more easily domesticated. Over half of known DTEs putatively function as transcription factors, suggesting that this is a relatively easy evolutionary transition. Other DTEs have diverse functions, such as transposition repression, translation regulation, nuclear import, mRNA splicing, chromatin regulation, DNA maintenance and repair, telomere maintenance, centromere formation, chromosome segregation, and recombination (Feschotte and Pritham 2007; Sinzelle et al. 2009). Many DTEs regulate development and thus have major phenotypic effects (e.g., Bundock and Hooykaas 2005). Furthermore, DTEs have often played vital roles in major evolutionary innovations, which should not be surprising, as exaptation is inherently innovative. For example, the vertebrate adaptive immune system includes both domesticated proteins (recombination activating genes) and binding sites (recombination signal sequences), and is essentially a domesticated, specifically regulated transposition system (Agrawal et al. 1998; Fugmann 2010). Another example is the mammalian placenta, which evolved by multiple exaptations of both proteins (e.g., the *Syncytins*, *ERV-3*, *Peg10,* and *Rtl1/Peg11*) and regulatory elements (Edwards et al. 2008; Rawn and Cross 2008; Cornelis et al. 2012). Other systems in which

TE exaptations play important roles include programmed genome rearrangements in protozoans (Baudry et al. 2009; Cheng et al. 2010), mating-type switching in yeast (Rusche and Rine 2010), and neural development in mammals and other vertebrates (Cao et al. 2006; Santangelo et al. 2007; Okada et al. 2010; Beck et al. 2011; Franchini et al. 2011). TEs and DTEs have also been repeatedly recruited for functions in centromeres (Casola et al. 2008) and telomeres (Levin and Moran 2011; Pardue and DeBaryshe 2011). Indeed, ancient TE exaptations may even have been responsible for the very origin of telomeres in early eukaryotes, an innovation that may have been vital to enable the evolution of meiosis and sexual reproduction (which may also have been driven by TEs), prerequisites for complex life (Nosek et al. 2006).

12.5 Other Types of Protein-Coding Exaptation

Thus far, we have focused on cases of full molecular domestication, i.e., TE exaptations that form whole new genes, because they are illustrative, interesting, and the best characterized in plants. However, additional types of TE coding sequence exaptation also occur. One type is that certain DNA transposons acquire, by unknown mechanisms, duplicated fragments of ordinary genes, called *transduplications*. It is not yet clear whether transduplications have phenotypic coding functions, but they may generate regulatory small RNAs (Le et al. 2000; Yu et al. 2000; Turcotte et al. 2001; Jiang et al. 2004a, 2011; Juretic et al. 2005; Lai et al. 2005; Hanada et al. 2009). Furthermore, in some cases it appears that transduplicated genes serve self-replicative functions, i.e., they may have been exapted in the opposite direction to molecular domestication, an ordinary gene converted to TE gene (Hoen et al. 2006; Sela et al. 2008). It is possible that these self-replicative transduplicates might sometimes be re-exapted to again encode ordinary proteins, forming a cycle of exaptations, but no such cases are known.

Another potential path to coding sequence exaptation is through exonization. A TE that inserts into or near a gene can be incorporated as a novel cassette exon. Exonization is especially prevalent in animals, where alternative splicing is common (Sorek 2007; Cordaux and Batzer 2009; Schmitz and Brosius 2011). For instance, thousands of ordinary human gene transcripts may contain TE-derived sequences (Nekrutenko and Li 2001; Britten 2006; Sela et al. 2007). However, most exonized TEs are expressed only as rare splice variants and are probably not translated into functional peptides (Gotea and Makalowski 2006; Lin et al. 2008a). Indeed, the vast majority is derived not from TE coding sequences, but from nonautonomous elements. For instance, roughly half of exonized human TEs are originated from *Alu* elements, the most abundant human TE at over one million copies, accounting for more than 10% of the genome (Nekrutenko and Li 2001; Sela et al. 2007). *Alu* elements are themselves derived from retrotransposed 7SL RNAs and thus have no inherent coding capacity, but do contain multiple splice signals, which perhaps make them amenable to exonization (Makałowski et al.

1994). It is not yet clear whether or how exonized *Alu* elements may contribute to phenotypic evolution (Cordaux and Batzer 2009).

Exonization also occurs in plants but is not as prevalent or well characterized (Barbazuk et al. 2008). In *A. thaliana*, more than 2,000 loci have transcribed exons derived from TEs, but it is not clear how many of these are ordinary genes and how many are transduplications, nor is it known what fraction, if any, are translated into functional proteins (Lockton and Gaut 2009). MITEs, high copy-number TEs prominent in plants but also found in other eukaryotes that frequently insert near genes (see below), may be good candidates for exonization (Marino-Ramirez et al. 2005). However, like *Alu* elements, MITEs usually have no inherent coding capacity and at least certain families very rarely insert into coding exons or become exonized (Oki et al. 2008; Naito et al. 2009).

Exonization may seem to be an easier path to exaptation than full molecular domestication, given that existing genes are already be stably regulated. Conversely, adding new sequence to an existing gene, and thus changing its function, would likely be deleterious, unless perhaps it was a recently duplicated gene, or unless the novel exon were included in only one isoform leaving the original gene function intact. However, this scenario would also require a concurrent change in regulatory control.

Although exonized TEs appear to usually be nonfunctional, some functional chimeras are known. For instance, the primate gene *SETMAR* (Robertson and Zumpano 1997; Cordaux et al. 2006), *a.k.a. Metnase* (Lee et al. 2005), arose when a *Tc1-Mariner* element inserted downstream of a SET gene. *Tc1-Mariner* elements are present in all eukaryotic kingdoms and undergo frequent horizontal transfer (Lohe et al. 1995). In humans, *Tc1-Mariner* is now extinct, like other DNA transposons, but 60–80 million years ago, during the primate radiation, *Tc1-Mariner* underwent a burst of transposition (Pace and Feschotte 2007). *SETMAR*, which formed shortly thereafter (Shaheen et al. 2010), is conserved in humans, apes, and monkeys (Lee et al. 2005) and, unlike other *Tc1-Mariner* fossils in the human genome, is evolving slowly with uninterrupted, expressed open reading frames (Robertson and Zumpano 1997).

SETMAR contains both a transposase domain, evolving under strong purifying selection, and a SET histone methyl transferase domain (Shaheen et al. 2010). It has maintained its ancestral DNA-binding specificity for a 19 base-pair ancestral *Tc1-Mariner* terminal sequence still present in thousands of copies in the human genome, suggesting that it may be involved in a large network (Cordaux et al. 2006). The DDE motif, normally required for transposition, is absent and SETMAR does not efficiently catalyze transposition, nor does it, like functional transposases such as *h*AT in maize (Zhang et al. 2009), mediate chromosomal translocation, but instead it represses translocation (Shaheen et al. 2010). Yet, SETMAR does maintain certain transposase functions, including limited DNA endonuclease activity. One activity of SETMAR is histone methylation, and thus it potentially functions in transcriptional regulation, DNA repair, DNA replication, and imprinting. It also interacts Pso4, a protein involved both in DNA double-stranded break repair and RNA splicing. It plays a role in double-stranded break repair by nonhomologous end joining, and possibly in other DNA repair processes, and in restarting stalled replication forks (Shaheen et al. 2010). Replication fork restart is also a function of another DTE, a CENP-B homolog in fission yeast (Zaratiegui et al. 2011).

SETMAR was formed through a unique series of mutations. After inserting downstream of the ancestral SET gene, a *Tc1-Mariner* was immobilized by an *Alu* insertion. The original stop codon of the SET gene was deleted, resulting in the exonization of a segment of 3′ UTR and creation of a novel intron, which fused the SET gene to the transposase. Curiously, the intron acceptor site is encoded by the ancestral *Tc1-Mariner* element itself, just three base-pairs upstream of the transposase start codon; furthermore, the ancestral element also encoded two putative branch sites (Cordaux et al. 2006). These features perhaps suggest that ancestral *Tc1-Mariners* may, like *Alu* elements, have evolved a capacity to exonize, and thus that TEs themselves can benefit from exonization. This might suggest an evolutionary feedback between self-replicative and phenotypic selection to increase the rate of exonization, which in turn may increase the rate of TE exaptation.

In any case, *SETMAR* demonstrates that transposase functions can be exapted not only through the formation of whole new genes but also by chimerization with existing ordinary genes. Although only a small fraction of exonized TEs may become exapted, over the long course of evolution they may have made a large contribution to the genome, especially in functions such as DNA binding, protein–protein interaction, and recombination.

12.6 Regulatory Exaptation: Coding and Noncoding

Complex multicellular life has evolved primarily not through changes to structural genes but by changes to how genes are regulated (King and Wilson 1975; Carroll 2005; Wray 2007). Transcription factors are a major component of gene regulation. While most genes are similar between different species, some transcription factor families evolve quickly (Nowick and Stubbs 2010). TEs contribute to regulatory evolution in various ways, including by creating new families of lineage-specific transcription factors through molecular domestication (Britten and Davidson 1971; Bourque et al. 2008; Feschotte 2008; Kunarso et al. 2010). DNA transposons, in particular, have innate characteristics that predispose them to exaptation as de novo regulatory networks, i.e., transcription factors and binding sites (Feschotte 2008). Transposases have DNA-binding domains that recognize motifs within TEs (Haren et al. 1999; Lisch and Jiang 2009). Some TEs preferentially insert upstream of genes, an ideal location for elements that regulate transcription (Bureau and Wessler 1992; Lisch and Jiang 2009; Naito et al. 2009). When the transposase is expressed, it binds to the TEs and can modify the expression of flanking genes. If the TEs happen to be distributed in a way that the coordinated expression change is phenotypically beneficial, then the transposase and the binding sites are favored by phenotypic selection and can be domesticated. Of eukaryotic DTEs derived from DNA transposons with putative or known functions, about half are transcription factors (Feschotte and Pritham 2007; Sinzelle et al. 2009). As we have seen, these probably include all the plant DTE families: *FHY3* (Lin et al. 2007), *DAYSLEEPER* (Bundock and Hooykaas 2005), and *MUG* (Z. Joly-Lopez, E. Forczek, D. Hoen, and T. Bureau, unpublished data).

Indeed, molecular domestication may have been the ancient foundation of several transcription factor families. We have already discussed THAP genes, many of which are transcription factors, which may have originated by *P* element domestication. Additional examples include the AP2/ERF and WRKY superfamilies of plant-specific transcription factors, which may also have arisen through the ancient domestication of DNA-binding domains (Magnani et al. 2004; Babu et al. 2006). However, as may be expected for ancient events, these gene families have limited similarity to extant TEs making it difficult to determine exactly how they arose. There may have been multiple domestication events, or single domestication events followed by exon shuffling (not mediated by TEs), or it is even possible, though unlikely, that the TEs co-opted these domains through transduplication, rather than the transcription factors arising by domestication.

TE exaptation also contributes to regulatory evolution in other ways. TEs contain binding sites for not only transposases, which are needed for transposition, but also for ordinary transcription factors, as well as promoters, which are needed for transposase expression. These binding sites and promoters can be exapted for phenotypic function, independently of transposase domestication, singly or in networks. Indeed, this type of exaptation appears to be far more prevalent than coding sequence domestication. In general, transcription factor binding sites and other functional noncoding sequences evolve rapidly and are frequently restricted to single species or narrow phyletic ranges (Ponting et al. 2011). While some may originate de novo (Eichenlaub and Ettwiller 2011), TEs are appropriate sources of lineage-specific regulatory elements, since TEs themselves evolve rapidly, both in sequence and genomic distribution (e.g., Hollister et al. 2011). Indeed, noncoding TE sequences are frequently exapted as short- and long-distance (Pi et al. 2010) enhancers, repressors, silencers, insulators, alternative transcription start sites, alternative polyadenylation sites (Lee et al. 2008), antisense transcripts, or alternative exons (Beauregard et al. 2008; Feschotte 2008; Shapiro 2010; Ponting et al. 2011; Studer et al. 2011). For instance, in humans, one-quarter to one-third of promoters and transcription factor binding sites may be derived from TEs (Jordan et al. 2003; Bourque et al. 2008; Kunarso et al. 2010), and thousands of conserved noncoding elements derived from TEs are especially enriched near genes involved in development and transcriptional regulation (Lowe et al. 2007). In plants, small nonautonomous DNA transposons known as miniature inverted-repeat transposable elements (MITEs) are especially abundant. MITEs insert preferentially into 5′ flanking regions of genes, can be exapted as various types of regulatory element, and can form de novo regulatory networks (Bureau et al. 1996; Jiang et al. 2004b; Oki et al. 2008; Naito et al. 2009).

In addition to transcription factors and *cis*-binding sites, TEs can also be exapted as various classes of RNAs (Brosius 1999). Of these, the most important for regulation are small RNAs (sRNAs). sRNAs of different types are used in several related RNAi pathways to target transcriptional and posttranscriptional silencing (Chapman and Carrington 2007; Malone and Hannon 2009). sRNAs are generated from double-stranded RNAs (dsRNAs), which ordinary genes do not normally produce, but which TEs produce in various ways, for example, by intermolecular hybridization of bidirectional transcripts or by intramolecular hybridization of read-through transcripts containing inverted repeats (Lisch 2009). In plants, transcriptional silencing of TEs by chromatin compaction is activated by sRNA-directed DNA methylation as well as repressive histone modifications. Secondary sRNAs spread DNA methylation to adjacent areas (Simon and Meyers 2011). DNA methylation is initiated and reinforced in plant embryos by sRNAs produced in the vegetative cell of pollen and the central cell of endosperm, cells which are hypomethylated to activate TEs, but which do not contribute genetic material to subsequent generations (Law and Jacobsen 2010; Calarco and Martienssen 2011). Once established, DNA methylation is epigenetically inherited (Feng and Jacobsen 2010); thus, degenerate TEs may eventually be desilenced making them more available for exaptation.

Epigenetic regulation of TEs can alter the expression of nearby ordinary genes, and *cis*-regulatory elements exapted from TEs can also be epigenetically regulated (Slotkin and Martienssen 2007; Weil and Martienssen 2008; Markljung et al. 2009). In fact, McClintock first discovered TEs by observing effects caused by the epigenetic regulation of maize *En/Spm (Enhancer/Suppressor Mutator)* elements (Fedoroff 1999). Methylation of TEs inserted near ordinary genes decreases their expression (Hollister and Gaut 2009), for instance by methylation spreading to *cis*-regulatory elements (Martin et al. 2009). Conversely, TE-derived regulatory elements may also be targeted by active histone modifications (Huda et al. 2011).

TE-derived sRNAs can regulate ordinary genes not only epigenetically, but also posttranscriptionally. For instance, transduplications can produce, via antisense transcription or possibly inverted duplication, sRNAs that posttranscriptionally downregulate parent gene expression in *trans* (Juretic et al. 2005; Slotkin et al. 2005; Hanada et al. 2009). Also, microRNA (miRNA) genes (*MIR*) can be derived from TEs (Smalheiser and Torvik 2005; Piriyapongsa et al. 2007; Li et al. 2011b). *MIR* transcripts form stem-loop dsRNAs, which are processed into sRNAs that target the mRNA transcripts of ordinary genes for cleavage or translation inhibition. While some plant *MIR* genes have ancient origin and are highly conserved, the majority are young and restricted to single species. Young *MIR* genes may originate by random inversion of coding sequence, or they may be derived from short inverted repeat TEs, such as MITEs (Voinnet 2009). However, many young *MIR* genes may be evolving neutrally (Fahlgren et al. 2010).

While some TE-derived miRNAs produce canonical stem-loop structures, many are atypical and more similar to ancestral TE configurations. This suggests a model (Piriyapongsa and Jordan 2007) of *MIR* exaptation similar in principle to the Frequent Birth model. Consider a specific TE locus that generates sRNAs targeting homologous TEs for silencing. As a side effect of silencing, a set of ordinary genes may also be downregulated, perhaps via a distributed set of exonized TEs targeted by the sRNAs. If the new regulatory network were phenotypically beneficial, the locus would come under phenotypic selection and could evolve to become a *MIR* gene. The main principle here is the same as that for de novo exaptation of transcription factor networks: a network of sequences derived from TEs randomly distributed near ordinary genes are targets of regulation, either by an ordinary transcription factor, a domesticated transcription factor, or an miRNA exapted from a TE. Indeed, miRNAs have been likened to posttranscriptional transcription factors (Malone and Hannon 2009). The first step in this process, the exaptation of a TE to produce repressive sRNAs is nicely illustrated by the maize Mu killer locus (*Muk*). *Muk* is an inverted duplication of part of a *Mutator* TE. It produces a stem-loop dsRNA transcript that is processed into sRNAs, which effectively silence, both posttranscriptionally and epigenetically, this otherwise highly active TE family (Slotkin et al. 2005). The large number of observed young *MIR* genes is consistent with a rapid birth–death cycle as predicted by the Frequent Birth model, and with a role for these miRNAs in lineage-specific diversification. Perhaps a similar phenomenon might be found for domesticated protein-coding genes, given sufficiently sensitive searches.

On a grander scale, the RNAi system itself may itself be an indirect exaptation of TE-genome coevolution. Perhaps originally evolved to regulate TEs (Lisch and Bennetzen 2011), epigenetic regulation is now used for various purposes such as genomic imprinting (Köhler and Weinhofer-Molisch 2010), gene body methylation (Saze and Kakutani 2011), developmental plasticity, and the buffering of developmental programs (Obbard et al. 2009; Martin and Bendahmane 2010; Feng and Jacobsen 2011; Mirouze and Paszkowski 2011; Simon and Meyers 2011). On an even larger scale, the modulation of epigenetic TE regulation may provide a mechanism for evolvability and may thus play a fundamental evolutionary role. Global epigenetic desilencing increases the rate of transposition, so it may enable periods of rapid evolution and, ultimately, punctuated equilibrium (Gould and Eldredge 1977; Zeh et al. 2009; Johnson and Tricker 2010; Okada et al. 2010; Oliver 2011; Werren 2011).

12.7 Concluding Remarks

We have reviewed the known DTE genes in plants. Although some results suggest that there may be few plant DTEs left to be discovered (Cowan et al. 2005; Jiao and Deng 2007), the question remains open. Indeed, other lines of evidence suggest that molecular domestication may be a relatively common phenomena: the convergent domestication of *pogo* transposases into CENP-B-like DTEs in mammals, fission yeast, and possibly insects and plants (Barbosa-Cisneros and Herrera-Esparza 2002; Casola et al. 2008; d'Alençon et al. 2011); the repeated and convergent domestication of placental genes (Rawn and Cross 2008); the recurrent domestication of *P* elements (Miller et al. 1999; Quesneville et al. 2005); and multiple domestication events during *FHY3* evolution (Lin et al. 2007). Thus, before we assess the frequency and global significance of molecular domestication, additional systematic genomic searches are needed. Regardless of how frequent TE coding sequence exaptation is, it has already become apparent that noncoding TE exaptations, such as transcription factor binding sites, contribute significantly to regulatory evolution.

It is common to describe TEs as selfish parasites, a viewpoint supposedly justified by the capacity of TEs to persist outside the need for phenotypic selection. Gould cautioned that an exclusive focus on immediate adaptation may lead to an inverted understanding of evolution, a focus on spandrels not arches, on paintings not architecture (Gould and Lewontin 1979). Modern genomics supports this view (Koonin 2009). We find it fascinating and beautiful that natural selection appears to have created a system of molecular evolution where TEs, which are not directly maintained in the short term by phenotypic selection but rather through self-replication, without which they would otherwise become extinct, may nonetheless be indispensible in the long term to the evolution of complex organisms. Indeed, by evolving mechanisms such as molecular domestication that increase the rate of adaptive mutation, TEs may be able to increase their own chances of survival and proliferation (Le Rouzic et al. 2007). Thus, rather than viewing TEs as parasites living off a genomic system in which ordinary genes and phenotypic selection are paramount, a more productive analogy may be to view TEs as important components of a more complex genomic ecosystem (Brookfield 2005). When Barbara McClintock first discovered transposable elements, she recognized their potential importance in gene regulation and genome evolution (McClintock 1950). In the end, she may well be proven right.

Box 12.1 Terminology

CHIMERIZATION—The fusion of two separate genes into a single gene.

EXAPTATION—A feature evolved for one role (or no role) that has been co-opted to perform a different role. Also, a feature evolved at one level of selection that produces effects at a different level of selection. Although the concept was understood at least as early as Darwin, the term was coined by Gould and Vrba. The original role of an exaptation can also be called, in retrospect, a preadaptation.

EXONIZATION—The incorporation into a gene of a new coding sequence not originally part of the gene, usually from a noncoding source.

HORIZONTAL TRANSFER—The transfer of a gene or TE between individuals other than from parent to offspring, also known as lateral transfer. In bacteria, horizontal transfer via plasmids of certain genes, such as those conferring antibiotic resistance, is common. In eukaryotes, TEs and genes are ordinarily passed vertically from parent to offspring; however, certain types of TEs, such as *P* elements, appear to undergo relatively frequent horizontal transfer by unknown mechanisms, which may be critical to counteract stochastic loss and permit long-term persistence.

TRANSDUPLICATION—A non-TE sequence that has been copied into a DNA transposon. Transduplication occurs frequently in certain plant TEs such as rice MULEs and maize Helitrons. MULEs with transduplications are also called Pack-MULEs. The mechanisms and evolutionary consequences of transduplication are not yet well established. Similarly, retrotransposons can mobilize non-TE DNA by a related but distinct process called transduction.

MOLECULAR DOMESTICATION—A process whereby a TE gene or other sequence is coopted to perform a nonmobility-related function.

ORDINARY GENE—A non-TE gene.

PHENOTYPIC SELECTION—Natural selection acting on the phenotypes of organisms. Ordinary genes (and other genetic elements) are replicated en-masse via the production of progeny organisms, which are the immediate object of selection. Ordinary genes persist by producing beneficial phenotypes; TEs can also experience selection at this level to decrease their deleterious effects.

SELF-REPLICATIVE SELECTION—Natural selection at the level of self-replicating DNA. TEs can persist without producing beneficial phenotypes, because they have multiple copies that sustain mutations independently. By replicating frequently enough in germ cells, at least one autonomous copy can escape any disabling mutations that would prevent further replication and be passed to the next generation.

Acknowledgement The authors would like to thank Thomas Eickbush for his comments. This work was supported by funds from the Natural Sciences and Engineering Research Council of Canada (NSERC), Genome Québec, and Genome Canada.

References

Agrawal A, Eastman QM, Schatz DG (1998) Transposition mediated by RAG1 and RAG2 and its implications for the evolution of the immune system. Nature 394:744–751

Agren JA, Wright SI (2011) Co-evolution between transposable elements and their hosts: a major factor in genome size evolution? Chromosome Res 19:777–786

Allen T, Koustenis A, Theodorou G, Somers DE, Kay SA, Whitelam GC, Devlin PF (2006) Arabidopsis FHY3 specifically gates phytochrome signaling to the circadian clock. Plant Cell 18:2506–2516

Aziz RK, Breitbart M, Edwards RA (2010) Transposases are the most abundant, most ubiquitous genes in nature. Nucleic Acids Res 38:4207–4217

Babu MM, Iyer LM, Balaji S, Aravind L (2006) The natural history of the WRKY-GCM1 zinc fingers and the relationship between transcription factors and transposons. Nucleic Acids Res 34:6505–6520

Bae G, Choi G (2008) Decoding of light signals by plant phytochromes and their interacting proteins. Annu Rev Plant Biol 59:281–311

Barbazuk WB, Fu Y, McGinnis KM (2008) Genome-wide analyses of alternative splicing in plants: opportunities and challenges. Genome Res 18:1381–1392

Barbosa-Cisneros O, Herrera-Esparza R (2002) CENP-B is a conserved gene among vegetal species. Genet Mol Res 1:241–245

Baudry C, Malinsky S, Restituito M, Kapusta A, Rosa S, Meyer E, Betermier M (2009) PiggyMac, a domesticated *piggyBac* transposase involved in programmed genome rearrangements in the ciliate *Paramecium tetraurelia*. Genes Dev 23:2478–2483

Beauregard A, Curcio MJ, Belfort M (2008) The take and give between retrotransposable elements and their hosts. Annu Rev Genet 42:587–617

Beck CR, Garcia-Perez JL, Badge RM, Moran JV (2011) LINE-1 elements in structural variation and disease. Annu Rev Genomics Hum Genet 12:187–215

Benjak A, Forneck A, Casacuberta JM (2008) Genome-wide analysis of the "cut-and-paste" transposons of grapevine. PLoS One 3:e3107

Biémont C (2010) A brief history of the status of transposable elements: from junk DNA to major players in evolution. Genetics 186:1085–1093

Bourque G, Leong B, Vega VB, Chen X, Lee YL, Srinivasan KG, Chew J-L, Ruan Y, Wei C-L, Ng HH, Liu ET (2008) Evolution of the mammalian transcription factor binding repertoire via transposable elements. Genome Res 18:1752–1762

Brennecke J, Aravin AA, Stark A, Dus M, Kellis M, Sachidanandam R, Hannon GJ (2007) Discrete small RNA-generating loci as master regulators of transposon activity in Drosophila. Cell 128:1089–1103

Britten R (2006) Transposable elements have contributed to thousands of human proteins. Proc Natl Acad Sci USA 103:1798–1803

Britten RJ, Davidson EH (1971) Repetitive and non-repetitive DNA sequences and a speculation on the origins of evolutionary novelty. Q Rev Biol 46:111–138

Brookfield JF (1982) Interspersed repetitive DNA sequences are unlikely to be parasitic. J Theor Biol 94:281–299

Brookfield JF (2005) The ecology of the genome—mobile DNA elements and their hosts. Nat Rev Genet 6:128–136

Brosius J (1999) RNAs from all categories generate retrosequences that may be exapted as novel genes or regulatory elements. Gene 238:115–134

Bundock P, Hooykaas P (2005) An Arabidopsis *hAT*-like transposase is essential for plant development. Nature 436:282–284

Bureau TE, Wessler SR (1992) Tourist: a large family of small inverted repeat elements frequently associated with maize genes. Plant Cell 4:1283–1294

Bureau TE, Ronald PC, Wessler SR (1996) A computer-based systematic survey reveals the predominance of small inverted-repeat elements in wild-type rice genes. Proc Natl Acad Sci USA 93:8524–8529

Calarco JP, Martienssen RA (2011) Genome reprogramming and small interfering RNA in the Arabidopsis germline. Curr Opin Genet Dev 21:134–139

Cam HP, K-i N, Ebina H, Levin HL, Grewal SIS (2008) Host genome surveillance for retrotransposons by transposon-derived proteins. Nature 451:431–436

Cao X, Yeo G, Muotri A, Kuwabara T (2006) Noncoding RNAs in the mammalian central nervous system. Annu Rev Neurosci 29:77–103

Carroll SB (2005) Evolution at two levels: on genes and form. PLoS Biol 3:e245

Casola C, Hucks D, Feschotte C (2008) Convergent domestication of *pogo*-like transposases into centromere-binding proteins in fission yeast and mammals. Mol Biol Evol 25:29–41

Chapman EJ, Carrington JC (2007) Specialization and evolution of endogenous small RNA pathways. Nat Rev Genet 8:884–896

Charlesworth B, Sniegowski P, Stephan W (1994) The evolutionary dynamics of repetitive DNA in eukaryotes. Nature 371:215–220

Chen M, Chory J (2011) Phytochrome signaling mechanisms and the control of plant development. Trends Cell Biol 21:664–671

Cheng C-Y, Vogt A, Mochizuki K, Yao M-C (2010) A domesticated *piggyBac* transposase plays key roles in heterochromatin dynamics and DNA cleavage during programmed DNA deletion in *Tetrahymena thermophila*. Mol Biol Cell 21:1753–1762

Clouaire T, Roussigne M, Ecochard V, Mathe C, Amalric F, Girard JP (2005) The THAP domain of THAP1 is a large C2CH module with zinc-dependent sequence-specific DNA-binding activity. Proc Natl Acad Sci USA 102:6907–6912

Conrad DF, Pinto D, Redon R, Feuk L, Gokcumen O, Zhang Y, Aerts J, Andrews TD, Barnes C, Campbell P, Fitzgerald T, Hu M, Ihm CH, Kristiansson K, Macarthur DG, Macdonald JR, Onyiah I, Pang AWC, Robson S, Stirrups K, Valsesia A, Walter K, Wei J, Consortium WTCC, Tyler-Smith C, Carter NP, Lee C, Scherer SW, Hurles ME (2010) Origins and functional impact of copy number variation in the human genome. Nature 464:704–712

Cordaux R, Batzer MA (2009) The impact of retrotransposons on human genome evolution. Nat Rev Genet 10:691–703

Cordaux R, Udit S, Batzer MA, Feschotte C (2006) Birth of a chimeric primate gene by capture of the transposase gene from a mobile element. Proc Natl Acad Sci USA 103:8101–8106

Cornelis G, Heidmann O, Bernard-Stoecklin S, Reynaud K, Véron G, Mulot B, Dupressoir A, Heidmann T (2012) From the Cover: Ancestral capture of syncytin-Car1, a fusogenic endogenous retroviral envelope gene involved in placentation and conserved in Carnivora. Proc Natl Acad Sci 109:E432–E441

Cowan RK, Hoen DR, Schoen DJ, Bureau TE (2005) *MUSTANG* is a novel family of domesticated transposase genes found in diverse angiosperms. Mol Biol Evol 22:2084–2089

d'Alençon E, Nègre N, Stanojcic S, Alassoeur B, Gimenez S, Léger A, Abd-Alla A, Juliant S, Fournier P (2011) Characterization of a CENP-B homolog in the holocentric Lepidoptera *Spodoptera frugiperda*. Gene 485:91–101

Daniels SB, Peterson KR, Strausbaugh LD, Kidwell MG, Chovnick A (1990) Evidence for horizontal transmission of the P transposable element between Drosophila species. Genetics 124:339–355

Darwin C (1876) The origin of species by means of natural selection, or the preservation of favoured races in the struggle for life, 6th edn. John Murray, London

Diao X, Freeling M, Lisch DR (2005) Horizontal transfer of a plant transposon. PLoS Biol 4:e5

Doolittle WF, Sapienza C (1980) Selfish genes, the phenotype paradigm and genome evolution. Nature 284:601–603

Dooner HK, Weil CF (2007) Give-and-take: interactions between DNA transposons and their host plant genomes. Curr Opin Genet Dev 17:486–492

Edwards CA, Mungall AJ, Matthews L, Ryder E, Gray DJ, Pask AJ, Shaw G, Graves JAM, Rogers J, Consortium S, Dunham I, Renfree MB, Ferguson-Smith AC (2008) The evolution of the DLK1-DIO3 imprinted domain in mammals. PLoS Biol 6:e135

Eichenlaub MP, Ettwiller L (2011) De novo genesis of enhancers in vertebrates. PLoS Biol 9:e1001188

Fahlgren N, Jogdeo S, Kasschau KD, Sullivan CM, Chapman EJ, Laubinger S, Smith LM, Dasenko M, Givan SA, Weigel D, Carrington JC (2010) MicroRNA gene evolution in *Arabidopsis lyrata* and *Arabidopsis thaliana*. Plant Cell 22:1074–1089

Fedoroff NV (1999) The suppressor-mutator element and the evolutionary riddle of transposons. Genes Cells 4:11–19

Feng S, Jacobsen S (2010) Epigenetic reprogramming in plant and animal development. Science 330:622–627

Feng S, Jacobsen SE (2011) Epigenetic modifications in plants: an evolutionary perspective. Curr Opin Plant Biol 14:179–186

Feschotte C (2008) Transposable elements and the evolution of regulatory networks. Nat Rev Genet 9:397–405

Feschotte C, Pritham EJ (2007) DNA transposons and the evolution of eukaryotic genomes. Annu Rev Genet 41:331–368

Flagel LE, Wendel JF (2009) Gene duplication and evolutionary novelty in plants. New Phytol 183:557–564

Franchini LF, López-Leal R, Nasif S, Beati P, Gelman DM, Low MJ, de Souza FJS, Rubinstein M (2011) Convergent evolution of two mammalian neuronal enhancers by sequential exaptation of unrelated retroposons. Proc Natl Acad Sci USA 108:15270–15275

Fugmann SD (2010) The origins of the Rag genes—from transposition to V(D)J recombination. Semin Immunol 22:10–16

Gaut BS, Wright SI, Rizzon C, Dvorak J, Anderson LK (2007) Recombination: an underappreciated factor in the evolution of plant genomes. Nat Rev Genet 8:77–84

Gotea V, Makalowski W (2006) Do transposable elements really contribute to proteomes? Trends Genet 22:260–267

Gould SJ, Eldredge N (1977) Punctuated equilibria: the tempo and mode of evolution reconsidered. Paleobiology 3:115–151

Gould SJ, Lewontin RC (1979) The spandrels of San Marco and the Panglossian paradigm: a critique of the adaptationist programme. Proc R Soc Lond B Biol Sci 205:581–598

Gould SJ, Lloyd EA (1999) Individuality and adaptation across levels of selection: how shall we name and generalize the unit of Darwinism? Proc Natl Acad Sci USA 96:11904–11909

Gould SJ, Vrba ES (1982) Exaptation-a missing term in the science of form. Paleobiology 8:4–15

Hammer SE (2005) Homologs of *Drosophila P* transposons were mobile in zebrafish but have been domesticated in a common ancestor of chicken and human. Mol Biol Evol 22:833–844

Hanada K, Vallejo V, Nobuta K, Slotkin RK, Lisch DR, Meyers BC, Shiu S-H, Jiang N (2009) The functional role of Pack-MULEs in rice inferred from purifying selection and expression profile. Plant Cell 21:25–38

Haren L, Ton-Hoang B, Chandler M (1999) Integrating DNA: transposases and retroviral integrases. Annu Rev Microbiol 53:245–281

Hickey DA (1982) Selfish DNA: a sexually-transmitted nuclear parasite. Genetics 101:519–531

Hoen DR, Park KC, Elrouby N, Yu Z, Mohabir N, Cowan RK, Bureau TE (2006) Transposon-mediated expansion and diversification of a family of *ULP*-like genes. Mol Biol Evol 23:1254–1268

Hollister JD, Gaut BS (2009) Epigenetic silencing of transposable elements: a trade-off between reduced transposition and deleterious effects on neighboring gene expression. Genome Res 19:1419–1428

Hollister JD, Smith LM, Guo Y-L, Ott F, Weigel D, Gaut BS (2011) Transposable elements and small RNAs contribute to gene expression divergence between *Arabidopsis thaliana* and *Arabidopsis lyrata*. Proc Natl Acad Sci USA 108:2322–2327

Hua-Van A, Le Rouzic A, Boutin TS, Filée J, Capy P (2011) The struggle for life of the genome's selfish architects. Biol Direct 6:19

Huda A, Bowen NJ, Conley AB, Jordan IK (2011) Epigenetic regulation of transposable element derived human gene promoters. Gene 475:39–48

Hudson ME, Ringli C, Boylan MT, Quail PH (1999) The *FAR1* locus encodes a novel nuclear protein specific to phytochrome A signaling. Genes Dev 13:2017–2027

Hudson ME, Lisch DR, Quail PH (2003) The *FHY3* and *FAR1* genes encode transposase-related proteins involved in regulation of gene expression by the phytochrome A-signaling pathway. Plant J 34:453–471

Jiang N, Bao Z, Zhang X, Eddy SR, Wessler SR (2004a) Pack-MULE transposable elements mediate gene evolution in plants. Nature 431:569–573

Jiang N, Feschotte C, Zhang X, Wessler SR (2004b) Using rice to understand the origin and amplification of miniature inverted repeat transposable elements (MITEs). Curr Opin Plant Biol 7:115–119

Jiang N, Ferguson AA, Slotkin RK, Lisch DR (2011) Pack-*Mutator*-like transposable elements (Pack-MULEs) induce directional modification of genes through biased insertion and DNA acquisition. Proc Natl Acad Sci USA 108:1537–1542

Jiao Y, Deng X (2007) A genome-wide transcriptional activity survey of rice transposable element-related genes. Genome Biol 8:R28

Jiao Y, Lau OS, Deng XW (2007) Light-regulated transcriptional networks in higher plants. Nat Rev Genet 8:217–230

Johnson LJ (2008) Selfish genetic elements favor the evolution of a distinction between soma and germline. Evolution 62:2122–2124

Johnson LJ, Tricker PJ (2010) Epigenomic plasticity within populations: its evolutionary significance and potential. Heredity 105:113–121

Joly-Lopez Z, Forczek E, Hoen DR, Juretic N, Bureau TE (2012) A gene family derived from transposable elements during early angiosperm evolution has reproductive fitness benefits in *Arabidopsis thaliana*. PLoS Genet 8(9): e1002931. doi:10.1371/journal.pgen.1002931

Jordan IK, Rogozin IB, Glazko GV, Koonin EV (2003) Origin of a substantial fraction of human regulatory sequences from transposable elements. Trends Genet 19:68–72

Juretic N, Hoen DR, Huynh ML, Harrison PM, Bureau TE (2005) The evolutionary fate of MULE-mediated duplications of host gene fragments in rice. Genome Res 15:1292–1297

Kaufman PD, Doll RF, Rio DC (1989) Drosophila P element transposase recognizes internal P element DNA sequences. Cell 59:359–371

Khurana JS, Wang J, Xu J, Koppetsch BS, Thomson TC, Nowosielska A, Li C, Zamore PD, Weng Z, Theurkauf WE (2011) Adaptation to *P* element transposon invasion in *Drosophila melanogaster*. Cell 147:1551–1563

Kidwell MG, Lisch DR (2001) Perspective: transposable elements, parasitic DNA, and genome evolution. Evolution 55:1–24

Kidwell MG, Kidwell JF, Sved JA (1977) Hybrid Dysgenesis in *DROSOPHILA MELANOGASTER*: a syndrome of aberrant traits including mutation, sterility and male recombination. Genetics 86:813–833

King DG, Kashi Y (2007) Mutability and evolvability: indirect selection for mutability. Heredity 99:123–124

King MC, Wilson AC (1975) Evolution at two levels in humans and chimpanzees. Science 188:107–116

Köhler C, Weinhofer-Molisch I (2010) Mechanisms and evolution of genomic imprinting in plants. Heredity 105:57–63

Koonin EV (2009) Darwinian evolution in the light of genomics. Nucleic Acids Res 37:1011–1034

Kunarso G, Chia N-Y, Jeyakani J, Hwang C, Lu X, Chan Y-S, Ng HH, Bourque G (2010) Transposable elements have rewired the core regulatory network of human embryonic stem cells. Nat Genet 42:631–634

Lai J, Li Y, Messing J, Dooner HK (2005) Gene movement by *Helitron* transposons contributes to the haplotype variability of maize. Proc Natl Acad Sci USA 102:9068–9073

Law JA, Jacobsen SE (2010) Establishing, maintaining and modifying DNA methylation patterns in plants and animals. Nat Rev Genet 11:204–220

Le Rouzic A, Deceliere G (2005) Models of the population genetics of transposable elements. Genet Res 85:171–181

Le Rouzic A, Boutin TS, Capy P (2007) Long-term evolution of transposable elements. Proc Natl Acad Sci USA 104:19375–19380

Le QH, Wright S, Yu Z, Bureau T (2000) Transposon diversity in *Arabidopsis thaliana*. Proc Natl Acad Sci USA 97:7376–7381

Lee S-H, Oshige M, Durant ST, Rasila KK, Williamson EA, Ramsey H, Kwan L, Nickoloff JA, Hromas R (2005) The SET domain protein Metnase mediates foreign DNA integration and links integration to nonhomologous end-joining repair. Proc Natl Acad Sci USA 102:18075–18080

Lee JY, Ji Z, Tian B (2008) Phylogenetic analysis of mRNA polyadenylation sites reveals a role of transposable elements in evolution of the 3'-end of genes. Nucleic Acids Res 36:5581–5590

Levin HL, Moran JV (2011) Dynamic interactions between transposable elements and their hosts. Nat Rev Genet 12:615–627

Li J, Li G, Gao S, Martinez C, He G, Zhou Z, Huang X, Lee JH, Zhang H, Shen Y, Wang H, Deng XW (2010) Arabidopsis transcription factor ELONGATED HYPOCOTYL5 plays a role in the feedback regulation of phytochrome A signaling. Plant Cell 22:3634–3649

Li G, Siddiqui H, Teng Y, Lin R, X-y W, Li J, Lau OS, Ouyang X, Dai M, Wan J, Devlin PF, Deng XW, Wang H (2011a) Coordinated transcriptional regulation underlying the circadian clock in Arabidopsis. Nat Cell Biol 13:616–622

Li Y, Li C, Xia J, Jin Y (2011b) Domestication of transposable elements into MicroRNA genes in plants. PLoS One 6:e19212

Lin R, Wang H (2004) Arabidopsis *FHY3/FAR1* gene family and distinct roles of its members in light control of Arabidopsis development. Plant Physiol 136:4010–4022

Lin R, Ding L, Casola C, Ripoll DR, Feschotte C, Wang H (2007) Transposase-derived transcription factors regulate light signaling in Arabidopsis. Science 318:1302–1305

Lin L, Shen S, Tye A, Cai JJ, Jiang P, Davidson BL, Xing Y (2008a) Diverse splicing patterns of exonized Alu elements in human tissues. PLoS Genet 4:e1000225

Lin R, Teng Y, Park H-J, Ding L, Black C, Fang P, Wang H (2008b) Discrete and essential roles of the multiple domains of Arabidopsis FHY3 in mediating phytochrome A signal transduction. Plant Physiol 148:981–992

Lisch DR (2009) Epigenetic regulation of transposable elements in plants. Annu Rev Plant Biol 60:43–66

Lisch DR, Bennetzen JL (2011) Transposable element origins of epigenetic gene regulation. Curr Opin Plant Biol 14:156–161

Lisch DR, Jiang N (2009) *Mutator* and MULE transposons. In: Bennetzen JL, Hake S (eds) Handbook of maize. Springer, New York, NY, pp 277–306

Lisch DR, Freeling M, Langham RJ, Choy MY (2001) Mutator transposase is widespread in the grasses. Plant Physiol 125:1293–1303

Lockton S, Gaut BS (2009) The contribution of transposable elements to expressed coding sequence in *Arabidopsis thaliana*. J Mol Evol 68:80–89

Lohe AR, Moriyama EN, Lidholm DA, Hartl DL (1995) Horizontal transmission, vertical inactivation, and stochastic loss of *Mariner*-like transposable elements. Mol Biol Evol 12:62–72

Lowe CB, Bejerano G, Haussler D (2007) Thousands of human mobile element fragments undergo strong purifying selection near developmental genes. Proc Natl Acad Sci USA 104:8005–8010

Lunyak VV, Atallah M (2011) Genomic relationship between SINE retrotransposons, Pol III-Pol II transcription, and chromatin organization: the journey from junk to jewel. Biochem Cell Biol 89:495–504

Magnani E, Sjölander K, Hake S (2004) From endonucleases to transcription factors: evolution of the AP2 DNA binding domain in plants. Plant Cell 16:2265–2277

Makałowski W, Mitchell GA, Labuda D (1994) Alu sequences in the coding regions of mRNA: a source of protein variability. Trends Genet 10:188–193

Malone CD, Hannon GJ (2009) Small RNAs as guardians of the genome. Cell 136:656–668

Marino-Ramirez L, Lewis KC, Landsman D, Jordan IK (2005) Transposable elements donate lineage-specific regulatory sequences to host genomes. Cytogenet Genome Res 110:333–341

Markljung E, Jiang L, Jaffe JD, Mikkelsen TS, Wallerman O, Larhammar M, Zhang X, Wang L, Saenz-Vash V, Gnirke A, Lindroth AM, Barres R, Yan J, Stromberg S, De S, Ponten F, Lander ES, Carr SA, Zierath JR, Kullander K, Wadelius C, Lindblad-Toh K, Andersson G, Hjalm G, Andersson L (2009) ZBED6, a novel transcription factor derived from a domesticated DNA transposon regulates IGF2 expression and muscle growth. PLoS Biol 7(12):e1000256

Martin A, Bendahmane A (2010) A blessing in disguise: transposable elements are more than parasites. Epigenetics 5:378–380

Martin A, Troadec C, Boualem A, Rajab M, Fernandez R, Morin H, Pitrat M, Dogimont C, Bendahmane A (2009) A transposon-induced epigenetic change leads to sex determination in melon. Nature 461:1135–1138

Mathews S (2006) Phytochrome-mediated development in land plants: red light sensing evolves to meet the challenges of changing light environments. Mol Ecol 15:3483–3503

McClintock B (1950) The origin and behavior of mutable loci in maize. Proc Natl Acad Sci USA 36:344–355

McClintock B (1984) The significance of responses of the genome to challenge. Science 226:792–801

Miller WJ, Hagemann S, Reiter E, Pinsker W (1992) *P*-element homologous sequences are tandemly repeated in the genome of Drosophila guanche. Proc Natl Acad Sci USA 89:4018–4022
Miller WJ, McDonald JF, Pinsker W (1997) Molecular domestication of mobile elements. Genetica 100:261–270
Miller WJ, McDonald JF, Nouaud D, Anxolabehere D (1999) Molecular domestication–more than a sporadic episode in evolution. Genetica 107:197–207
Mirouze M, Paszkowski J (2011) Epigenetic contribution to stress adaptation in plants. Curr Opin Plant Biol 14:267–274
Muehlbauer GJ, Bhau BS, Syed NH, Heinen S, Cho S, Marshall D, Pateyron S, Buisine N, Chalhoub B, Flavell AJ (2006) A *hAT* superfamily transposase recruited by the cereal grass genome. Mol Genet Genomics 275:553–563
Naito K, Zhang F, Tsukiyama T, Saito H, Hancock CN, Richardson AO, Okumoto Y, Tanisaka T, Wessler SR (2009) Unexpected consequences of a sudden and massive transposon amplification on rice gene expression. Nature 461:1130–1134
Nekrutenko A, Li WH (2001) Transposable elements are found in a large number of human protein-coding genes. Trends Genet 17:619–621
Nosek J, Kosa P, Tomaska L (2006) On the origin of telomeres: a glimpse at the pre-telomerase world. Bioessays 28:182–190
Nouaud D, Anxolabehere D (1997) *P* element domestication: a stationary truncated P element may encode a 66-kDa repressor-like protein in the *Drosophila montium* species subgroup. Mol Biol Evol 14:1132–1144
Nowick K, Stubbs L (2010) Lineage-specific transcription factors and the evolution of gene regulatory networks. Brief Funct Genomics 9:65–78
Obbard DJ, Gordon KHJ, Buck AH, Jiggins FM (2009) The evolution of RNAi as a defence against viruses and transposable elements. Philos Trans R Soc Lond B Biol Sci 364:99–115
Ohno S (1972) So much "junk" DNA in our genome. Brookhaven Symp Biol 23:366–370
Okada N, Sasaki T, Shimogori T, Nishihara H (2010) Emergence of mammals by emergency: exaptation. Genes Cells 15:801–812
Oki N, Yano K, Okumoto Y, Tsukiyama T, Teraishi M, Tanisaka T (2008) A genome-wide view of miniature inverted-repeat transposable elements (MITEs) in rice, Oryza sativa ssp. japonica. Genes Genet Syst 83:321–329
Oliver KR (2011) Mobile DNA and the TE-Thrust hypothesis: supporting evidence from the primates. Mob DNA 2:8
Orgel LE, Crick FHC (1980) Selfish DNA: the ultimate parasite. Nature 284:604–607
Ouyang X, Li J, Li G, Li B, Chen B, Shen H, Huang X, Mo X, Wan X, Lin R, Li S, Wang H, Deng XW (2011) Genome-wide binding site analysis of FAR-RED ELONGATED HYPOCOTYL3 reveals its novel function in Arabidopsis development. Plant Cell 23:2514–2535
Pace JK, Feschotte C (2007) The evolutionary history of human DNA transposons: evidence for intense activity in the primate lineage. Genome Res 17:422–432
Pardue M-L, DeBaryshe PG (2011) Retrotransposons that maintain chromosome ends. Proc Natl Acad Sci USA 108(51):20317–20324
Paricio N, Perez-Alonso M, Martinez-Sebastian MJ, de Frutos R (1991) P sequences of *Drosophila subobscura* lack exon 3 and may encode a 66 kd repressor-like protein. Nucleic Acids Res 19:6713–6718
Parisod C, Alix K, Just J, Petit M, Sarilar V, Mhiri C, Ainouche M, Chalhoub B, Grandbastien M-A (2010) Impact of transposable elements on the organization and function of allopolyploid genomes. New Phytol 186:37–45
Pi W, Zhu X, Wu M, Wang Y, Fulzele S, Eroglu A, Ling J, Tuan D (2010) Long-range function of an intergenic retrotransposon. PNAS 107:12992–12997
Pinsker W, Haring E, Hagemann S, Miller WJ (2001) The evolutionary life history of *P* transposons: from horizontal invaders to domesticated neogenes. Chromosoma 110:148–158
Piriyapongsa J, Jordan IK (2007) A family of human microRNA genes from miniature inverted-repeat transposable elements. PLoS One 2:e203

Piriyapongsa J, Marino-Ramirez L, Jordan IK (2007) Origin and evolution of human microRNAs from transposable elements. Genetics 176:1323–1337

Ponting CP, Nellåker C, Meader S (2011) Rapid turnover of functional sequence in human and other genomes. Annu Rev Genomics Hum Genet 12:275–299

Pritham EJ (2009) Transposable elements and factors influencing their success in eukaryotes. J Hered 100:648–655

Quesneville H, Nouaud D, Anxolabehere D (2005) Recurrent recruitment of the THAP DNA-binding domain and molecular domestication of the *P*-transposable element. Mol Biol Evol 22:741–746

Raizada MN, Benito MI, Walbot V (2001) The MuDR transposon terminal inverted repeat contains a complex plant promoter directing distinct somatic and germinal programs. Plant J 25:79–91

Rawn SM, Cross JC (2008) The evolution, regulation, and function of placenta-specific genes. Annu Rev Cell Dev Biol 24:159–181

Rebollo R, Horard B, Hubert B, Vieira C (2010) Jumping genes and epigenetics: towards new species. Gene 454:1–7

Reiss D, Nouaud D, Ronsseray S, Anxolabéhère D (2005) Domesticated *P* elements in the *Drosophila montium* species subgroup have a new function related to a DNA binding property. J Mol Evol 61:470–480

Rio DC (1990) Molecular mechanisms regulating *Drosophila* P element transposition. Annu Rev Genet 24:543–578

Robertson HM, Zumpano KL (1997) Molecular evolution of an ancient mariner transposon, Hsmar1, in the human genome. Gene 205:203–217

Rose MR, Oakley TH (2007) The new biology: beyond the modern synthesis. Biol Direct 2:30

Roussigne M, Kossida S, Lavigne AC, Clouaire T, Ecochard V, Glories A, Amalric F, Girard JP (2003) The THAP domain: a novel protein motif with similarity to the DNA-binding domain of *P* element transposase. Trends Biochem Sci 28:66–69

Rusche LN, Rine J (2010) Switching the mechanism of mating type switching: a domesticated transposase supplants a domesticated homing endonuclease. Genes Dev 24:10–14

Sabogal A, Lyubimov AY, Corn JE, Berger JM, Rio DC (2010) THAP proteins target specific DNA sites through bipartite recognition of adjacent major and minor grooves. Nat Struct Mol Biol 17:117–123

Saccaro NL Jr, Van Sluys M-A, de Mello VA, Rossi M (2007) MudrA-like sequences from rice and sugarcane cluster as two bona fide transposon clades and two domesticated transposases. Gene 392:117–125

Saijo Y, Zhu D, Li J, Rubio V, Zhou Z, Shen Y, Hoecker U, Wang H, Deng XW (2008) Arabidopsis COP1/SPA1 complex and FHY1/FHY3 associate with distinct phosphorylated forms of phytochrome A in balancing light signaling. Mol Cell 31:607–613

Sánchez-Gracia A, Maside X, Charlesworth B (2005) High rate of horizontal transfer of transposable elements in *Drosophila*. Trends Genet 21:200–203

Santangelo AM, de Souza FS, Franchini LF, Bumaschny VF, Low MJ, Rubinstein M (2007) Ancient exaptation of a CORE-SINE retroposon into a highly conserved mammalian neuronal enhancer of the proopiomelanocortin gene. PLoS Genet 3:1813–1826

Saze H, Kakutani T (2011) Differentiation of epigenetic modifications between transposons and genes. Curr Opin Plant Biol 14:81–87

Schaack S, Gilbert C, Feschotte C (2010) Promiscuous DNA: horizontal transfer of transposable elements and why it matters for eukaryotic evolution. Trends Ecol Evol 25:537–546

Schmitz J, Brosius J (2011) Exonization of transposed elements: a challenge and opportunity for evolution. Biochimie 93:1928–1934

Sela N, Mersch B, Gal-Mark N, Lev-Maor G, Hotz-Wagenblatt A, Ast G (2007) Comparative analysis of transposed element insertion within human and mouse genomes reveals Alu's unique role in shaping the human transcriptome. Genome Biol 8:R127

Sela N, Stern A, Makalowski W, Pupko T, Ast G (2008) Transduplication resulted in the incorporation of two protein-coding sequences into the *turmoil-1* transposable element of *C. elegans*. Biol Direct 3:41

Shaheen M, Williamson E, Nickoloff J, Lee S-H, Hromas R (2010) Metnase/SETMAR: a domesticated primate transposase that enhances DNA repair, replication, and decatenation. Genetica 138:559–566
Shapiro JA (2010) Mobile DNA and evolution in the 21st century. Mob DNA 1:4
Simon SA, Meyers BC (2011) Small RNA-mediated epigenetic modifications in plants. Curr Opin Plant Biol 14:148–155
Sinzelle L, Izsvák Z, Ivics Z (2009) Molecular domestication of transposable elements: from detrimental parasites to useful host genes. Cell Mol Life Sci 66:1073–1093
Slotkin RK, Martienssen RA (2007) Transposable elements and the epigenetic regulation of the genome. Nat Rev Genet 8:272–285
Slotkin RK, Freeling M, Lisch DR (2005) Heritable transposon silencing initiated by a naturally occurring transposon inverted duplication. Nat Genet 37:641–644
Smalheiser NR, Torvik VI (2005) Mammalian microRNAs derived from genomic repeats. Trends Genet 21:322–326
Smith JJ, Sumiyama K, Amemiya CT (2011) A living fossil in the genome of a living fossil: *Harbinger* transposons in the coelacanth genome. Mol Biol Evol 29(3):985–993
Sorek R (2007) The birth of new exons: mechanisms and evolutionary consequences. RNA 13:1603–1608
Studer A, Zhao Q, Ross-Ibarra J, Doebley J (2011) Identification of a functional transposon insertion in the maize domestication gene tb1. Nat Genet 43(11):1160–1163
Turcotte K, Srinivasan S, Bureau T (2001) Survey of transposable elements from rice genomic sequences. Plant J 25:169–179
Voinnet O (2009) Origin, biogenesis, and activity of plant microRNAs. Cell 136:669–687
Volff J-N (2006) Turning junk into gold: domestication of transposable elements and the creation of new genes in eukaryotes. Bioessays 28:913–922
Wang H, Deng XW (2002) Arabidopsis FHY3 defines a key phytochrome A signaling component directly interacting with its homologous partner FAR1. EMBO J 21:1339–1349
Weil CF, Martienssen RA (2008) Epigenetic interactions between transposons and genes: lessons from plants. Curr Opin Genet Dev 18:188–192
Werren JH (2011) Selfish genetic elements, genetic conflict, and evolutionary innovation. Proc Natl Acad Sci USA 108(Suppl 2):10863–10870
Whitelam GC, Johnson E, Peng J, Carol P, Anderson ML, Cowl JS, Harberd NP (1993) Phytochrome A null mutants of Arabidopsis display a wild-type phenotype in white light. Plant Cell 5:757–768
Wicker T, Sabot F, Hua-Van A, Bennetzen JL, Capy P, Chalhoub B, Flavell A, Leroy P, Morgante M, Panaud O, Paux E, SanMiguel P, Schulman AH (2007) A unified classification system for eukaryotic transposable elements. Nat Rev Genet 8:973–982
Wray GA (2007) The evolutionary significance of cis-regulatory mutations. Nat Rev Genet 8:206–216
Xing J, Zhang Y, Han K, Salem AH, Sen SK, Huff CD, Zhou Q, Kirkness EF, Levy S, Batzer MA, Jorde LB (2009) Mobile elements create structural variation: analysis of a complete human genome. Genome Res 19:1516–1526
Yanovsky MJ, Whitelam GC, Casal JJ (2000) *fhy3-1* retains inductive responses of phytochrome A. Plant Physiol 123:235–242
Yu Z, Wright SI, Bureau TE (2000) *Mutator*-like elements in *Arabidopsis thaliana*. Structure, diversity and evolution. Genetics 156:2019–2031
Zaratiegui M, Vaughn MW, Irvine DV, Goto D, Watt S, Bähler J, Arcangioli B, Martienssen RA (2011) CENP-B preserves genome integrity at replication forks paused by retrotransposon LTR. Nature 469:112–115
Zeh DW, Zeh JA, Ishida Y (2009) Transposable elements and an epigenetic basis for punctuated equilibria. Bioessays 31:715–726
Zhang J, Yu C, Pulletikurti V, Lamb J, Danilova T, Weber DF, Birchler J, Peterson T (2009) Alternative *Ac/Ds* transposition induces major chromosomal rearrangements in maize. Genes Dev 23:755–765

Chapter 13
SINE Exaptation as Cellular Regulators Occurred Numerous Times During Eukaryote Evolution

Jean-Marc Deragon

Abstract Exaptation is defined as a successful adaptation by acquiring a new function from previously useless DNA sequences. I review here recent evidences suggesting that Short INterspersed Elements (SINEs) were exapted numerous times during eukaryote evolution. I propose that the ubiquitous presence of SINEs in eukaryotes depends mainly on their parasitic nature coupled to the facility by which their RNA can be exapted and preserved long after the SINE family responsible for its biogenesis stopped amplifying. Therefore, exapted SINE RNAs (and loci responsible for their production) represent a reservoir to initiate new rounds of retroposition (and new SINE families) that in turn generate new materials for further exaptation events. While most examples of SINE exaptation come from animals, the ubiquitous nature of SINEs in plants coupled with the recent availability of a significant number of fully sequenced plant genomes should soon reveal new examples of plant SINE exaptation.

Keywords Transposable elements • SINEs • LINEs • Retrotransposons • Exaptation

13.1 Introduction

13.1.1 SINE Structure and Evolution

SINEs are nonautonomous retroelements, transcribed by the polymerase III (polIII) machinery (Chu et al. 1995; Roy et al. 2000; Arnaud et al. 2001; Nikitina et al. 2011) and propagated by a process called retroposition (Rogers 1985; Kramerov and

J.-M. Deragon (✉)
LGDP, Université de Perpignan, UMR5096, 58 av. Paul Alduy, 66860 Perpignan, France

LGDP, CNRS, UMR5096, 58 av. Paul Alduy, 66860 Perpignan, France
e-mail: jean-marc.deragon@univ-perp.fr

M.-A. Grandbastien and J.M. Casacuberta (eds.), *Plant Transposable Elements*, Topics in Current Genetics 24, DOI 10.1007/978-3-642-31842-9_13,

Vassetzky 2005). SINEs are molecular parasites of autonomous LINEs (Long INterspersed Elements) (Eickbush and Jamburuthugoda 2008) that code for essential enzymes involved in retroposition (Ohshima et al. 1996; Boeke 1997; Kajikawa and Okada 2002; Dewannieux et al. 2003). SINEs emerged de novo several times in eukaryote evolution, reusing biologically functional "modules" possessing a polIII promoter such as tRNA, 7SL RNA, and 5S RNA [for a recent review on SINE origin and evolution see Kramerov and Vassetzky (2011)]. SINE organization can become more complex when the initial module (for example the tRNA-related sequence) multimerize to form either homodimer or homotrimer SINEs or fuse with other functional modules (5S RNA or 7SL RNA) to form heterodimer and heterotrimer SINEs. The initial module can also fuse with an unrelated sequence (often of unknown origin). These composite SINEs can also form subsequently homo- and hetero-multimeric versions. Finally, all SINEs possess a 3′-terminal "tail" of variable length important for LINE protein interactions (Ohshima et al. 1996; Boeke 1997; Kajikawa and Okada 2002; Dewannieux et al. 2003). This region usually ends with more or less degenerated simple repeats (Kramerov and Vassetzky 2005).

13.1.2 SINE Distribution in Eukaryotes

SINEs are widespread among eukaryotes (Kramerov and Vassetzky 2005; Wenke et al. 2011). They are found in most metazoans with the exception of certain species such as *Caenorhabditis elegans* and a few terrestrial insects particularly of the Drosophila and Apis genders. SINEs are nevertheless present in many Isoptera, Blattoidea, Orthoptera, Lepidoptera, and Diptera species (Wilson et al. 1988; Liao et al. 1998; Tu 1999; Feschotte et al. 2001; Luchetti and Mantovani 2011). SINEs have a patchy distribution in fungi. They are absent from several species (including *Saccharomyces cerevisiae*) but are present in many others (Rasmussen et al. 1993; Kachroo et al. 1995; Mes et al. 2000). SINEs have been found in the genome of many protists (such as *Schistosoma japonicum* (Laha et al. 2000), *Entamoeba histolytica* (Irmer et al. 2010), and *Phytophtora infestans* (Whisson et al. 2005)) but data are lacking to evaluate their presence in many others. SINEs are ubiquitous in the "green" lineage being present in unicellular green algae (Cognat et al. 2008) and in most (if not all) gymnosperms and angiosperms (Wenke et al. 2011).

13.1.3 How to Explain the Ubiquitous Presence of SINEs in Genomes

The ubiquitous presence of SINEs in eukaryotes is usually proposed to result from their parasitic nature, their expansion by retroposition being faster than their rate of removal from genomes (Rogers 1985). This hypothesis is apparently supported by the general view that SINEs are very different even between closely related species (i.e., SINEs are evolving in a concerted manner). This "parasitic only" model of SINE survival in genomes is now being challenged by new data. First, in plants, the general rate of SINE turnover has been shown to be extremely high (Lenoir et al. 2005; Baucom et al. 2009), yet SINE families are ubiquitous in modern plant genomes (Wenke et al. 2011). This implies that, using a pure parasitic model for SINE expansion, the rate of SINE amplification would had to be exceptionally high in all plant lineages to ensure SINE survival over hundred of million years. Based on the low level of expression and retroposition of known autonomous plant LINE partners (Noma et al. 1999, 2000), this scenario seems unlikely. Another possibility is that new plant SINE families are created de novo from functional RNA module (i.e., tRNAs) at a very high rate compensating for the regular extinction of older SINE families, but in that case we should be able to retrace easily the (recent) tRNA origin of plant SINEs and this is not what is generally observed (Deragon and Zhang 2006; Wenke et al. 2011). Second, several examples or highly conserved SINE families have now being found in animals and plants strongly suggesting that at least some SINE loci are evolving under purifying selection (Gilbert and Labuda 1999; Ogiwara et al. 2002; Bejerano et al. 2006; Fawcett et al. 2006; Akasaki et al. 2010; Piskurek and Jackson 2011). Also, most tRNA-related SINEs that are not conserved at the primary sequence level have a related RNA structure (Sun et al. 2007), suggesting that selection can also operate at this level. These new data suggest that the ubiquitous distribution of SINEs in eukaryote may not rely solely on the parasitic nature of these elements but that selection may come into play from time to time in a process called exaptation (Brosius and Gould 1992).

Exaptation was originally defined as a process in which an existing trait is acquiring a new function different from a previous one (Gould and Vrba 1982). An example of exaptation concerns bird feathers that evolved initially as insulation but were exapted eventually for bird flight. Latter this concept was extended to junk DNA that can eventually get a "useful" function (Brosius and Gould 1992). Apparently SINEs are very good material to be exapted (see below for why this could be the case) as several cases have been published lately. I will summarize below the data supporting putative cases of SINEs exaptation in eukaryotes. Table 13.1 summarizes the strongest evidence of SINE exaptation described in the paper.

13.2 Exaptation of polIII-Specific SINE RNAs

13.2.1 Conservation of SINE RNA Structures

Most SINEs originate from functional RNAs (tRNA, 5S RNA, or 7SL RNA) (Kramerov and Vassetzky 2005). In the process of becoming a SINE, the postulant SINE RNA must first evade the ancestral RNA structure to avoid being treated as the ancestral RNA by some cellular processing factors (Rozhdestvensky et al. 2001;

Table 13.1 Summary of the main exaptation events involving SINEs

SINE (SINE-derived)	Level of exaptation	Level of regulation	Species	Main references
Alu/B2	RNA (polIII-specific transcript)	PolII-transcription (heat shock)	Human/ mouse	Espinoza et al. (2004), Yakovchuk et al. (2009)
Alu	RNA (polIII-specific transcript)	Translation (viral and other stresses)	Human	Schmid (1998), Rubin et al. (2002)
BC200/BC1	RNA (polIII-specific transcript)	Translation (neuron)	Human/ mouse	Kondrashov et al. (2005)
SB1	RNA (polIII-specific transcript)	miRNAs pathway	Arabidopsis	Pouch-Pelissier et al. (2008)
(NMD29)	RNA (polIII Alu-related transcript)	Cell cycle and differentiation	Human	Castelnuovo et al. (2010)
(SnaR)	RNA (polIII Alu-related transcript)	Translation (testis, pituitary gland)	Primates	Parrott et al. (2011)
Alu	RNA (polII co-transcript)	Staufen1-mediated mRNA decay	Human	Gong and Maquat (2011a, b)
ID	RNA (polII co-transcript)	Targeting of mRNAs to dendrites	Rodents	Buckley et al. (2011)
CORE-SINE	DNA	Enhancer function	Mammals	Santangelo et al. (2007)
LF-SINE	DNA	Enhancer function	Tetrapods	Bejerano et al. (2006)
Nin-DC-SINE	DNA	Enhancer function brain development	Metazoans	Piskurek and Jackson (2011), Sasaki et al. (2008)
B2	DNA	Boundary element	Mouse	Lunyak et al. (2007)
B1	DNA	Boundary element	Mouse	Roman et al. (2011)

Sun et al. 2007). For example, tRNA-related SINE must avoid recognition by tRNA-processing enzymes, in particular the 3′ endonuclease (RNase Z) (Morl and Marchfelder 2001) that by cleaving the 3′-end of SINE RNA would prevent selection by the LINE machinery (Kajikawa and Okada 2002; Dewannieux et al. 2003). To generate a SINE family, the "new" SINE RNA structure must also be competent to interact efficiently with LINE products. LINEs are evolving fast to evade host-repressing factors (Eickbush and Jamburuthugoda 2008). If the SINE/LINE interaction is the major force shaping SINE RNA structures in evolution, then these structures should not be conserved over long evolutionary periods. Sun and collaborators analyzed the RNA secondary structure of many eukaryotes (tRNA-related) SINEs (Sun et al. 2007). They observed that, as expected, the typical tRNA cloverleaf structure is not apparent for most SINE consensus RNAs but that surprisingly common secondary structural motifs are nevertheless present.

Using a cladistic method where RNA structural components were coded as polarized and ordered multistate characters, they were able to show that indeed related structural motifs are present in most SINE RNAs from mammals, fishes, and plants suggesting common selective constraints are imposed on SINEs at the RNA structural level (Sun et al. 2007). One possibility to explain this result is that before being able to interact with LINE products, SINE RNAs must survive the many degradation pathways present in the host. To do so, SINE polIII transcripts are likely to interact with a limited number of conserved host factors usually involved in stabilizing functional cellular RNAs. Although the nature of these host factors is mostly unknown, the La-autoantigen and the Poly(A)-binding protein may be involved in assembling the SINE ribonucleoprotein complex (RNP) (Goodier and Maraia 1998; Kremerskothen et al. 1998; Fleurdepine et al. 2007). The necessity for SINE RNA to interact with conserved stabilizing (RNA-binding) host factors is likely responsible for the observed selective constraints imposed on SINEs at the RNA structural level. Of course, a new SINE family can emerge only if these stabilized polIII SINE RNAs can also interact efficiently with LINE factors and behave as retroposition intermediates. Therefore, to generate a burst of SINE retroposition, a compromise must be reached between the RNA structure needed to evade cellular degradation pathways and the RNA structure allowing LINE-mediated retroposition. This compromise is unlikely to last since the SINE RNA structure imposed by the binding of conserved host factors is unlikely to stay compatible with fast evolving LINE products over long evolutionary periods. Nevertheless, during the period of SINE retroposition, this compromise will be amplified, generating a population of slightly different stable noncoding RNAs in the host. These stable noncoding RNAs represent interesting material to be exapted to perform a biologically function unrelated with retroposition. Indeed these slightly different SINE RNPs contain conserved core host functional proteins that likely can interact with a network of other functional cellular proteins possibly involved in RNA metabolism (transport, stability, transcription, translation etc.). SINE RNPs may, therefore, be particularly prone to evolve as regulators of RNA metabolism. This type of exaptation event may be hard to detect by comparative genome analysis as selection at the RNA structure level allows SINE primary sequence to accumulate conservative changes, so that SINE under this type of selection are not necessary highly conserved at the DNA level even among closely related species. Also the exapted SINE RNAs can come from many distinct loci limiting again the action of selection at the primary sequence level. Nevertheless, the direct consequence of SINE RNA exaptation is that the corresponding (SINE-like) RNA secondary structure can survive long evolutionary period even if the SINE family that generated this noncoding RNA dies out. Subsequently, following the introduction of a new LINE partner for example, transcripts from exapted SINE loci can be used to reach a new compromise leading to the emergence of a different SINE family. In that scenario, exapted SINE RNAs (and the loci responsible for their production) represent a reservoir to initiate new round of retroposition that in turn generate new materials for further exaptation events. Such a mechanism could explain the ubiquitous presence of numerous SINE families from protist to human. If this scenario is correct, we should be able to document cases were SINE RNAs are involved in regulating important cellular functions. Also the biological impact of different SINE RNAs is likely to vary since it results from independent exaptation events. A few cases of putative SINE RNA exaptation supporting this model have been described recently.

13.2.2 Exaptation of Animal SINE RNA

Rodent tRNA-derived B2 SINE RNAs have been implicated as regulator of polII transcription during heat shock (Allen et al. 2004; Espinoza et al. 2004, 2007). B2 polIII-specific transcripts accumulate during heat shock and bind specifically to the polII complex blocking transcription of most genes except the ones coding for heat shock response factors. B2 RNA is proposed to repress transcription by preventing polII from properly engaging the DNA after assembling into complexes with promoter-associated general transcription factors (Yakovchuk et al. 2009). Surprisingly, the 7SL-derived human Alu SINE RNA (but not B1, its rodent equivalent) may also act in a similar way (Mariner et al. 2008; Yakovchuk et al. 2009). Human Alu-specific transcripts were also proposed to regulate translation during viral infection (as well as other stresses) either by binding to PKR (double stranded RNA-activated protein kinase R) or by an uncharacterized PKR-independent mechanism (Schmid 1998; Williams 1999; Rubin et al. 2002; Hasler and Strub 2006). Alu RNA accumulates following cell stress and viral infection (Liu et al. 1995; Wick et al. 2003) and modulates PKR activity as does some viral noncoding (polIII) RNAs such as VA1 and 2 from adenovirus and EBER1 and 2 for Epstein–Barr virus (Maran and Mathews 1988; Kitagawa et al. 2000). PKR can phosphorylate the eIF2α translation factor leading to a general downregulation of translation (Williams 1999). PKR plays also a role in regulating cell apoptosis (Williams 1999) and Alu RNA could be involved in regulating this process as well. The tRNA-derived BC1 SINE RNA in rodent also regulates translation in neuron but using a different mechanism (Martignetti and Brosius 1993b; Kondrashov et al. 2005; Wang et al. 2005; Lin et al. 2008). BC1 RNAs bind to eIF4A and the poly(A) binding protein (PABP) sequestering locally key translational factor and preventing 48S preinitiation complex formation in neurons. BC1 is also a "founder locus" responsible for the amplification of the ID SINE family in rodent (Kim et al. 1994). This represents a clear example of an exapted SINE locus that have maintained retroposition activity. A recent secondary exaptation of rat ID sequences was also found (Buckley et al. 2011 and see below) illustrating the potential of successive exaptation/retroposition cycles.

13.2.3 Exaptation of Plant SINE RNAs

A plant tRNA-related SINE (named SB1) is also possibly involved in regulating the microRNA pathway (Pouch-Pelissier et al. 2008). SB1 SINE RNAs can interact with a double stranded RNA-binding protein (DRB1) present in association with the Dicer-like protein 1 (DCL1), in the major dicer complex involved in microRNA biogenesis (Mallory and Vaucheret 2006). In *Arabidopsis thaliana*, the binding of SB1 RNAs to DRB1 involves the first stem–loop that mimics the structure of a microRNA precursor, the natural substrate of DRB1 (Pouch-Pelissier et al. 2008). Following binding, the SINE RNA is cleaved in small 21 nucleotides SB1-related small RNAs by DCL1. This whole process competes with the cleavage of microRNA precursors by DCL1, and plants expressing constitutively SB1 RNAs have a reduced amount of most mature miRNAs and present serious developmental defects (Pouch-Pelissier et al. 2008). SINE transcription in *Arabidopsis* is developmentally regulated, as SINE RNAs are only present in roots and flowers (Clavel and Deragon, unpublished result). One possibility is that plant SINE RNAs are able to finely tune microRNA production at critical step of plant development. It is interesting to note that the first stem–loop in the SINE RNA secondary structure is one of the most conserved features of eukaryote tRNA-derived SINEs (Sun et al. 2007), and competition between SINE RNAs and dicer complexes may take place in many eukaryotes.

13.2.4 Reorganization of SINE RNAs Leading to New Riboregulators

Apart from these SINE exaptation events, where the whole polIII SINE transcript is directly exapted, a number of events where part of a SINE polIII transcript was reorganized to yield a new functional polIII noncoding RNA have been documented. The BC200 RNA, a neuron-specific noncoding RNA present in a single locus in all primates, is a reorganized ancestral SINE Alu sequence (FLAMC) (Watson and Sutcliffe 1987; Martignetti and Brosius 1993a; Tiedge et al. 1993; Khanam et al. 2007). Curiously, the 7SL-related BC200 was shown to be a functional equivalent of the rodent-specific (tRNA-related) BC1 RNA discussed above (Kondrashov et al. 2005). This example of convergent evolution support the idea that retroposition is a powerful diversity generating device providing raw material (in that case, in the form of a population of stable small noncoding RNAs) that can be adapted to cellular needs by selection. The Alu RNA sequence was also recycled by retroposition into a noncoding polIII transcript named NDM29 (Castelnuovo et al. 2010). In human, NMD29 is strongly increased in differentiating cells (where it leads to a slowdown of the cell cycle) and is markedly reduced in malignant highly proliferating cells. Indeed, this small noncoding RNA was shown to promote cell differentiation and reduce malignancy of human neuroblastoma cells (Castelnuovo et al. 2010). In this case, Alu retroposition (and its associated shuffling of sequences) generated a new noncoding RNA that was apparently exapted to control cell cycle. Another very interesting case where portions of SINE RNA were exapted concerns the snaR family of small noncoding RNAs in primates (Parrott et al. 2011). SnaR RNAs are polIII transcripts found only in human and chimpanzee that bind ribosomes and likely regulate translation in their tissues of expression (mainly testis and pituitary gland). SnaR RNAs are produced from multiple loci that arose by segmental duplications. SnaR RNAs originate from the rearrangement of two families of SINE-like retroposons; the ASR family present in all monkeys and apes and the related CAS family present only in Old World Monkey and apes (Parrott et al. 2011). Interestingly both ASR and CAS evolved from the left monomer of the SINE Alu family. This represent an illustration of the retroposition/exaptation cycle discussed above. In the common ancestor of primates, a well-established retroposing SINE family (Alu) gives rise by retroposition to a copy that, after diverging, was subsequently amplified in a new retroposition burst to generate the ASR and latter the CAS families. These sequences continue to amplify and diversify by retroposition until a copy is rearranged into the first snaR, in the common ancestor of human and chimpanzee. In that particular case, this snaR copy apparently lost its capacity to retropose and diversified latter by segmental duplications, but one can imagine that if the first exapted snaR copy had kept its capacity to retropose, it could have generated a new SINE family and possibly further exaptation events.

13.2.5 Toxic Effects of SINE RNA Deregulation

Two independent studies, one in animals and one in plants revealed that miss expressing SINE RNAs can be toxic. The deregulated accumulation of full length polIII-specific Alu or B1/B2 SINE RNAs in retinal pigmented epithelium (RPE) cells is responsible for the development of an advanced form of age-related macular degeneration in human and mice, respectively (Kaneko et al. 2011). The accumulation of SINE RNAs is linked to a depletion of the DICER1 RNase in RPE. Subretinal injection of Alu RNAs induced RPE degeneration in wild-type mice, supporting the assignment of disease causality, while injection of tRNA or 7SL RNA did not (Kaneko et al. 2011). Therefore, Alu RNA-induced RPE degeneration cannot be attributed solely to its double-stranded nature but more likely to its precise RNA structure. The mechanism leading to RPE degeneration following Alu RNA accumulation is not known but results obtained in plants may provide a first hint. As mentioned above, the deregulated expression of SB1 SINE RNAs in *Arabidopsis* is leading to the miss-expression of several microRNAs and to a very severe developmental phenotype (Pouch-Pelissier et al. 2008). However, only part of this aberrant phenotype can be linked to the deregulation of the microRNA pathway. Using a double-stranded RNA binding domain from *Xenopus laevis* ADAR1, these authors have shown that the plant SB1 RNA is fit to bind highly divergent double-stranded RNA-binding protein (dsRBP). Therefore many dsRBPs could be affected by SINE RNA expression and not only those involved in microRNA biogenesis or RNAi. This result suggest that SINE RNAs can potentially affect many basic cellular processes involving dsRBPs and highlights the requirement for the host to strictly control the expression of these elements, although they are nonautonomous for retroposition.

13.3 Exaptation of SINE Sequences Present in Messenger RNAs

SINE RNA can also be exapted following the integration of a SINE copy into a transcribed genic region. It that case, the SINE RNA is produced as part of a messenger RNA by the polII machinery (polII co-transcription). This type of SINE RNA exaptation is also potentially very important, but it will not contribute to the retroposition/exaptation cycle described above. SINE RNAs embedded in polII transcript have been shown to affect mRNA expression in numerous ways (reviewed in Ponicsan et al. 2010). However one must distinguish the general impact SINE sequences may have on mRNA splicing, stability, polyadenylation, edition, nuclear export, protein production, etc., from precise exaptation events, where a SINE fragment in a given mRNA has been selected to perform a function important for host survival. A very convincing example of SINE sequence exaptation in mRNAs involves the regulation of the Staufen 1 (STAU1)-mediated messenger RNA decay (SMD) pathway in mammals (Kim et al. 2007; Gong and Maquat 2011a, b). SMD is a decay pathway that targets a significant amount of mRNAs in various human tissues (for example, 1.6 % of protein-coding transcripts are targets of SMD in human epithelial cells). SMD is in competition with the more general nonsense-mediated mRNA decay (NMD) pathway and this competition contributes to differentiation of myoblasts to myotubes (Gong et al. 2009). STAU1, as for *Arabidopsis* DRB1 and human PKR, possesses two double stranded RNA-binding domains and can bind structured RNA motifs in the 3′UTR of mRNA targets. Gong and Maquat observed that 13 % of all SMD targets possess a single Alu SINE element in their 3′ UTR (this number is only 4 % for other mRNAs) showing that Alu elements are enriched in SMD targets relative to the bulk of cellular mRNAs (Gong and Maquat 2011b). They also show that a STAU1-binding site can be formed in trans by base-pairing between an Alu element in the 3′UTR of a SMD mRNA target and another Alu in a cytoplasmic, polyadenylated (polII-generated) long noncoding RNAs (lncRNA). They identified 378 human lncRNAs containing a single Alu element and experimentally confirmed that at least four of them can regulate Alu-containing mRNAs in a STAU1-dependent manner. Each Alu-containing lncRNA has different binding properties and therefore different abilities to transactivate STAU1 binding to Alu-containing mRNAs. These finding strongly suggest that Alu sequences were exapted to create a mRNA regulatory network (Gong and Maquat 2011b). Another type of Alu-regulated mRNA network was also proposed to influence cancer development (Moolhuijzen et al. 2010). Alu-containing mRNAs in cancerous tissues are significantly underrepresented in comparison to the situation in normal tissues. In parallel Alu-siRNAs are increased in the same tissues suggesting that Alu-siRNAs could be involved in downregulating Alu-containing mRNAs in cancer cells (Moolhuijzen et al. 2010). However, for the moment, direct experimental evidences are lacking to support the existence of such a network. Another interesting, but more recent case of exaptation of SINE sequences in mRNAs, concerns the use of rat SINE ID sequences for the targeting of different mRNAs to the dendrites (Buckley et al. 2011). As mentioned above, ID elements are derived retroposed copies of the BC1 RNA (Kim et al. 1994), a regulator of neuron translation in rodent. ID elements are highly abundant in rats with approximately 150,000 copies while the mouse genome contains only around 1,000 copies (Buckley et al. 2011). Buckley and collaborators observed that introns are retained in a number of rat dentritically targeted mRNAs and that many of the retained introns contain ID elements. A portion of these SINEs was shown to be essential for targeting different essential mRNAs to the dendrites and by doing so to alter the distribution of endogenous proteins in neurons. The ID sequence was, therefore, exapted to work as a common dendritic-targeting element across multiple RNAs. This exaptation event is recent (less than 20 million years), and the very high copy number of ID elements in rats, compared to other rodents, is likely to have favored this rapid functionalization.

No clear example of the functional use of SINE sequences as part of plant mRNAs have been described yet. However the situation in maize is intriguing. In this plant, 1,190 out of 1,991 SINE copies (60 %) are found within transcribed genic regions that represent less than 15 % of the maize genome (Baucom et al. 2009). Furthermore, the presence of SINEs in transcribed maize region is apparently affecting splicing. Indeed, 88 % of the 767 maize genes with an intronic SINE possesses alternative mature mRNAs, compared to 33 % for the bulk of maize genes (Personal unpublished results). Reasons for these strong biases are unknown, but it should be interesting to investigate whether or not maize SINE sequences have been exapted to modulate splicing in this species.

13.4 Exaptation of SINE-Related DNA Modules: The Ultraconserved SINEs

The recent discovery that a few SINE families were conserved for several hundred million years was completely unexpected. SINEs are usually presented as noncoding and nonfunctional pieces of DNA and are expected to drift rapidly as for any nonfunctional DNA sequences. As discussed above, even in the hypothesis that SINE RNAs may evolve under some type of selection, this should not lead to a high conservation of SINE primary sequences over long evolutionary periods. Consequently SINEs from different species (even closely related species) are usually completely different (Kramerov and Vassetzky 2005, 2011; Deragon and Zhang 2006; Wenke et al. 2011). This "concerted" mode of evolution is used as a strong argument against the possibility that SINEs may perform any type of biologically important function. However, the discovery of ultraconserved SINE families revealed that a least some SINE DNA modules must evolve under strong purifying selection.

13.4.1 CORE-SINEs

The first ultraconserved SINE family discovered the CORE-SINE family (Gilbert and Labuda 1999, 2000) is present in the genomes of mammals, reptiles, birds, and mollusk and, therefore, survived for more than 550 millions years. This element gave rise to a number of different SINE subfamilies such as the mammalian-wide interspersed repeats (MIRs) that all share a common 65-bp "core sequence." Although CORE-SINEs are no-longer retropositionally active in eutherians, they are in marsupials and monotremes (Kirby et al. 2007; Munemasa et al. 2008). The reason for the high conservation of the core sequence was thought initially to be selfish. This core region could stabilize SINE RNA and its capacity to interact with its LINE partner (Gilbert and Labuda 1999). However this scenario appears unlikely as LINE partner in the various species are evolving at a very fast rate (see above), and the core domain should do the same and diverge rapidly. Another possibility is that the core domain strongly enhanced the probability of fusing de novo tRNA sequence with a 3′-tail composed of simple repeats, generating over and over new SINEs with the same internal domain (Gilbert and Labuda 1999, 2000). In theory, high recombination activity of the core domain could explain its conservation but it is mechanically very difficult to see how this short motif could perform this task and why core sequence should be conserved over long evolutionary period once retroposition has stopped in a given lineage (as for CORE-SINE in eutherians for example). A third possibility is that the core domain possesses an independent function in genomes and that its presence in multiple copies may be advantageous for host survival. This could explain why the core domain was conserved for long evolutionary periods even in lineages where the CORE-SINE family is no longer retropositionally active. Recently, this third hypothesis was supported by the observation that a key neuronal enhancer of the proopiomelanocortin gene originated from the exaptation of a CORE-SINE retroposon in the common ancestor of all mammals 170 million years ago (Santangelo et al. 2007). The same authors observed that several highly conserved exonic, intronic, and intergenic sequences in mammalian genomes also originated from the exaptation of CORE-SINE. Therefore, the core domain is likely to be under selection for its capacity to act as an enhancer (or contribute otherwise to gene expression).

13.4.2 LF-SINEs

A similar situation to the CORE-SINE was found latter for two other SINE families, LF-SINEs (Bejerano et al. 2006) and Nin-DC-SINEs (Piskurek and Jackson 2011). The LF-SINE family was active in the common ancestor of lobe-finned fishes and terrestrial vertebrates at least 410 million years ago. It is no longer retropositionally active in tetrapodes but is still active in the modern genome of the coelacanth. The consensus LF-SINE, made using the recent copies found in the coelacanth genome, is 481 bp and many human LF-SINEs are 80 % identical over nearly their entire length to the coelacanth consensus, strongly suggesting that the whole LF-SINE sequence is under selection. Supporting this view is the observation that hundreds of LF-SINEs detected in tetrapodes form orthology groups, in which each orthologue is in the same relative location with respect to the surrounding genes in all tetrapodes where it is present (Bejerano et al. 2006). Also examination of the orthology groups revealed that they all evolved significantly more slowly than would be expected assuming neutrality indicting that in most (if not all) instances, LF-SINEs in tetrapods have been exapted into cellular roles benefiting the host. Experimental work confirms that one LF-SINE copy is used in human (and likely in all tetrapods) as enhancer for the neuro-developmental gene ISL1 (Bejerano et al. 2006). Another copy codes for an ultra-conserved 31-amino-acid residue alternatively spliced exon of the messenger RNA processing gene PCBP2 (Bejerano et al. 2006).

13.4.3 Nin-DC SINEs

The Nin-DC SINE family emerged in the common ancestor of all metazoan more than 600 million years ago (this family was formerly named Deu-SINE since it was first found in many Deuterostomian species) (Piskurek and Jackson 2011). All Nin-DC SINEs possess a central conserved domain of around 300 bp named the Nin-domain (formerly the Deu-domain). Nin-DC-SINEs are found in representatives of Deuterostomia, Lophotrochozoa, Ecdysozoa, and Cnidaria making this family the most phylogenetically widespread SINE currently known. All Nin-DC-SINEs possess, in addition to the central Nin-domain, a 5′ tRNA-related polIII promoter region and a variable 3′ tail repeat sequence both unique to each metazoan. Therefore different metazoan lineages possess different subfamilies of Nin-DC-SINEs (the same is true for CORE-SINEs). Nin-DC SINEs are probably retropositionally active in metazoan, except in amniotes were they lost this capacity (Piskurek and Jackson 2011). Clear examples of Nin-DC-SINE exaptation were found by looking closely at mammalian genomes. Two copies of AmnSINEs (the amniote-specific variant of Nin-DC-SINE) were shown experimentally to be responsible for the correct expression of two genes (FGF8 and SATB2) both involved in mammalian brain development (Nishihara et al. 2006; Sasaki et al. 2008; Okada et al. 2010). Since more than 100 AmnSINE loci are highly conserved in all mammals, it is proposed that the mammalian-specific exaptation of AmnSINEs have contributed significantly to the morphological evolution of the mammalian-specific characters (Okada et al. 2010). By analogy to the mammalian situation, it is also possible that exaptation events of Nin-DC-SINEs in other metazoan lineages contributed to their evolution.

13.4.4 Other Ultraconserved SINEs

Three other ultraconserved SINE families have been identified: the Ceph-SINEs (Akasaki et al. 2010), the V-SINEs (Ogiwara et al. 2002), and the plant-specific Au-SINE (Fawcett et al. 2006). The Ceph-SINEs amplified in the common ancestor of cephalopods (around 500 million years ago). The V-SINE apparently amplified in the common ancestor of all vertebrates (around 550 million years ago) but was subsequently lost in many lineages (i.e., in all amniotes and in the Salmonidei).

Ceph-SINEs and V-SINEs possess a central conserved domain and more variable 5′ and 3′-region much like CORE-SINEs and Nin-DC-SINEs. The plant-specific Au-SINE amplified in a common ancestor of gymnosperm and angiosperm, more than 320 million years ago (Fawcett et al. 2006; Yagi et al. 2011). Like for V-SINEs, Au-SINEs present a patchy distribution. Copies of Au-SINEs are found in a few gymnosperms, in a basal angiosperm (*Asimina triloba*) and in many Gramineae, Solanaceae, and Fabaceae species but not in rice and *Arabidopsis thaliana* (Fawcett et al. 2006; Yagi et al. 2011). A phylogenetic tree constructed using Au-SINE sequences is fully compatible with a vertical transmission of these sequences and do not support horizontal transfer as a mechanism likely to explain the patchy distribution observed (Fawcett et al. 2006). As for LF-SINE, Au-SINE consensus sequences are more than 80 % identical over nearly their entire length among various plants species. Although SINEs are widespread in plants (Wenke et al. 2011), the Au-SINE is for the moment the only case of SINE conservation for long evolutionary time in that kingdom. It is likely that Ceph-SINE, V-SINE, and Au-SINE survival for several hundred of million years involved exaptation and selection at some point, although no case of exaptation has been described for them yet. It would be interesting to investigate the functional impact of these sequences and to determine, as for CORE-SINEs, LF-SINEs, and Nin-DC-SINEs, if they form clear orthology groups.

13.4.5 Sequence Conservation Does Not Necessarily Implies Retroposition

It is intriguing to see that SINE sequence conservation does not necessarily implies conservation of retroposition capacity for the corresponding family. For example, the CORE-SINE family lost its retroposition capacity in eutherians but kept amplifying in marsupials and monotremes (Kirby et al. 2007; Munemasa et al. 2008). LF-SINEs kept amplifying in the coelacanth lineage (for more than 410 million years) while their retropositition capacities was lost in tetrapodes (Bejerano et al. 2006). Nin-DC-SINEs kept amplifying in all metazoans except in amniotes (Piskurek and Jackson 2011). Could the retroposition of an ultraconserved SINE family be under selection? What could be the advantage of keeping the retroposition capacity of a given SINE family over long evolutionary period? How can this be achieved? Obviously, amplifying a functional SINE DNA module that possesses gene regulating properties may lead, following selection, to major innovations. However, this putative long-term advantage of retroposition is not accessible to selection and the reason why some SINE families kept their retroposition capacities for very long evolutionary periods must depend on more immediate constraints. If a given SINE locus is under selection for both generating a functional RNA and acting as a regulatory DNA module, then evading retroposition competency by mutation may be difficult without altering any of these two functions. This double selection hypothesis is compatible with the fact that these ultraconserved SINEs are part of the most evolutionarily constrained regions in eukaryotes (Bejerano et al. 2006). Keeping ultraconserved SINE retroposition competency may result in conserving a SINE copy that is both critical for host survival and retropositionally competent. In that scenario, the lineage that lost retroposition competency managed to destroy this amplification capacity without altering the two cellular functions. However, lineages that did not evolve this solution may have a higher potential for variability on the long term. It still remains difficult to understand how LINE partners managed to stay in tune with the same SINE sequence over such long evolutionary periods. One possibility is that at least part of the LINE sequence (coding for factors involved in RNA selection?) was also, in the same time period, stabilized by selection. Putative examples of LINE sequence exaptation have been described (reviewed in Kazazian 2004). In conclusion, I suggest that the very strong evolutionary constraints imposed simultaneously on ultraconserved SINEs at the DNA and RNA levels are responsible for their exceptionally long evolutionary retroposition period. This, in turn, may have been key in achieving some critical adaptations during eukaryote evolution (Okada et al. 2010).

13.5 Recent Cases of SINE-Related DNA Module Exaptation

Other more recent cases of SINE-related DNA module exaptation, not involving ultraconserved SINE, have been described recently. A B2 SINE was shown to act in chromatin as a boundary element regulating the mouse growth hormone (GH) gene in a developmental and tissue-specific manner (Lunyak et al. 2007). The bidirectional transcription of this B2 element (by polIII for the sense transcript and polII for the antisense transcript) is essential to reposition the GH gene from a heterochromatic region to a more permissive euchromatic environment during pituitary development leading to its activation. How bidirectional transcription of the B2 SINE results in boundary element function is not yet understood. A subfamily of the mouse SINE B1 (called B1-X35S) was also shown recently to have potent intrinsic boundary activity in cultured cells and live animals (Roman et al. 2011). Here again the B1 element needs to be transcribed by polIII to act as boundary, but this activity is strongly enhanced by the binding of two transcription factors (AHR and SLUG) and the engagement of polII to generate a second transcript on the same strand. Since a copy of B1-X35S is found in the promoter region of more than 1,300 mouse genes, it suggests that this SINE subfamily as a widespread impact on gene expression that may be important during normal development as well as in pathological conditions.

13.6 Conclusions

Why SINEs are so prone to exaptation compared to other type of repeats? SINEs are noncoding elements so that a given SINE RNA cannot depend on its own translation product to be stabilized and must rely on host factors for survival.

Most SINEs originate ancestrally from a highly conserved functional RNAs (tRNA, 7SL RNA or 5S RNA) and are transcribed by the polIII machinery. This situation likely favors the recruitment of similar host stabilizing factors on any given SINE RNA. Resulting SINE RNPs may be prone to exaptation since they contain core functional proteins that can interact with a network of other functional cellular proteins possibly involved in RNA metabolism. Also, retroposition of small SINEs is better tolerated in genomes (compared to the amplification of larger transposable elements) so that SINEs can sometime amplify to very high copy numbers, like in the mammalian lineage. The diversity generated by retroposition (that involves point mutations, insertions, deletions, and sequence shuffling), the high copy numbers of some SINE families, associated to the fact that SINEs possess an internal polIII promoter that can bind key transcription factors and chromatin regulators (Nikitina et al. 2011), makes secondary exaptation of SINE DNA modules likely. Therefore, SINE RNA and SINE DNA have both an intrinsically high potential for exaptation and retroposition generates the diversity on which selection can act to functionalize these sequences. I propose that SINE survival in genomes depends mainly on their parasitic nature coupled to the capacity of SINE RNAs to be exapted over relatively short evolutionary period. The capacity of SINEs to amplify to high copy numbers and to diversify can lead, in some cases, to secondary exaptation events where the SINE primary sequence is also recruited to perform a function. The two levels of selection (RNA and DNA) can possibly work in parallel leading to the conservation of SINE sequence over long evolutionary periods (i.e., ultraconserved SINEs). However, I suggest that in most cases selection at the RNA level is the major force driving SINE evolution leaving the SINE primary sequence (except for the internal polIII promoter) relatively free to change as long as key secondary RNA motifs are maintained. The number of documented cases of SINE RNA exaptation is at the moment too limited to confirm this retroposition/exaptation model, but if more cases were to be found it could provide an explanation to a long-standing question: how these so-called parasitic SINEs managed to colonize most eukaryote genomes so efficiently?

Acknowledgments I thank Cecile Bousquet-Antonelli, Nicolas Gilbert and Damian Labuda for critical reading of the manuscript.

References

Akasaki T, Nikaido M, Nishihara H, Tsuchiya K, Segawa S, Okada N (2010) Characterization of a novel SINE superfamily from invertebrates: "Ceph-SINEs" from the genomes of squids and cuttlefish. Gene 454:8–19

Allen TA, Von Kaenel S, Goodrich JA, Kugel JF (2004) The SINE-encoded mouse B2 RNA represses mRNA transcription in response to heat shock. Nat Struct Mol Biol 11:816–821

Arnaud P, Yukawa Y, Lavie L, Pélissier T, Sugiura M, Deragon JM (2001) Analysis of the SINE S1 Pol III promoter from Brassica; impact of methylation and influence of external sequences. Plant J 26:295–305

Baucom RS, Estill JC, Chaparro C, Upshaw N, Jogi A, Deragon JM, Westerman RP, Sanmiguel PJ, Bennetzen JL (2009) Exceptional diversity, non-random distribution, and rapid evolution of retroelements in the B73 maize genome. PLoS Genet 5:e1000732

Bejerano G, Lowe CB, Ahituv N, King B, Siepel A, Salama SR, Rubin EM, Kent WJ, Haussler D (2006) A distal enhancer and an ultraconserved exon are derived from a novel retroposon. Nature 441:87–90

Boeke JD (1997) LINEs and Alus–the polyA connection. Nat Genet 16:6–7

Brosius J, Gould SJ (1992) On "genomenclature": a comprehensive (and respectful) taxonomy for pseudogenes and other "junk DNA". Proc Natl Acad Sci USA 89:10706–10710

Buckley PT, Lee MT, Sul JY, Miyashiro KY, Bell TJ, Fisher SA, Kim J, Eberwine J (2011) Cytoplasmic intron sequence-retaining transcripts can be dendritically targeted via ID element retrotransposons. Neuron 69:877–884

Castelnuovo M, Massone S, Tasso R, Fiorino G, Gatti M, Robello M, Gatta E, Berger A, Strub K, Florio T, Dieci G, Cancedda R, Pagano A (2010) An Alu-like RNA promotes cell differentiation and reduces malignancy of human neuroblastoma cells. FASEB J 24:4033–4046

Chu WM, Liu WM, Schmid CW (1995) RNA polymerase III promoter and terminator elements affect Alu RNA expression. Nucleic Acids Res 23:1750–1757

Cognat V, Deragon JM, Vinogradova E, Salinas T, Remacle C, Marechal-Drouard L (2008) On the evolution and expression of *Chlamydomonas reinhardtii* nucleus-encoded transfer RNA genes. Genetics 179:113–123

Deragon JM, Zhang X (2006) Short interspersed elements (SINEs) in plants: origin, classification, and use as phylogenetic markers. Syst Biol 55:949–956

Dewannieux M, Esnault C, Heidmann T (2003) LINE-mediated retrotransposition of marked Alu sequences. Nat Genet 35:41–48

Eickbush TH, Jamburuthugoda VK (2008) The diversity of retrotransposons and the properties of their reverse transcriptases. Virus Res 134:221–234

Espinoza CA, Allen TA, Hieb AR, Kugel JF, Goodrich JA (2004) B2 RNA binds directly to RNA polymerase II to repress transcript synthesis. Nat Struct Mol Biol 11:822–829

Espinoza CA, Goodrich JA, Kugel JF (2007) Characterization of the structure, function, and mechanism of B2 RNA, an ncRNA repressor of RNA polymerase II transcription. RNA 13:583–596

Fawcett JA, Kawahara T, Watanabe H, Yasui Y (2006) A SINE family widely distributed in the plant kingdom and its evolutionary history. Plant Mol Biol 61:505–514

Feschotte C, Fourrier N, Desmons I, Mouches C (2001) Birth of a retroposon: the Twin SINE family from the vector mosquito *Culex pipiens* may have originated from a dimeric tRNA precursor. Mol Biol Evol 18:74–84

Fleurdepine S, Deragon JM, Devic M, Guilleminot J, Bousquet-Antonelli C (2007) A bona fide La protein is required for embryogenesis in *Arabidopsis thaliana*. Nucleic Acids Res 35:3306–3321

Gilbert N, Labuda D (1999) CORE-SINEs: eukaryotic short interspersed retroposing elements with common sequence motifs. Proc Natl Acad Sci USA 96:2869–2874

Gilbert N, Labuda D (2000) Evolutionary inventions and continuity of CORE-SINEs in mammals. J Mol Biol 298:365–377

Gong C, Maquat LE (2011a) "Alu"strious long ncRNAs and their role in shortening mRNA half-lives. Cell Cycle 10:1882–1883

Gong C, Maquat LE (2011b) lncRNAs transactivate STAU1-mediated mRNA decay by duplexing with 3′ UTRs via Alu elements. Nature 470:284–288

Gong C, Kim YK, Woeller CF, Tang Y, Maquat LE (2009) SMD and NMD are competitive pathways that contribute to myogenesis: effects on PAX3 and myogenin mRNAs. Genes Dev 23:54–66

Goodier JL, Maraia RJ (1998) Terminator-specific recycling of a B1-Alu transcription complex by RNA polymerase III is mediated by the RNA terminus-binding protein La. J Biol Chem 273:26110–26116

Gould SJ, Vrba ES (1982) Exaptation; a missing term in the science of form. Paleobiology 8:4–15

Hasler J, Strub K (2006) Alu RNP and Alu RNA regulate translation initiation in vitro. Nucleic Acids Res 34:2374–2385

Irmer H, Hennings I, Bruchhaus I, Tannich E (2010) tRNA gene sequences are required for transcriptional silencing in *Entamoeba histolytica*. Eukaryot Cell 9:306–314

Kachroo P, Leong SA, Chattoo BB (1995) Mg-SINE: a short interspersed nuclear element from the rice blast fungus, *Magnaporthe grisea*. Proc Natl Acad Sci USA 92:11125–11129

Kajikawa M, Okada N (2002) LINEs mobilize SINEs in the eel through a shared 3′ sequence. Cell 111:433–444

Kaneko H, Dridi S, Tarallo V, Gelfand BD, Fowler BJ, Cho WG, Kleinman ME, Ponicsan SL, Hauswirth WW, Chiodo VA, Kariko K, Yoo JW, Lee DK, Hadziahmetovic M, Song Y, Misra S, Chaudhuri G, Buaas FW, Braun RE, Hinton DR, Zhang Q, Grossniklaus HE, Provis JM, Madigan MC, Milam AH, Justice NL, Albuquerque RJ, Blandford AD, Bogdanovich S, Hirano Y, Witta J, Fuchs E, Littman DR, Ambati BK, Rudin CM, Chong MM, Provost P, Kugel JF, Goodrich JA, Dunaief JL, Baffi JZ, Ambati J (2011) DICER1 deficit induces Alu RNA toxicity in age-related macular degeneration. Nature 471:325–330

Kazazian HH Jr (2004) Mobile elements: drivers of genome evolution. Science 303:1626–1632

Khanam T, Rozhdestvensky TS, Bundman M, Galiveti CR, Handel S, Sukonina V, Jordan U, Brosius J, Skryabin BV (2007) Two primate-specific small non-protein-coding RNAs in transgenic mice: neuronal expression, subcellular localization and binding partners. Nucleic Acids Res 35:529–539

Kim J, Martignetti JA, Shen MR, Brosius J, Deininger P (1994) Rodent BC1 RNA gene as a master gene for ID element amplification. Proc Natl Acad Sci USA 91:3607–3611

Kim YK, Furic L, Parisien M, Major F, DesGroseillers L, Maquat LE (2007) Staufen1 regulates diverse classes of mammalian transcripts. EMBO J 26:2670–2681

Kirby PJ, Greaves IK, Koina E, Waters PD, Marshall Graves JA (2007) Core-SINE blocks comprise a large fraction of monotreme genomes; implications for vertebrate chromosome evolution. Chromosome Res 15:975–984

Kitagawa N, Goto M, Kurozumi K, Maruo S, Fukayama M, Naoe T, Yasukawa M, Hino K, Suzuki T, Todo S, Takada K (2000) Epstein-Barr virus-encoded poly(A)(−) RNA supports Burkitt's lymphoma growth through interleukin-10 induction. EMBO J 19:6742–6750

Kondrashov AV, Kiefmann M, Ebnet K, Khanam T, Muddashetty RS, Brosius J (2005) Inhibitory effect of naked neural BC1 RNA or BC200 RNA on eukaryotic in vitro translation systems is reversed by poly(A)-binding protein (PABP). J Mol Biol 353:88–103

Kramerov DA, Vassetzky NS (2005) Short retroposons in eukaryotic genomes. Int Rev Cytol 247:165–221

Kramerov DA, Vassetzky NS (2011) Origin and evolution of SINEs in eukaryotic genomes. Heredity 107:487–495

Kremerskothen J, Nettermann M, op de Bekke A, Bachmann M, Brosius J (1998) Identification of human autoantigen La/SS-B as BC1/BC200 RNA-binding protein. DNA Cell Biol 17:751–759

Laha T, McManus DP, Loukas A, Brindley PJ (2000) Sjalpha elements, short interspersed element-like retroposons bearing a hammerhead ribozyme motif from the genome of the oriental blood fluke *Schistosoma japonicum*. Biochim Biophys Acta 1492:477–482

Lenoir A, Pelissier T, Bousquet-Antonelli C, Deragon JM (2005) Comparative evolution history of SINEs in *Arabidopsis thaliana* and *Brassica oleracea*: evidence for a high rate of SINE loss. Cytogenet Genome Res 110:441–447

Liao C, Rovira C, He H, Edstrom JE (1998) Site-specific insertion of a SINE-like element, Cp1, into centromeric tandem repeats from *Chironomus pallidivittatus*. J Mol Biol 280:811–821

Lin D, Pestova TV, Hellen CU, Tiedge H (2008) Translational control by a small RNA: dendritic BC1 RNA targets the eukaryotic initiation factor 4A helicase mechanism. Mol Cell Biol 28:3008–3019

Liu WM, Chu WM, Choudary PV, Schmid CW (1995) Cell stress and translational inhibitors transiently increase the abundance of mammalian SINE transcripts. Nucleic Acids Res 23:1758–1765

Luchetti A, Mantovani B (2011) Molecular characterization, genomic distribution and evolutionary dynamics of Short INterspersed Elements in the termite genome. Mol Genet Genomics 285:175–184

Lunyak VV, Prefontaine GG, Nunez E, Cramer T, Ju BG, Ohgi KA, Hutt K, Roy R, Garcia-Diaz A, Zhu X, Yung Y, Montoliu L, Glass CK, Rosenfeld MG (2007) Developmentally regulated activation of a SINE B2 repeat as a domain boundary in organogenesis. Science 317:248–251

Mallory AC, Vaucheret H (2006) Functions of microRNAs and related small RNAs in plants. Nat Genet 38(Suppl):S31–S36

Maran A, Mathews MB (1988) Characterization of the double-stranded RNA implicated in the inhibition of protein synthesis in cells infected with a mutant adenovirus defective for VA RNA. Virology 164:106–113

Mariner PD, Walters RD, Espinoza CA, Drullinger LF, Wagner SD, Kugel JF, Goodrich JA (2008) Human Alu RNA is a modular transacting repressor of mRNA transcription during heat shock. Mol Cell 29:499–509

Martignetti JA, Brosius J (1993a) BC200 RNA: a neural RNA polymerase III product encoded by a monomeric Alu element. Proc Natl Acad Sci USA 90:11563–11567

Martignetti JA, Brosius J (1993b) Neural BC1 RNA as an evolutionary marker: guinea pig remains a rodent. Proc Natl Acad Sci USA 90:9698–9702

Mes JJ, Haring MA, Cornelissen BJ (2000) Foxy: an active family of short interspersed nuclear elements from *Fusarium oxysporum*. Mol Gen Genet 263:271–280

Moolhuijzen P, Kulski JK, Dunn DS, Schibeci D, Barrero R, Gojobori T, Bellgard M (2010) The transcript repeat element: the human Alu sequence as a component of gene networks influencing cancer. Funct Integr Genomics 10:307–319

Morl M, Marchfelder A (2001) The final cut: the importance of tRNA 3′-end processing. EMBO Rep 2:17–20

Munemasa M, Nikaido M, Nishihara H, Donnellan S, Austin CC, Okada N (2008) Newly discovered young CORE-SINEs in marsupial genomes. Gene 407:176–185

Nikitina TV, Tischenko LI, Schulz WA (2011) Recent insights into regulation of transcription by RNA polymerase III and the cellular functions of its transcripts. Biol Chem 392:395–404

Nishihara H, Smit AF, Okada N (2006) Functional noncoding sequences derived from SINEs in the mammalian genome. Genome Res 16:864–874

Noma K, Ohtsubo E, Ohtsubo H (1999) Non-LTR retrotransposons (LINEs) as ubiquitous components of plant genomes. Mol Gen Genet 261:71–79

Noma K, Ohtsubo H, Ohtsubo E (2000) ATLN elements, LINEs from *Arabidopsis thaliana*: identification and characterization. DNA Res 7:291–303

Ogiwara I, Miya M, Ohshima K, Okada N (2002) V-SINEs: a new superfamily of vertebrate SINEs that are widespread in vertebrate genomes and retain a strongly conserved segment within each repetitive unit. Genome Res 12:316–324

Ohshima K, Hamada M, Terai Y, Okada N (1996) The 3′ ends of tRNA-derived short interspersed repetitive elements are derived from the 3′ ends of long interspersed repetitive elements. Mol Cell Biol 16:3756–3764

Okada N, Sasaki T, Shimogori T, Nishihara H (2010) Emergence of mammals by emergency: exaptation. Genes Cells 15:801–812

Parrott AM, Tsai M, Batchu P, Ryan K, Ozer HL, Tian B, Mathews MB (2011) The evolution and expression of the snaR family of small non-coding RNAs. Nucleic Acids Res 39:1485–1500

Piskurek O, Jackson DJ (2011) Tracking the ancestry of a deeply conserved eumetazoan SINE domain. Mol Biol Evol 28:2727–2730

Ponicsan SL, Kugel JF, Goodrich JA (2010) Genomic gems: SINE RNAs regulate mRNA production. Curr Opin Genet Dev 20:149–155

Pouch-Pelissier MN, Pelissier T, Elmayan T, Vaucheret H, Boko D, Jantsch MF, Deragon JM (2008) SINE RNA induces severe developmental defects in *Arabidopsis thaliana* and interacts with HYL1 (DRB1), a key member of the DCL1 complex. PLoS Genet 4:e1000096

Rasmussen M, Rossen L, Giese H (1993) SINE-like properties of a highly repetitive element in the genome of the obligate parasitic fungus *Erysiphe graminis* f.sp. hordei. Mol Gen Genet 239:298–303

Rogers JH (1985) The origin and evolution of retroposons. Int Rev Cytol 93:187–279
Roman AC, Gonzalez-Rico FJ, Molto E, Hernando H, Neto A, Vicente-Garcia C, Ballestar E, Gomez-Skarmeta JL, Vavrova-Anderson J, White RJ, Montoliu L, Fernandez-Salguero PM (2011) Dioxin receptor and SLUG transcription factors regulate the insulator activity of B1 SINE retrotransposons via an RNA polymerase switch. Genome Res 21:422–432
Roy AM, West NC, Rao A, Adhikari P, Aleman C, Barnes AP, Deininger PL (2000) Upstream flanking sequences and transcription of SINEs. J Mol Biol 302:17–25
Rozhdestvensky TS, Kopylov AM, Brosius J, Huttenhofer A (2001) Neuronal BC1 RNA structure: evolutionary conversion of a tRNA(Ala) domain into an extended stem-loop structure. RNA 7:722–730
Rubin CM, Kimura RH, Schmid CW (2002) Selective stimulation of translational expression by Alu RNA. Nucleic Acids Res 30:3253–3261
Santangelo AM, de Souza FS, Franchini LF, Bumaschny VF, Low MJ, Rubinstein M (2007) Ancient exaptation of a CORE-SINE retroposon into a highly conserved mammalian neuronal enhancer of the proopiomelanocortin gene. PLoS Genet 3:1813–1826
Sasaki T, Nishihara H, Hirakawa M, Fujimura K, Tanaka M, Kokubo N, Kimura-Yoshida C, Matsuo I, Sumiyama K, Saitou N, Shimogori T, Okada N (2008) Possible involvement of SINEs in mammalian-specific brain formation. Proc Natl Acad Sci USA 105:4220–4225
Schmid CW (1998) Does SINE evolution preclude Alu function? Nucleic Acids Res 26:4541–4550
Sun FJ, Fleurdepine S, Bousquet-Antonelli C, Caetano-Anolles G, Deragon JM (2007) Common evolutionary trends for SINE RNA structures. Trends Genet 23:26–33
Tiedge H, Chen W, Brosius J (1993) Primary structure, neural-specific expression, and dendritic location of human BC200 RNA. J Neurosci 13:2382–2390
Tu Z (1999) Genomic and evolutionary analysis of Feilai, a diverse family of highly reiterated SINEs in the yellow fever mosquito, *Aedes aegypti*. Mol Biol Evol 16:760–772
Wang H, Iacoangeli A, Lin D, Williams K, Denman RB, Hellen CU, Tiedge H (2005) Dendritic BC1 RNA in translational control mechanisms. J Cell Biol 171:811–821
Watson JB, Sutcliffe JG (1987) Primate brain-specific cytoplasmic transcript of the Alu repeat family. Mol Cell Biol 7:3324–3327
Wenke T, Döbel T, Rosleff ST, Junghans H, Weisshaar B, Schimdt T (2011) Targeted identification os SINE (Short Interspersed Nuclear Element) families show their wide-spread existence and extreme heterogeneity in plant genomes. Plant Cell 23:3117–3128
Whisson SC, Avrova AO, Lavrova O, Pritchard L (2005) Families of short interspersed elements in the genome of the oomycete plant pathogen, *Phytophthora infestans*. Fungal Genet Biol 42:351–365
Wick N, Luedemann S, Vietor I, Cotten M, Wildpaner M, Schneider G, Eisenhaber F, Huber LA (2003) Induction of short interspersed nuclear repeat-containing transcripts in epithelial cells upon infection with a chicken adenovirus. J Mol Biol 328:779–790
Williams BR (1999) PKR; a sentinel kinase for cellular stress. Oncogene 18:6112–6120
Wilson ET, Condliffe DP, Sprague KU (1988) Transcriptional properties of BmX, a moderately repetitive silkworm gene that is an RNA polymerase III template. Mol Cell Biol 8:624–631
Yagi E, Akita T, Kawahara T (2011) A novel Au SINE sequence found in a gymnosperm. Genes Genet Syst 86:19–25
Yakovchuk P, Goodrich JA, Kugel JF (2009) B2 RNA and Alu RNA repress transcription by disrupting contacts between RNA polymerase II and promoter DNA within assembled complexes. Proc Natl Acad Sci USA 106:5569–5574

Chapter 14
LTR Retrotransposons as Controlling Elements of Genome Response to Stress?

Quynh Trang Bui and Marie-Angèle Grandbastien

Abstract Transposable elements can impact gene expression and regulatory patterns. This is particularly true for LTR retrotransposons, whose Long Terminal Repeats (LTRs) promoter/regulatory capsules are present at both ends of the element and make them particularly prone to influencing adjacent genes. LTRs can act as promoters, as sources of regulatory sequences, or initiate antisense transcripts regulating gene expression. As a consequence, LTR responses to specific stimuli can influence adjacent host genes and contribute to the organism's response to these stimuli. Most plant LTR retrotransposons are activated in response to stress or environmental changes, and in this review, we will update current data on this stress response. After a short journey across the animal kingdom, where the regulatory impact of LTRs is well documented, we will present recent reports suggesting that LTRs may also play a role in the modulation of gene expression and in the generation of phenotypic plasticity in plants.

Keywords Retrotransposon • Retroviral • LTR • Long Terminal Repeat • Stress • Cotranscrit • Expression • Host gene

Abbreviations

ERV	Endogenous Retroviral Element
LINE	Long INterspersed Nuclear Element
LTR	Long Terminal Repeat
SINE	Short INterspersed Nuclear Element
TE	Transposable Element
TSS	Transcription Start Site

Q.T. Bui • M.-A. Grandbastien (✉)
Institut Jean Pierre Bourgin, UMR 1318 INRA/AgroParisTech, INRA-Versailles, 78026 Versailles Cedex, France
e-mail: gbastien@versailles.inra.fr

M.-A. Grandbastien and J.M. Casacuberta (eds.), *Plant Transposable Elements*, Topics in Current Genetics 24, DOI 10.1007/978-3-642-31842-9_14,

14.1 Transposable Elements: A "Functionalist" Perspective

In spite of their abundance and role in genome restructuring and fluidity, transposable elements (TEs) have for a long time been considered as parasitic junk DNA and at best as "mortar" elements of the chromosomal structure. Nevertheless, TEs were originally named "Controlling Elements" by Barbara McClintock in the 1940s, the official terminology "Transposable Elements" arising only decades later. More important than the concept of DNA mobility that led to a Nobel Prize in 1983, McClintock actually considered the ability of these mobile elements to modify gene expression as their fundamental characteristic, being convinced from the beginning that they were involved in regulating cellular differentiation during development: "*The real point is control. The real secret of all of this is control. It is not transposition*" (McClintock cited in Comfort 1999, an excellent review on the early evolution of concepts on TEs).

Discarded for a long time, this prescient view has received support in the past decade, where experimental evidence for a central role of TEs in the diversification and modulation of genic functions has accumulated. Upon insertion in or next to coding regions, TEs impact gene expression and function in various ways. Besides disrupting gene function, TEs can be exapted in coding or noncoding regions, a process that leads to the creation of splicing variants and new proteins. More importantly, TEs themselves are subject to transcriptional and epigenetic regulations in response to developmental cues and external stimuli. As a consequence, host genes can be placed under the control of these TE responses, either under direct control of neighboring TE promoter/regulatory sequences or via RNAi pathways. The ability of TEs to respond to specific signals, combined with their repetitive and widespread nature, is thus expected to be fundamental to the fine-tuning of gene expression and function. These TE-generated variations may significantly expand the functional potentialities of genes and the diversification of their activities, bearing important consequences for the generation of phenotypic diversity. There is now a growing interest by the scientific community in this "functionalist" view of mobile elements, by which TEs can be considered as "distributed genomic control modules" (Shapiro 2005) at the core of regulatory networks, leading to reprogramming of batteries of genes as part of the organism's response to specific stimuli.

This is particularly true for a specific type of TE, the LTR (Long Terminal Repeat) retrotransposons, or retroviral-like elements, whose LTRs can act as promoter and/or regulator of the expression of adjacent cellular genes. The role of retroviral LTRs in driving the large-scale coordinated regulation of host cellular genes and in shaping regulatory networks is now well documented in mammalian models. In plants, a growing body of evidence suggests their potential involvement in the modulation of gene expression in response to several stimuli, notably stresses and external challenges, the prevailing conditions of activation for plant retrotransposons. In this review, we will update current data on the stress response of plant LTR retrotransposons, and on recent reports suggesting that this response may play a role in the modulation of host gene expression.

14.2 Plant LTR Retrotransposons and Stress

14.2.1 LTR Retrotransposon Life Cycle

LTR retrotransposons are the predominant class of TEs in plant genomes and can represent over 80% of the DNA of cereals with large genomes. Overlying polyploidy, they are primary agents of genome size differences (for a review, see Chap. 3). LTR retrotransposons are found in a variety of diverse types (Wicker et al. 2007; see Chap. 1), that all have common features (and a common origin) with vertebrate retroviruses, hence their frequent designation as retroviral-like elements.

Like all retroelements, the amplification of LTR retrotransposons involves reverse transcription of an RNA template into a daughter DNA copy subsequently inserted into the genome (for a detailed description of the LTR retrotransposon life cycle, see Chap. 5). LTR retrotransposons are bounded by two Long Terminal Repeats (LTRs) that are identical in newly inserted copies. The proteins required for the retrotransposition cycle are encoded between the two LTRs, and transcription of the full length LTR-to-LTR template RNA is initiated in the 5′ LTR and ends in the 3′ LTR. The LTRs contain the functional signals required for transcription (promoter, transcriptional start, transcriptional end), as well as a significant part of the regulatory sequences that determine expression patterns. As a consequence the functional integrity of the LTR is a key feature of the element's life cycle and of its amplification patterns.

14.2.2 Plant LTR Retrotransposon Response to Stress

With few exceptions, most LTR plant retrotransposons are inactive under normal plant development and are frequently activated under stress conditions or in response to environmental changes. Transcriptional activation, and sometimes mobilization, of plant LTR retrotransposons has been documented after in vitro tissue culture, a process that involves cellular dedifferentiation and activation of plant defense responses, and in response to a variety of biotic and abiotic environmental challenges. The response of LTR retrotransposons to genome shocks such as interspecific crosses and allopolyploidy has also been documented and will not be reviewed here, as it is presented in Chap. 9.

LTR retrotransposon stress responses were particularly well studied in tobacco, where a tight connection has been established since the 1990s between expression of the two best known plant LTR retrotransposons, Tnt1A (Grandbastien et al. 1989) and Tto1 (Hirochika 1993), and stress response pathways (reviewed in Grandbastien 1998; Grandbastien et al. 2005). Tnt1A was originally detected in plants regenerated from protoplast-derived cell cultures and its expression is strongly activated by biotic stresses such as pathogen inoculations and microbial factors. Tto1 is similarly activated by various biotic stresses, as well as by tissue culture (see Takeda et al. 2001 and references therein), a stimulus that only poorly activates Tnt1A in its original host, indicating subtle differences between the two elements in their stress response. Tnt1A and Tto1 expression is also activated by wounding and by intermediates in the plant defense responses such as salicylic acid and methyl jasmonate, and is detected in roots of healthy plants, a tissue in which stress responses are known to be activated.

Activation in various stress conditions has also been reported for a large number of other LTR retrotransposons, and a current update of elements for which differential expression in stress conditions has been formally reported is presented in Table 14.1. Stress-related expression of many other LTR retrotransposon sequences was also detected through genome-wide analysis such as nonspecific RT-PCR targeting of reverse transcriptase domains, differential display, production of EST collections, or microarray analyses (Table 14.2). These global studies were generally not associated with further evaluation of the stress response of each individual element; they, however, frequently point out an increase in the frequency of LTR retrotransposon sequences in transcriptome data obtained in stress conditions. High-throughput analyses such as LTR retrotransposon tiling arrays (Picault et al. 2009) or next-generation resequencing (Sabot et al. 2011; Miyao et al. 2012; see Chap. 4) have been recently successfully developed in rice to demonstrate expression of retrotransposons in tissue culture, as well as their amplification in plants regenerated from tissue culture.

14.2.3 LTRs as Autonomous Promoter/Regulatory Capsules

Studies of structural features involved in LTR retrotransposon stress-response all demonstrate the involvement of LTR sequences in this regulation, and the striking similarities of their regulatory regions, notably the U3 region located upstream of the transcription start (see Fig. 14.1a), with those of plant stress response genes. For instance, Tnt1A expression features involve several U3 *cis*-acting elements similar to well-characterized motifs involved in the activation of defense genes, such as a G-box and repeated H-boxes, and parallels tightly the expression of host defense genes (detailed in Grandbastien et al. 2005). Tnt1-related elements present in tomato and related species, Retrolyc1/TLC1, also display stress-related expression mediated by repeated U3 regulatory motifs similar to those plant defense genes (Tapia et al. 2005; Salazar et al. 2007). Activation of the Tto1 tobacco retrotransposon also parallels the expression of host defense genes and involves tandemly repeated U3 sequences carrying H-boxes, and activation of Tto1 is mediated *via* binding to these U3 motifs of a stress-inducible transcription factor NtMYB2 that is also involved in activation of the PAL defense gene (see Takeda et al. 2001 and references therein).

LTR involvement in the response to environmental changes has been shown for many other elements, such as for the rice Tos17 element (Hirochika et al. 1996), and BARE-1 of barley, whose LTR is involved in expression in calli and contains U3

Table 14.1 Plant LTR retrotransposons differentially expressed in response to external challenges (listed by chronological order of first report)

LTR-RT	Species	Induction of expression (E) or amplification (A)	References
Bs1	*Zea mays*	Virus infection (A)	Johns et al. (1985)
Tnt1	*Nicotiana tabacum*	Protoplasts, microbial factors (E, A), elicitins, wounding, pathogen inoculations, JA, SA (E)	Grandbastien et al. (1989)
Tto1	*Nicotiana tabacum*	Protoplasts, cell and tissue cultures (E, A), microbial factors, wounding, pathogen inoculations, JA, SA (E)	Hirochika (1993)
BARE-1	*Hordeum vulgare*	Protoplasts, tissue culture (E)	Suoniemi et al. (1996), Chang and Schulman (2008)
Tos17	*Oryza sativa*	Tissue culture (E, A)	Hirochika et al. (1996)
PsrA, PsrB, PsrC	*Pisum sativum*	Protoplasts and/or fungal elicitor (E)	Kato et al. (1999)
BARE-1	*Hordeum spontaneum*	Microclimatic changes (A)	Kalendar et al. (2000)
OARE1	*Avena sativa*	UV light, wounding, fungal inoculation, JA, SA (E)	Kimura et al. (2001)
MCIRE	*Medicago sativa*	Cold stress (E)	Ivashuta et al. (2002)
ZmMI1	*Zea mays*	Cold stress (E)	Steward et al. (2002)
Rtsp-1	*Ipomoea batatas*	Tissue culture (E, A)	Tahara et al. (2004)
TLC1 (Retrolyc1)	*Solanum chilense*	Ethylene, ABA, JA, SA, H_2O_2 (E)	Tapia et al. (2005), Salazar et al. (2007)
Hopscotch-like	*Saccharum officinarum*	Tissue culture, endophytic bacterial inoculation (E)	Araujo et al. (2005)
Morgane	*Triticum aestivum*	Nitrogen stress, fungal infection (E)	Sabot et al. (2006)
CIRE1	*Citrus sinensis*	Wounding, phytohormones (E)	Rico-Cabanas and Martínez-Izquierdo (2007)
Reme1	*Cucumis melo*	UV light (E)	Ramallo et al. (2008)
CLCoy1	*Citrus limon*	Wounding, salt stress, cell culture (E)	De Felice et al. (2009)
Lullaby	*Oryza sativa*	Tissue culture (E, A)	Picault et al. (2009)
MERE1	*Medicago truncatula*	Tissue culture (E)	Rakocevic et al. (2009)
Osr23, Osr36, Osr42	*Oryza sativa*	Space flight (= multiple environmental factors) (A)	Long et al. (2009)
FaRE1	*Fragaria x ananassa*	Phytohormones, including ABA (E)	He et al. (2010)
Ttd1a	*Triticum durum*	Salt and light stress (E, A)	Woodrow et al. (2011)
ONSEN	*Arabidopsis thaliana*	Heat (E)	Ito et al. (2011)
Tcs1, Tcs2	*Citrus sinensis*	Cold (E)	Butelli et al. (2012)

JA jasmonic acid, *SA* salicylic acid

ABA response elements typical of water stress-induced genes (Suoniemi et al. 1996; Chang and Schulman 2008). Similarly, the LTR of the ABA-responsive FaRE1 element of strawberry also contains regulatory motifs associated with response to ABA and stress (He et al. 2010), and the U3 of the cold-responsive MCIRE element of *Medicago sativa* contains a LTRE-low temperature responsive regulatory element (Ivashuta et al. 2002). U3 *cis*-acting motifs of the stress-responsive Tdt1 element of

Table 14.2 Other LTR retrotransposons detected in stress-related conditions by genome-wide analyses (listed by chronological order of first report)

Species	Strategies	Activation conditions	References
Expression			
Oryza sativa	RT-PCR of pol domain	Cell culture-derived protoplasts	Hirochika et al. (1996)
Solanum tuberosum	RT-PCR of pol domain	Protoplasts	Pearce et al. (1996)
Nicotiana tabacum	RT-PCR of pol domain	BY2 cell culture-derived protoplasts	Hirochika (1993)
Nicotiana attenuata	cRNA differential display	Herbivorous insect leaf damage	Hermsmeier et al. (2001)
Avena sativa	RT-PCR of pol domain	Tissue culture	Kimura et al. (2001)
Gramineae spp.	EST collections	Cell culture, stress-induced plants	Vicient et al. (2001)
Triticeae spp.	EST collections	Various biotic and abiotic stresses	Echenique et al. (2002)
Sorghum bicolor	RT-PCR of pol domain	Tissue culture, protoplasts	Muthukumar and Bennetzen (2004)
Saccharum officinarum	EST collections + macroarrays	Tissue culture	Araujo et al. (2005)
Arabidopsis thaliana	Microarrays	Calli derived from habituated cell culture	Pischke et al. (2006)
Agrostis spp.	cDNA libraries	Fungal disease (?)	Rotter et al. (2007)
Triticum aestivum	cDNA differential display	Mycotoxin	Ansari et al. (2007)
Hordeum vulgare	cDNA differential display	Senescence	Ay et al. (2008)
Coffea spp.	EST collections	Calli, cell cultures treated with fongicide or salt, parasitic infections, water stress	Lopes et al. (2008)
Arabidopsis thaliana	Microarrays	Tissue culture, suspension cells	Tanurdzic et al. (2008)
Oryza sativa	Tiling arrays	Tissue culture	Picault et al. (2009)
Arabidopsis thaliana	Microarrays	Salt, osmotic and cold stress, ABA treatment	Zeller et al. (2009)
Triticum aestivum	Microarrays	Water stress	Aprile et al. (2009)
Zea mays	EST collections	Cell culture	Vicient (2010)
Zea mays	Microarrays	Water stress	Lu et al. (2011)
Vitis vinifera	Microarrays	Ripening process (oxydative stress)	Fortes et al. (2011)
Pissodes strobi	Microarrays	Downregulated in weevil resistant genotypes	Verne et al. (2011)
Oryza sativa	RNA-seq	Atrazine (herbicide)	Zhang et al. (2012)
Mobility			
Oryza sativa	NGS genome resequencing	Tissue culture-derived plant	Sabot et al. (2011)
Oryza sativa	NGS genome resequencing	Tissue culture-derived plants	Miyao et al. (2012)

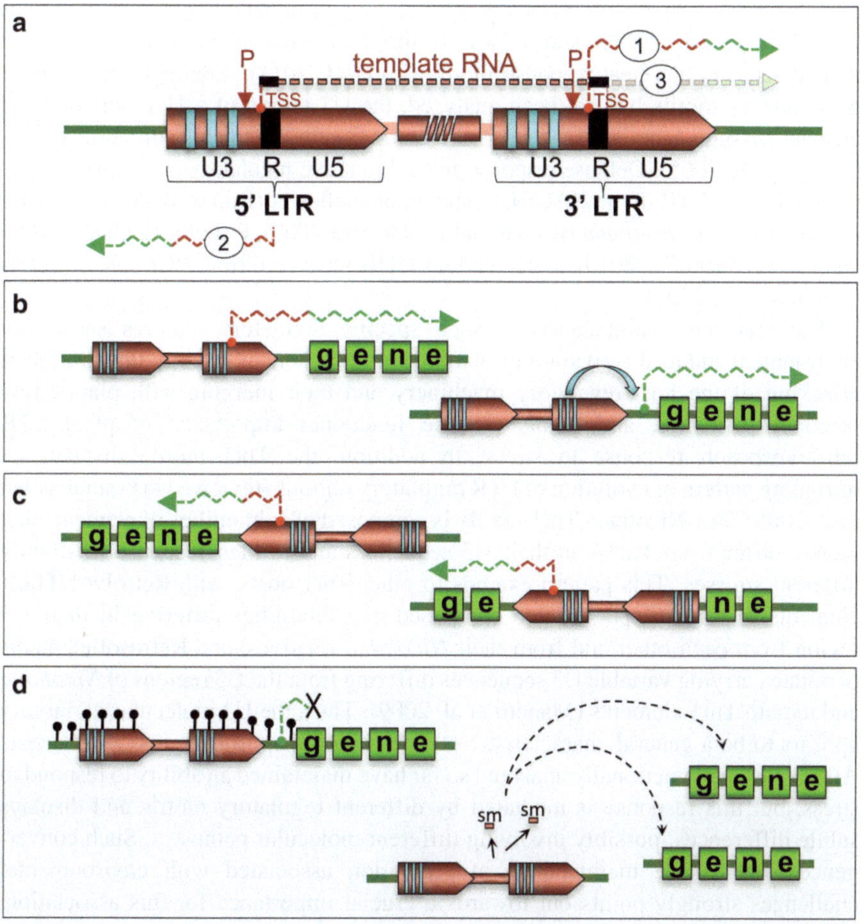

Fig. 14.1 Examples of LTR-mediated gene regulation mechanisms. (**a**) Structure and transcriptional features of LTR retrotransposons: the element is bounded by two Long Terminal Repeats (LTRs) that are identical in recently transposed copies. LTRs carry promoter (P), transcriptional start (TSS), and regulatory sequences (*blue boxes*), and the RNA template used for amplification is initiated at the U3/R boundary in the 5′ LTR and ends at the R/U5 boundary in the 3′ LTR. (1) The 3′ LTR also contains promoter, TSS, and regulatory sequences, and can drive the readout cotranscription of downstream adjacent sequences. (2) LTRs can also carry cryptic antisense promoters driving the readout cotranscription of upstream adjacent sequences from the 5′ LTR. (3) Template RNAs sometimes fail to terminate in the 3′ LTR and can extend readthrough transcripts in downstream sequences. (**b**) When inserted upstream from genes, 3′ LTR can act as promoter by initiating transcription or provide *cis*-regulatory sequences such as binding sites for transcription factors. (**c**) When inserted in antisense to the gene (or using cryptic antisense promoters), LTRs can initiate antisense transcripts that may downregulate the target gene. (**d**) LTR retrotransposons can also transfer epigenetic regulations such as DNA or histone methylation to adjacent genes or be source of small RNAs (sm) that can regulate distant genes. These a few examples of the multiple possibilities of genic impact of LTRs that vary depending on their orientation regarding the adjacent genes and on their position in the genic sequence

durum wheat are involved in DNA-protein binding in salt and light stress conditions (Woodrow et al. 2011), and LTRs of the heat-responsive Onsen element of Arabidopsis contain heat response motifs (Ito et al. 2011). Although the presence of regulatory motifs has not been analyzed, the U3 region of a Hopscotch-related element of sugarcane is also able to drive expression specifically in callus tissues (Araujo et al. 2005). Defense response-related putative regulatory *cis*-elements were reported in the LTR of the HACRE1 element of sunflower (Buti et al. 2009), CARE1 element of *Cicer arietinum* (Rajput and Upadhyaya 2009), Cotzilla element of *Beta vulgaris* (Weber et al. 2010), and in various LTRs recovered from *Phaseolus vulgaris* (Galindo et al. 2004).

The recurrent maintenance of such specific regulatory features show that environment-induced activation of plant retrotransposons is directly linked to their hijacking of the host regulatory machinery and their merging with plant stress response pathways, and points out the functional importance of plant LTR retrotransposon response to stress. In addition, the Tnt1 family displays an intriguing pattern of evolution of LTR regulatory regions (reviewed in Grandbastien et al. 2005). The Nicotiana Tnt1 family is composed of subfamilies of elements that mostly differ from Tnt1A in their U3 sequences and in their response to slightly different stresses. This pattern extends to other Tnt1 hosts, with Retrolyc1/TLC1 elements of tomato species also composed of subfamilies differing in their U3 region from each other and from their *Nicotiana* relatives, and Retrosol elements of potato carrying variable U3 sequences differing from the U3 regions of *Nicotiana* and tomato Tnt1 elements (Manetti et al. 2009). Thus, the U3 molecular variability appears to be a general characteristic of Tnt1 retrotransposons across Solanaceae. All U3 variants functionally analyzed so far have maintained an ability to respond to stress, but this response is mediated by different regulatory motifs and displays subtle differences, possibly involving different molecular pathways. Such convergence towards the maintenance of regulation associated with environmental challenges strongly points out towards a crucial importance for this association, whether for the survival of elements or for some benefit to their hosts.

14.3 A Structural Impact of Retrotransposon Stress Response?

In spite of the tight correlation between the expression features of many LTR retrotransposons and plant stress responses, direct evidence of stress-induced amplification is to this day mostly restricted to artificial systems such as tissue culture or plants regenerated from tissue culture (Table 14.1) that very poorly reproduce naturally in planta natural environmental challenges. Similarly, the amplification of Tnt1A in response to microbial factors was monitored using an in vitro experimental system, and Tnt1A amplification was not demonstrated in response to in planta infections. A few exceptions include the mobilization of rice retrotransposons (as well as other TEs) in plants derived from seeds submitted to spaceflight, an environment characterized by multiple stress factors (Long et al. 2009), and a recent report of mobilization of Ttd1a in durum wheat submitted to light and salt stresses (see Woodrow et al. 2011). The mobility of Bs1 was also initially detected in progeny of virus-infected maize plants (Johns et al. 1985), although a direct link between Bs1 mobility and viral infection remains to be confirmed.

From such scattered evidence, a significant role of LTR retrotransposons in host genome restructuring in response to external challenges cannot be really be inferred in plants so far, let alone any possible adaptive role of these changes. Nevertheless, some significant, albeit indirect, examples of potential large-scale impact of retrotransposon mobilization by stress have been reported. The most notable is certainly a seminal study that showed that the genomic BARE-1 content of natural wild barley populations increased linearly (up to 25%) with increasing altitude and aridity (Kalendar et al. 2000), an observation that correlates with the presence of ABA-response elements in BARE-1 LTR. The proliferation of LTR retrotransposons was also reported in hybrid sunflower species that evolved in extreme conditions such as a desert environment or saline marshes (Ungerer et al. 2006).

The paucity of data relative to LTR retrotransposon mobilization by stress may simply be due to experimental limitations preventing easy detection of somatic stress-induced transpositions in natural stress conditions. It may as well be due to possible restrictions in the transmission of new transpositions to the progeny, especially in the cases of pathogen-related stress, that usually affect the host plant somatically.

14.4 Functional Impact of LTR Retroelements

Upon insertion in or next to genic regions, TEs can create mutant phenotypes. This ability to modulate gene expression and function was at the origin of their discovery by B. McClintock. It was thus unsurprising, yet elating, to discover much later on that the wrinkled-seed pea character upon which Gregor Mendel established the basic laws of modern genetics was actually due to a transposon insertion into a starch-branching enzyme (Bhattacharyya et al. 1990). Farsighted early reports pointed out the potential importance of TEs in functional variations (McDonald 1990; Robins and Samuelson 1992; White et al. 1994; Britten 1996; Kidwell and Lisch 1997), however, TE influence on host gene expression and function was for a long time largely regarded as a circumstantial consequence of their insertional mutagenic activity.

The major importance of TE exaptation for regulatory functions was fully recognized when large mammalian genome and transcriptome sequence data demonstrated that TEs played an extremely important role in the regulation of host gene expression. This influence can range from the local supply of promoters and/or *cis*-regulatory elements, to the creation of alternative splicing or premature termination mediated by signals carried by insertions. TE influence on host gene expression also results from the transfer of TE-targeted epigenetic regulation, such as local spreading of chromatin modifications or long distance impact of small interfering RNAs (smRNAs). These processes are mediated by many TE types and are well documented in mammals (see excellent reviews such as Van de Lagemaat et al. 2003; Medstrand et al. 2005; Feschotte 2008; Gogvadze and Buzdin 2009; Kines and Belancio 2012).

LTR retroelements, however, display specific structural features that make them particularly prone to influencing adjacent genes, notably the presence of promoter/regulatory sequences at both extremities. As described above, the expression of LTR retroelements is under control of promoter/regulatory sequences that are generally located in LTRs, and LTRs thus represent small independent promoter/regulatory capsules of a few hundred base pairs that contain transcriptional start sites (TSS) and maintain their regulation features at different genomic positions. As LTRs are found at both ends of the retrotransposon, 3′ LTRs also possess promoter and regulatory abilities and can drive the readout cotranscription of adjacent sequences that can in turn exert a profound effect on the expression of neighboring genes. Depending on their orientation regarding the adjacent genes and on their position in the genic sequence, LTR-driven transcription has multiple and antagonistic effects on target genes. When inserted in the same orientation, LTRs in upstream regions can activate genes that normally are not expressed under the same condition and act as alternative or primary promoters driving readout transcripts (Fig. 14.1b). When inserted in opposite orientation to the gene, LTRs may repress gene expression by producing antisense readout transcripts (Fig. 14.1c). LTRs have also been shown to carry cryptic antisense promoters, and can also simply act as enhancer/repressor modules, providing regulatory sequences such as binding sites for transcription factors to neighboring genes (Fig. 14.1a).

In this review, we will mostly focus on how LTR retrotransposons contribute to host gene regulation, and how these processes may be crucial for plant phenotypic plasticity. But we will first make a short journey across the animal kingdom, where fascinating examples of the regulatory impact of LTRs have accumulated over the last years in mammalian models.

14.4.1 LTRs and Mammalian Regulatory Networks?

In contrast to plants, LTR retroelements, such as LTR retrotransposons or endogenous retroviral elements (ERVs), i.e., remnants from ancient retroviral infections, are moderately represented in mammalian genomes, with current estimates of 8% in human. The most abundant cohorts of mammalian TEs are represented by non-LTR retrotransposons, such as LINEs and SINEs, that also play important roles in host gene expression regulation but will not be reviewed here (for a recent review see Kines and Belancio 2012: see also Chap. 13).

From early works on Drosophila Adh, human amylase, and mouse Slp genes, McDonald (1990) and Robins and Samuelson (1992) were among the very first ones to point out the potential importance of regulatory changes mediated by retroviral-like insertions. But the global importance of TE involvement in the control of host cellular genes was fully unveiled in the early 2000s, with reports that 5′ upstream promoter/regulatory sequences of nearly 20% of human and mouse genes contained TE insertions, and that the transcription of many human genes started within a 5′ TE insertion, including a large number of cases involving LTRs (Jordan et al. 2003; Van de Lagemaat et al. 2003).

14.4.1.1 LTR-Derived Gene Promoters

The potential for LTRs to drive expression has been best illustrated by large-scale analyses of transcript ends that have shown that at least 50% of the human HERV-K LTRs possessed promoter activity (Buzdin et al. 2006) and that dozens of thousands of TSS are derived from LTRs in human and mouse (Conley et al. 2008b; Faulkner et al. 2009), with a large number of LTRs mapping within transcriptional units of human genes and driving alternative tissue-specific expression of adjacent genes (Conley et al. 2008b). Interestingly, LTRs can also act as bidirectional promoters, as shown for a human ERV1 LTR that drives transcription in similar tissues of two head-to-tail adjacent genes from two TSS closely positioned in the LTR (Dunn et al. 2006). In mouse and human, antisense TSS were found to represent, respectively, 47% and 56% of the TSS present on LTRs (Faulkner et al. 2009).

Mammalian LTRs appear to have been recruited in several major biological processes such as embryo development or reproductive biology. For instance, various LTRs act as alternative promoters for many genes during embryonic development in mouse (Peaston et al. 2004), and a cell-stage specific activation of the MuERV-L leads to numerous LTR-driven readout transcripts with adjacent genes in mouse embryonic stem cells (Macfarlan et al. 2012), pointing out the importance of these processes in the early embryo regulatory network. Epigenetic derepression of ERV elements by histone demethylation leads to upregulation of various genes via LTR-driven readout transcripts in mouse embryonic stem cells, indicating the complementary role of epigenetic regulation in these processes (Karimi et al. 2011). Human ERV LTRs have been recruited to drive placenta-specific expression of several genes (Cohen et al. 2009), and a solo-LTR acting as an alternative promoter redirects pituitary prolactin production to the human endometrium (Gerlo et al. 2006).

Striking examples of independent recruitment of different LTRs and other TEs for similar promoter functions have been described. In addition to the human solo-LTR redirecting prolactin production in the endometrium described above, a different LTR and a non-LTR retrotrotransposon have also been independently recruited in rodents and elephant, respectively, to act as alternative promoters for endometrial prolactin production (Emera et al. 2012), and regulatory motifs derived from a hAT DNA transposon have contributed to the establishment of an endometrial gene regulatory network dedicated to pregnancy in placental mammals (Lynch et al. 2011). All together, these data indicate that host cellular genes have repetitively recruited TEs for insuring crucial functions in the reproduction of placental mammals. Other examples include the mammalian anti-apoptotic Neuronal Apoptosis Inhibitory Protein (NAIP) locus that plays a role in neuronal survival. Different combinations of LTRs have been independently domesticated in human and rodents to insure similar promoter functions at the orthologous NAIP loci (Romanish et al. 2007). Interestingly, LTRs (and SINEs) are globally overrepresented in upstream regions of human and mouse inhibitor of apoptosis genes, with no shared insertions between the two species, reinforcing the evolutionary importance of this process.

14.4.1.2 LTR as Sources of Regulatory Sequences and Regulatory RNAs

In addition to acting as promoters, LTRs contribute extensively to host *cis*-regulatory sequences and constitute a large fraction of transcription factor binding sites identified in embryogenic stem cells and cancer cell lines (Bourque et al. 2008; Kunarso et al. 2010). One clear example is the involvement of several ERVs in the transcriptional network of the human protein p53 involved in DNA damage-triggered apoptosis, with a large number of LTRs containing p53 binding sites (Wang et al. 2007). Binding sites for the NF-kappaB, a transcription factor regulating the immune response, are provided by LTRs (and a SINE) upstream of the human antiviral IFN-λ 1 gene (Thomson et al. 2009). LTRs also form a significant fraction of the TE-derived c-Myc regulatory subnetwork by providing binding sites to a number of genes co-regulated with c-Myc and modulated in cancer cell lines (Wang et al. 2009).

More globally, LTRs were shown to provide nearly 20% of human TE-derived regulatory sequences driving gene expression in immunity-related $CD4^+$ T lymphocyte cells, identified as DNaseI-hypersensitive sites (Mariño-Ramírez and Jordan 2006), and a survey of all human *bona fide* TF binding sites shows that nearly 10% are derived from TEs, 18% of which from LTRs (Polavarapu et al. 2008). LTR-derived binding sites are the most conserved and are more prevalent than expected based on their genome frequencies, confirming that LTRs are particularly prone to donating regulatory sequences to the human genome.

More complex long-distance LTR functions have also been reported, such as the looping of a human ERV9 LTR transcription complex with far downstream globin promoters, resulting in the transfer of LTR-bound transcription factors to these promoters in immature red blood cells (Pi et al. 2010).

When inserted in opposite orientation to genes, LTRs may repress adjacent gene expression by producing transcripts antisense to the genes, and nearly 10,000 such *cis*-natural antisense transcripts (*cis*-NATs) to human genes were found to initiate in LTRs, mostly located at 3′ ends of genes (Conley et al. 2008a). Such antisense transcripts were shown to decrease target gene expression in several cases (Gogvadze et al. 2009). LTR retroelements also transfer epigenetic regulation to adjacent genes, for instance DNA methylation patterns as originally shown in the Agouti mouse (Michaud et al. 1994), or histone modifications involved in cell-type specific expression, such as upregulation in cancer cell lines (Huda et al. 2011).

14.4.1.3 LTR-Mediated Evolution of Regulatory Networks?

Taken together, these studies point out the global recruitment of LTRs in mammalian host regulatory functions. However, a detailed evaluation of experimentally confirmed LTRs acting as alternative or primary promoters suggests that, with the notable exception of their action on placental gene expression, LTRs so far appear to drive expression similar to that of the native promoter(s) and to contribute to changes in expression levels, rather than leading to strikingly novel expression patterns (Cohen et al. 2009). A major role in rewiring host regulatory networks during development may thus not be the LTR's primary impact. Interspecies comparisons have pointed out early on that LTRs may instead be major factors in the evolution of regulatory networks, leading to species-specific expression differences depending on the presence or absence of insertions at orthologous loci (Van de Lagemaat et al. 2003). For instance, recent species-specific insertions, mostly LTRs and SINEs, account for 20% of all expression profile divergence between mouse and rat across various tissues (Pereira et al. 2009), and several human–rodent genome-wide comparisons of regulatory binding sites have revealed a large fraction of species-specific LTR-derived binding sites (as well as binding sites derived from other TEs), resulting in the rewiring of genes into species-specific regulatory networks (Bourque et al. 2008; Kunarso et al. 2010).

In contrast to protein-coding genes, TE populations are indeed markedly dynamic, with waves of insertions creating species-specific TE lineages and insertion pools. Lineage-specific waves of SINE populations have for instance recently been shown to contribute to the diversification of regulatory binding sites in different mammals (Schmidt et al. 2012). The variability of LTR retroelements and other TE insertions, associated with their role in dispersing regulatory motifs and expression specificities, could thus make them essential agents of gene expression evolutionary plasticity.

14.4.2 A Functional Impact of Plant LTR Retrotransposons on Stress Response?

Few large-scale analysis such as those reported in mammalian models have been reported in plants; however, examples are now accumulating showing their association with the regulation of plant genes, and suggesting that, like their mammalian counterparts, they may play an important role in expanding the repertoire of host gene regulation and of regulatory sequences.

14.4.2.1 A Frequent Impact of LTRs on Adjacent Plant Genes

The first clues that LTR retrotransposons could supply promoter/regulatory sequences to plant genes were reported by White et al. (1994), who identified a number of insertions flanking plant genes, including several examples of LTRs with potential for playing a role in expression of adjacent genes. These include a Tnt1-related LTR upstream from the tomato pectate lyase LAT59 gene (Twell et al. 1991) and LTRs of stress-responsive pea PsrC elements upstream from two defense response genes (Kato et al. 1999). Although the direct involvement of LTRs in the modulation of these genes has not been reported, it is interesting to note that PsrC elements and downstream defense response genes respond to similar microbial stimuli. Most interestingly, White et al. (1994) also reported the embedding in LTR sequences of promoters and TSS of several members of the zein multigene family of maize and of transcriptional repressors of pea rbcS alleles.

Since then, numerous phenotypic changes associated with the presence of adjacent LTR retrotransposons have been reported in plants. The first striking example was undoubtedly the white color of grape berry due to an insertion of the Gret1 element upstream of a *Myb*-related gene that regulates anthocyanin biosynthesis (Kobayashi et al. 2004). Although the mechanism by which the upstream Gret1 insertion represses the gene remains to be established, the phenotype is partially reversed to a red berry color after internal recombination of Gret1 leaving a solo-LTR at the insertion site. This indicates that solo-LTRs, frequent recombination derivatives of LTR retrotransposon insertions, may exert regulatory impacts that differ from those of complete elements. Other examples include the much prized "hose-in-hose" primrose flower phenotype, due to upregulation of the PvGlo MADS box gene caused by a retrotransposon insertion in its promoter (Li et al. 2010), and the over-expression of the Auxin-binding protein 1 (ABP1) gene in teosinte, likely due to the additive effect of *cis*-acting regulatory sequences present in several transposon insertions in its promoter, including a solo LTR (Elrouby and Bureau 2012). Insertions of small LTR retrotransposon derivatives termed SMARTs in 5′ and 3′ ends of rice genes resulted in increased expression in specific tissues, while an intronic insertion had little effect, suggesting that SMART sequences act as enhancers (Gao et al. 2012).

Changes in response to environmental conditions were also associated with the presence of adjacent LTR retrotransposons. Insertion of a retrotransposon in the promoter of the Vrn-B1 vernalization gene of *Triticum turgidum* results in expression of the gene without vernalization, conferring spring growth habit (Chu et al. 2011), and insertions of the heat-responsive ONSEN element of Arabidopsis confer heat responsiveness to nearby genes (Ito et al. 2011). The rose continuous flowering phenotype (blooming in all seasons) is linked to an intronic insertion resulting in splicing failure of the KSN gene controlling flower transition, a characteristic under photoperiodic and thermal control (Iwata et al. 2012). Recombination of the retrotransposon to form a solo-LTR restores correct splicing, yet the resulting phenotype is not the wild-type phenotype (spring blooming), but a climbing phenotype (occasional reblooming in autumn), indicating that the intronic solo-LTR exerts a more subtle regulatory effect on the KSN gene.

14.4.2.2 Plant LTR Retrotransposons as Mediators of Epigenetic Regulations

In most insertions reported above, the molecular mechanisms resulting in gene expression changes were not formally reported, and it is not known whether these LTRs act as promoters or as providers of regulatory sequences or epigenetic modulations. TE regulation by epigenetic mechanisms (reviewed in Chap. 8) has been particularly well studied in plants and can influence host gene expression by various mechanisms (see Slotkin and Martienssen 2007), that range from local impacts, such as production of readout transcripts antisense to the gene or local spreading of chromatin modifications from insertions, to distant impact of smRNAs (see Fig. 14.1d).

For instance, the barley Brittle Stem mutation is due to an antisense intronic Sasandra solo-LTR downregulating a cellulase synthase gene, and Sasandra itself appears upregulated in the mutant line, suggesting that LTR-driven transcription antisense to the gene may be involved in the phenotype (Burton et al. 2010). The activation of Tos17 in rice tissue culture is correlated with its demethylation, a process that extends into some flanking genomic regions (Liu et al. 2004), and LTR demethylation in Arabidopsis mutants is correlated with transcriptional upregulation of neighboring genes (Huettel et al. 2006). More recently, it was shown that TEs targeted by smRNAs are globally associated with reduced expression of neighboring genes in *Arabidopsis thaliana* and *A. lyrata* (Hollister et al. 2011).

Distant insertions can also play a role in redirecting TE epigenetic regulation to host genes, as shown for the Arabidopsis *UPB1b* gene, repressed under the influence of specific smRNAs produced from distant Athila retrotransposons (Arteaga-Vázquez et al. 2006; McCue et al. 2012). UPB1b regulates cellular stress and Athila-derived smRNAs target UPB1b 3′UTR sequences, leading to a stress-sensitive phenotype. Similar mechanisms may be involved in the upregulation of the maize tb1 gene, leading to increase in apical dominance associated with maize domestication from teosinte and mediated via a Hopscotch insertion acting as a long distance (ca. 60 kb) enhancer (Studer et al. 2011), although a looping transfer of LTR-bound transcription factors to the tb1 promoter, such as the one reported above for human ERV9 LTR at globin promoters cannot be excluded.

This clearly points to the possibility of genome-wide gene regulation directed by retrotransposon-derived smRNAs, provided smRNAs short recognition sites are present on targeted genes. Stress-induced changes in retrotransposon epigenetic status may thus exert a global influence on the plant stress response, with many reports of retrotransposon-derived, or more generally TE-derived, smRNA production in stress conditions (see Slotkin and Martienssen 2007; Tanurdzic et al. 2008; Mirouze and Paszkowski 2011; Ito 2012).

14.4.3 LTR-Derived Promoters in Plants

The first clear example of LTRs initiating transcription of surrounding sequences in plants was the identification of readout transcripts driven from LTRs of Wis2 following its transcriptional activation in synthetic wheat allopolyploids (Kashkush et al. 2003). The production of these readout transcripts is associated with the modulation of the genes, depending on their orientation relative to the readout transcript. The production of readout transcripts from Dasheng 3′ LTR was also documented in rice, with tissue- and subspecies-specific LTR methylation correlating with the expression of adjacent genes (Kashkush and Khasdan 2007). It is intriguing to note that in both wheat and rice studies, most readout transcripts were produced in opposite orientation to the gene, LTR activation thus resulting in silencing of adjacent genes. In the wheat Wis2 study, readout cotranscripts were also produced in antisense from cryptic promoters in the 5′ LTR. Sense and antisense cotranscripts produced from derepressed LTRs were also detected in Arabidopsis methylation mutants (Huettel et al. 2006), confirming that LTRs can act as bidirectional promoters/enhancers in plants, as shown for ERVs. Recent studies performed in our laboratory also detected a high number of LTR-driven readout transcripts produced from tobacco retrotransposon insertions in various genes (unpublished data). These readout transcripts are produced in stress conditions such as microbial factors or wounding, and their production parallels element expression patterns. As in the case of the wheat Wis2 element, they are produced mostly from 3′ LTRs, but also in antisense from the 5′ LTR, and are often in opposite orientation to the gene.

Transcriptional activation of 5′ LTRs can also produce template readthrough transcripts that fail to terminate in the 3′ LTR and extends in adjacent sequences, with potentially similar impacts on adjacent gene expression (see Fig. 14.1b). In tobacco, a Tnt1 insertion within an NBS-LRR disease-resistance gene was shown to produce such readthrough cotranscripts, and antisense transcripts extending from the gene into the LTR were also identified, suggesting that the Tnt1/NBS-LRR structure may be involved in epigenetic regulation of tobacco resistance genes (Hernández-Pinzón et al. 2009).

In addition, several examples of phenotypic changes that have been experimentally associated to LTR activity as promoter/regulatory units were recently reported. The Cg1-R (corngrass) mutation of maize, that results in large developmental changes, is due to transcriptional initiation of a miRNA locus within a neighboring Stonor element in meristem and lateral organs, resulting in downregulation of several developmental genes targeted by the overexpressed miRNA (Chuck et al. 2007). The Pit disease resistance gene is transcriptionally reactivated in a resistant rice cultivar, due to exaptation of 3′ regions of the Renovator element as promoter (Hayashi and Yoshida 2009). This results in Pit upregulation in response to fungal inoculations, and, interestingly, methylation levels are lower in the 3′ LTR compared to the 5′ LTR, indicating differential targeting of the two LTRs by silencing pathways.

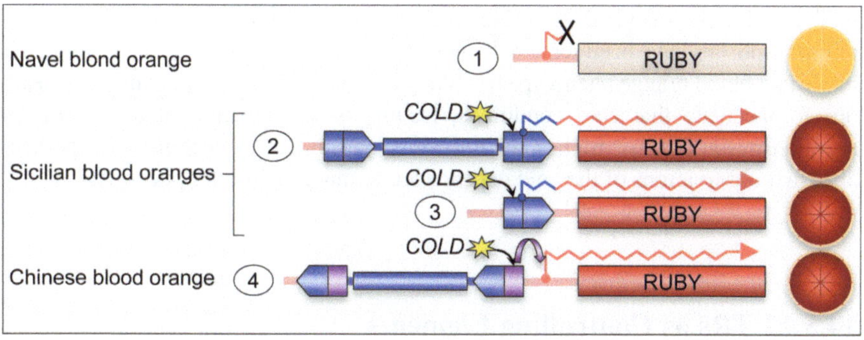

Fig. 14.2 LTR-mediated control of the blood orange phenotype (freely inspired from Butelli et al. 2012). The Ruby gene is inactive in Navel blond oranges (1) and reactivated in fruit-specific and cold-dependent manner in blood oranges as a consequence of a LTR retrotransposon inserted in its promoter. In Sicilan blood oranges, the Tcs1 3′ LTR (2) or the solo-LTR in some accessions (3) provides Ruby transcriptional start and regulatory sequences. In a Chinese blood orange accession (4), an upstream insertion of the closely related Tsc2 element, inserted in opposite orientation, supplies regulatory sequences

Finally, the most exciting example to this day is the blood orange fruit trait, due to LTR-driven transcriptional activation of the Ruby Myb gene, an activator of anthocyanin synthesis (Butelli et al. 2012; Fig. 14.2). Ruby appears inactive in sweet blond oranges, and is expressed in a fruit-specific and cold-dependent manner in Sicilian blood oranges, due to the insertion of the Tcs1 retrotransposon in its promoter. The Tcs1 3′ LTR provides Ruby transcriptional start and regulation, as Tcs1 transcription is also fruit-specific and cold-dependent. Furthermore, Ruby expression specificities are maintained in some Sicilian blood orange accessions where Tcs1 has recombined to form a solo-LTR, confirming that regulatory sequences contained within the LTR capsule are sufficient to insure Ruby specific expression. Very remarkably, another blood orange variety of Chinese origin contains an upstream insertion of Tsc2, another copy of the same retrotransposon. Tsc2 is, however, inserted in reverse orientation to Ruby, indicating that Tcs1 and Tcs2 insertion were unrelated events. Interestingly, Tcs1 and Tcs2 are very closely related, except in their U3 region, a pattern of LTR evolution similar to the one observed for Tnt1 elements (see Sect. 14.2.3). They nevertheless maintain and redirect similar patterns of fruit-specific cold-dependent expression, suggesting that regulatory motifs either have been preserved in the U3 or are carried by the U5. Ruby expression in the Chinese blood orange accession is initiated outside of Tcs2, indicating that in the Chinese blood orange, the LTR capsule only supplies regulatory sequences (Fig. 14.2).

The LTR-mediated blood orange fruit coloration is a very spectacular and exciting example, in the sense that it is a perfect textbook case for various molecular characteristics associated with LTR-mediated impact on host genes. Furthermore, it implies two parallel, yet independent, LTR recruitments to perform similar functions, a situation very reminiscent to those observed for the mammalian prolactin and inhibitor of apoptosis genes. The most puzzling observation, however, is that similar retrotransposons have been separately recruited, an unusual situation, even though it is likely that blood orange phenotype has likely been selected for by humans. This leads one to speculate that cold conditions, leading to mobilization of this particular retrotransposon family, might have been involved in early selection steps, perhaps after first observations of the cold dependance of this sought-after phenotype.

14.5 LTRs as Controlling Elements

The importance of retrotransposon LTRs in plant regulatory networks remains to be fully grasped, especially in comparison with mammalian models that currently lead the game. However, it is quite likely that, with the increasing availability of plant genome sequences, the gap will be rapidly bridged, revealing that, like their mammalian counterparts, plant LTRs play an important role in expanding the repertoire of host gene regulation and of regulatory sequences, and in the evolution of this repertoire. From this perspective, the maintenance of LTR regulatory features allows diverse possibilities of activation from LTRs, among which the fundamental function of the production of the RNA template needed for amplification would represent only the tip of the iceberg.

Current examples of LTR-mediated phenotypic changes or LTR-driven readout transcripts in plants suggest an involvement in regulatory changes in response to both developmental and environmental cues, and illustrate the role of LTRs as intermediate "sensors" of various stimuli as well as their ability to translate and redirect these messages towards adjacent cellular functions. Most plant LTR retrotransposons studied so far, however, carry complex regulatory features that all converge towards a response to various stresses and environmental challenges. Whether these regulatory features lead plant retrotransposons to play a major role in the reprogramming of host cellular genes in response to external cues remain to be established, but may be of crucial importance for plants that cannot escape stress and have evolved complex and highly coordinated responses to biotic and abiotic challenges.

LTR retrotransposons, that are by far the most abundant TEs in higher plants, are likely bound to reveal themselves as particularly efficient examples of the Controlling Elements described by Barbara McClintock. This new and open field of research is still largely uncharted in plants and will undoubtedly represent one of the most fascinating yet rewarding challenges, bearing important consequences for understanding the mechanisms involved in plant phenotypic plasticity.

Acknowledgment We are very thankful to Prof. Howard Laten for critical reading of the manuscript.

References

Ansari KI, Walter S, Brennan JM, Lemmens M, Kessans S, McGahern A, Egan D, Doohan FM (2007) Retrotransposon and gene activation in wheat in response to mycotoxigenic and non-mycotoxigenic-associated *Fusarium* stress. Theor Appl Genet 114:927–937

Aprile A, Mastrangelo AM, De Leonardis AM, Galiba G, Roncaglia E, Ferrari F, De Bellis L, Turchi L, Giuliano G, Cattivelli L (2009) Transcriptional profiling in response to terminal drought stress reveals differential responses along the wheat genome. BMC Genomics 10:279

Araujo PG, Rossi M, de Jesus EM, Saccaro NL Jr, Kajihara D, Massa R, de Felix JM, Drummond RD, Falco MC, Chabregas SM, Ulian EC, Menossi M, Van Sluys M-A (2005) Transcriptionally active transposable elements in recent hybrid sugarcane. Plant J 44:707–717

Arteaga-Vázquez M, Caballero-Pérez J, Vielle-Calzada JP (2006) A family of microRNAs present in plants and animals. Plant Cell 18:3355–3369

Ay N, Clauss K, Barth O, Humbeck K (2008) Identification and characterization of novel senescence-associated genes from barley (*Hordeum vulgare*) primary leaves. Plant Biol (Stuttg) 10:121–135

Bhattacharyya MK, Smith AM, Ellis TH, Hedley C, Martin C (1990) The wrinkled-seed character of pea described by Mendel is caused by a transposon-like insertion in a gene encoding starch-branching enzyme. Cell 60:115–122

Bourque G, Leong B, Vega VB, Chen X, Lee YL, Srinivasan KG, Chew JL, Ruan Y, Wei CL, Ng HH, Liu ET (2008) Evolution of the mammalian transcription factor binding repertoire via transposable elements. Genome Res 18:1752–1762

Britten RJ (1996) DNA sequence insertion and evolutionary variation in gene regulation. Proc Natl Acad Sci USA 93:9374–9377

Burton RA, Ma G, Baumann U, Harvey AJ, Shirley NJ, Taylor J, Pettolino F, Bacic A, Beatty M, Simmons CR, Dhugga KS, Rafalski JA, Tingey SV, Fincher GB (2010) A customized gene expression microarray reveals that the brittle stem phenotype fs2 of barley is attributable to a retroelement in the HvCesA4 cellulose synthase gene. Plant Physiol 153:1716–1728

Butelli E, Licciardello C, Zhang Y, Liu J, Mackay S, Bailey P, Reforgiato-Recupero G, Martin C (2012) Retrotransposons control fruit-specific, cold-dependent accumulation of anthocyanins in blood oranges. Plant Cell 24:1242–1255

Buti M, Giordani T, Vukich M, Gentzbittel L, Pistelli L, Cattonaro F, Morgante M, Cavallini A, Natali L (2009) HACRE1, a recently inserted copia-like retrotransposon of sunflower (*Helianthus annuus* L.). Genome 52:904–911

Buzdin A, Kovalskaya-Alexandrova E, Gogvadze E, Sverdlov E (2006) At least 50% of human-specific HERV-K (HML-2) long terminal repeats serve in vivo as active promoters for host nonrepetitive DNA transcription. J Virol 80:10752–10762

Chang W, Schulman AH (2008) BARE retrotransposons produce multiple groups of rarely polyadenylated transcripts from two differentially regulated promoters. Plant J 56:40–50

Chu CG, Tan CT, Yu GT, Zhong S, Xu SS, Yan L (2011) A novel retrotransposon inserted in the dominant Vrn-B1 allele confers spring growth habit in tetraploid wheat (*Triticum turgidum* L.). G3 (Bethesda) 1:637–645

Chuck G, Cigan AM, Saeteurn K, Hake S (2007) The heterochronic maize mutant Corngrass1 results from overexpression of a tandem microRNA. Nat Genet 39:544–549

Cohen CJ, Lock WM, Mager DL (2009) Endogenous retroviral LTRs as promoters for human genes: a critical assessment. Gene 448:105–114

Comfort NC (1999) "The real point is control": the reception of Barbara McClintock's controlling elements. J Hist Biol 32:133–162

Conley AB, Miller WJ, Jordan IK (2008a) Human *cis* natural antisense transcripts initiated by transposable elements. Trends Genet 24:53–56

Conley AB, Piriyapongsa J, Jordan IK (2008b) Retroviral promoters in the human genome. Bioinformatics 24:1563–1567

De Felice B, Wilson RR, Argenziano C, Kafantaris I, Conicella C (2009) A transcriptionally active copia-like retroelement in *Citrus limon*. Cell Mol Biol Lett 14:289–304

Dunn CA, Romanish MT, Gutierrez LE, van de Lagemaat LN, Mager DL (2006) Transcription of two human genes from a bidirectional endogenous retrovirus promoter. Gene 366:335–342

Echenique V, Stamova B, Wolters P, Lazo G, Carollo L, Dubcovsky J (2002) Frequencies of Ty1-copia and Ty3-gypsy retroelements within the *Triticeae* EST databases. Theor Appl Genet 104:840–844

Elrouby N, Bureau TE (2012) Modulation of auxin-binding protein 1 gene expression in maize and the teosintes by transposon insertions in its promoter. Mol Genet Genomics 287:143–153

Emera D, Casola C, Lynch VJ, Wildman DE, Agnew D, Wagner GP (2012) Convergent evolution of endometrial prolactin expression in primates, mice, and elephants through the independent recruitment of transposable elements. Mol Biol Evol 29:239–247

Faulkner GJ, Kimura Y, Daub CO, Wani S, Plessy C, Irvine KM, Schroder K, Cloonan N, Steptoe AL, Lassmann T, Waki K, Hornig N, Arakawa T, Takahashi H, Kawai J, Forrest AR, Suzuki H, Hayashizaki Y, Hume DA, Orlando V, Grimmond SM, Carninci P (2009) The regulated retrotransposon transcriptome of mammalian cells. Nat Genet 41:563–571

Feschotte C (2008) Transposable elements and the evolution of regulatory networks. Nat Rev Genet 9:397–405

Fortes AM, Agudelo-Romero P, Silva MS, Ali K, Sousa L, Maltese F, Choi YH, Grimplet J, Martinez-Zapater JM, Verpoorte R, Pais MS (2011) Transcript and metabolite analysis in Trincadeira cultivar reveals novel information regarding the dynamics of grape ripening. BMC Plant Biol 11:149

Galindo LM, Gaitán-Solís E, Baccam P, Tohme J (2004) Isolation and characterization of RNase LTR sequences of Ty1-copia retrotransposons in common bean (*Phaseolus vulgaris* L). Genome 47:84–95

Gao D, Chen J, Chen M, Meyers BC, Jackson S (2012) A highly conserved, small LTR retrotransposon that preferentially targets genes in grass genomes. PLoS One 7:e32010

Gerlo S, Davis JR, Mager DL, Kooijman R (2006) Prolactin in man: a tale of two promoters. Bioessays 28:1051–1055

Gogvadze E, Buzdin A (2009) Retroelements and their impact on genome evolution and functioning. Cell Mol Life Sci 66:3727–3742

Gogvadze E, Stukacheva E, Buzdin A, Sverdlov E (2009) Human-specific modulation of transcriptional activity provided by endogenous retroviral insertions. J Virol 83:6098–6105

Grandbastien M-A (1998) Activation of plant retrotransposons under stress conditions. Trends Plant Sci 3:181–187

Grandbastien M-A, Spielmann A, Caboche M (1989) Tnt1, a mobile retroviral-like transposable element of tobacco isolated by plant cell genetics. Nature 337:376–380

Grandbastien M-A, Audeon C, Bonnivard E, Casacuberta JM, Chalhoub B, Costa APP, Le QH, Melayah D, Petit M, Poncet C, Tam SM, Van Sluys M-A, Mhiri C (2005) Stress activation and genomic impact of Tnt1 retrotransposons in *Solanaceae*. Cytogenet Genome Res 110:229–241

Hayashi K, Yoshida H (2009) Refunctionalization of the ancient rice blast disease resistance gene Pit by the recruitment of a retrotransposon as a promoter. Plant J 57:413–425

He P, Ma Y, Zhao G, Dai H, Li H, Chang L, Zhang Z (2010) FaRE1: a transcriptionally active Ty1-copia retrotransposon in strawberry. J Plant Res 123:707–714

Hermsmeier D, Schittko U, Baldwin IT (2001) Molecular interactions between the specialist herbivore *Manduca sexta* (Lepidoptera, Sphingidae) and its natural host *Nicotiana attenuata*. I. Large-scale changes in the accumulation of growth- and defense-related plant mRNAs. Plant Physiol 125:683–700

Hernández-Pinzón I, Jesús E, Santiago N, Casacuberta JM (2009) The frequent transcriptional readthrough of the tobacco Tnt1 retrotransposon and its possible implications for the control of resistance genes. J Mol Evol 68:269–278

Hirochika H (1993) Activation of tobacco retrotransposons during tissue culture. EMBO J 12:2521–2528

Hirochika H, Sugimoto K, Otsuki Y, Tsugawa H, Kanda M (1996) Retrotransposons of rice involved in mutations induced by tissue culture. Proc Natl Acad Sci USA 93:7783–7788

Hollister JD, Smith LM, Guo YL, Ott F, Weigel D, Gaut BS (2011) Transposable elements and small RNAs contribute to gene expression divergence between *Arabidopsis thaliana* and *Arabidopsis lyrata*. Proc Natl Acad Sci USA 108:2322–2327

Huda A, Bowen NJ, Conley AB, Jordan IK (2011) Epigenetic regulation of transposable element derived human gene promoters. Gene 475:39–48

Huettel B, Kanno T, Daxinger L, Aufsatz W, Matzke AJ, Matzke M (2006) Endogenous targets of RNA-directed DNA methylation and Pol IV in Arabidopsis. EMBO J 25:2828–2836

Ito H (2012) Small RNAs and transposon silencing in plants. Dev Growth Differ 54:100–107

Ito H, Gaubert H, Bucher E, Mirouze M, Vaillant I, Paszkowski J (2011) An siRNA pathway prevents transgenerational retrotransposition in plants subjected to stress. Nature 472:115–119

Ivashuta S, Naumkina M, Gau M, Uchiyama K, Isobe S, Mizukami Y, Shimamoto Y (2002) Genotype-dependent transcriptional activation of novel repetitive elements during cold acclimation of alfalfa (*Medicago sativa*). Plant J 31:615–627

Iwata H, Gaston A, Remay A, Thouroude T, Jeauffre J, Kawamura K, Oyant LH, Araki T, Denoyes B, Foucher F (2012) The TFL1 homologue KSN is a regulator of continuous flowering in rose and strawberry. Plant J 69:116–125

Johns MA, Mottinger J, Freeling M (1985) A low copy number, copia-like transposon in maize. EMBO J 4:1093–1101

Jordan IK, Rogozin IB, Glazko GV, Koonin EV (2003) Origin of a substantial fraction of human regulatory sequences from transposable elements. Trends Genet 19:68–72

Kalendar R, Tanskanen J, Immonen S, Nevo E, Schulman AH (2000) Genome evolution of wild barley (*Hordeum spontaneum*) by BARE-1 retrotransposon dynamics in response to sharp microclimatic divergence. Proc Natl Acad Sci USA 97:6603–6607

Karimi MM, Goyal P, Maksakova IA, Bilenky M, Leung D, Tang JX, Shinkai Y, Mager DL, Jones S, Hirst M, Lorincz MC (2011) DNA methylation and SETDB1/H3K9me3 regulate predominantly distinct sets of genes, retroelements, and chimeric transcripts in mESCs. Cell Stem Cell 8:676–687

Kashkush K, Khasdan V (2007) Large-scale survey of cytosine methylation of retrotransposons and the impact of readout transcription from long terminal repeats on expression of adjacent rice genes. Genetics 177:1975–1985

Kashkush K, Feldman M, Levy AA (2003) Transcriptional activation of retrotransposons alters the expression of adjacent genes in wheat. Nat Genet 33:102–106

Kato H, Sriprasertsak P, Seki H, Ichinose Y, Shiraishi T, Yamada T (1999) Functional analysis of retrotransposons in pea. Plant Cell Physiol 40:933–941

Kidwell MG, Lisch D (1997) Transposable elements as sources of variation in animal and plants. Proc Natl Acad Sci USA 94:7704–7711

Kimura Y, Tosa Y, Shimada S, Sogo R, Kusaba M, Sunaga T, Betsuyaku S, Eto Y, Nakayashiki H, Mayama S (2001) OARE-1, a Ty1-copia retrotransposon in oat activated by abiotic and biotic stresses. Plant Cell Physiol 42:1345–1354

Kines KJ, Belancio VP (2012) Expressing genes do not forget their LINEs: transposable elements and gene expression. Front Biosci 17:1329–1344

Kobayashi S, Goto-Yamamoto N, Hirochika H (2004) Retrotransposon-induced mutations in grape skin color. Science 304:982

Kunarso G, Chia NY, Jeyakani J, Hwang C, Lu X, Chan YS, Ng HH, Bourque G (2010) Transposable elements have rewired the core regulatory network of human embryonic stem cells. Nat Genet 42:631–634

Li J, Dudas B, Webster MA, Cook HE, Davies BH, Gilmartin PM (2010) Hose in Hose, an S locus-linked mutant of *Primula vulgaris*, is caused by an unstable mutation at the *Globosa* locus. Proc Natl Acad Sci USA 107:5664–5668

Liu ZL, Han FP, Tan M, Shan XH, Dong YZ, Wang XZ, Fedak G, Hao S, Liu B (2004) Activation of a rice endogenous retrotransposon Tos17 in tissue culture is accompanied by cytosine demethylation and causes heritable alteration in methylation pattern of flanking genomic regions. Theor Appl Genet 109:200–209

Long L, Ou X, Liu J, Lin X, Sheng L, Liu B (2009) The spaceflight environment can induce transpositional activation of multiple endogenous transposable elements in a genotype-dependent manner in rice. J Plant Physiol 166:2035–2045

Lopes FR, Carazzolle MF, Pereira GA, Colombo CA, Carareto CM (2008) Transposable elements in *Coffea (Gentianales: Rubiacea)* transcripts and their role in the origin of protein diversity in flowering plants. Mol Genet Genomics 279:385–401

Lu HF, Dong HT, Sun CB, Qing DJ, Li N, Wu ZK, Wang ZQ, Li YZ (2011) The panorama of physiological responses and gene expression of whole plant of maize inbred line YQ7-96 at the three-leaf stage under water deficit and re-watering. Theor Appl Genet 123:943–958

Lynch VJ, Leclerc RD, May G, Wagner GP (2011) Transposon-mediated rewiring of gene regulatory networks contributed to the evolution of pregnancy in mammals. Nat Genet 43:1154–1159

Macfarlan TS, Gifford WD, Driscoll S, Lettieri K, Rowe HM, Bonanomi D, Firth A, Singer O, Trono D, Pfaff SL (2012) Embryonic stem cell potency fluctuates with endogenous retrovirus activity. Nature 487:57–63

Manetti ME, Rossi M, Nakabashi M, Grandbastien M-A, Van Sluys M-A (2009) The Tnt1 family member Retrosol copy number and structure disclose retrotransposon diversification in different *Solanum* species. Mol Genet Genomics 281:261–271

Mariño-Ramírez L, Jordan IK (2006) Transposable element derived DNaseI-hypersensitive sites in the human genome. Biol Direct 1:20

McCue AD, Nuthikattu S, Reeder SH, Slotkin RK (2012) Gene expression and stress response mediated by the epigenetic regulation of a transposable element small RNA. PLoS Genet 8: e1002474

McDonald JF (1990) Macroevolution and retroviral elements. Bioscience 40:183–191

Medstrand P, van de Lagemaat LN, Dunn CA, Landry JR, Svenback D, Mager DL (2005) Impact of transposable elements on the evolution of mammalian gene regulation. Cytogenet Genome Res 110:342–352

Michaud EJ, van Vugt MJ, Bultman SJ, Sweet HO, Davisson MT, Woychik RP (1994) Differential expression of a new dominant agouti allele (A^{iapy}) is correlated with methylation state and is influenced by parental lineage. Genes Dev 8:1463–1472

Mirouze M, Paszkowski J (2011) Epigenetic contribution to stress adaptation in plants. Curr Opin Plant Biol 14:267–274

Miyao A, Nakagome M, Ohnuma T, Yamagata H, Kanamori H, Katayose Y, Takahashi A, Matsumoto T, Hirochika H (2012) Molecular spectrum of somaclonal variation in regenerated rice revealed by whole-genome sequencing. Plant Cell Physiol 53:256–264

Muthukumar B, Bennetzen JL (2004) Isolation and characterization of genomic and transcribed retrotransposon sequences from sorghum. Mol Genet Genomics 271:308–316

Pearce SR, Kumar A, Flavell AJ (1996) Activation of the Ty1-copia group retrotransposons of potato (*Solanum tuberosum*) during protoplast isolation. Plant Cell Rep 16:949–953

Peaston AE, Evsikov AV, Graber JH, de Vries WN, Holbrook AE, Solter D, Knowles BB (2004) Retrotransposons regulate host genes in mouse oocytes and preimplantation embryos. Dev Cell 7:597–606

Pereira V, Enard D, Eyre-Walker A (2009) The effect of transposable element insertions on gene expression evolution in rodents. PLoS One 4:e4321

Pi W, Zhu X, Wu M, Wang Y, Fulzele S, Eroglu A, Ling J, Tuan D (2010) Long-range function of an intergenic retrotransposon. Proc Natl Acad Sci USA 107:12992–12997

Picault N, Chaparro C, Piegu B, Stenger W, Formey D, Llauro C, Descombin J, Sabot F, Lasserre E, Meynard D, Guiderdoni E, Panaud O (2009) Identification of an active LTR retrotransposon in rice. Plant J 58:754–765

Pischke MS, Huttlin EL, Hegeman AD, Sussman MR (2006) A transcriptome-based characterization of habituation in plant tissue culture. Plant Physiol 140:1255–1278

Polavarapu N, Mariño-Ramírez L, Landsman D, McDonald JF, Jordan IK (2008) Evolutionary rates and patterns for human transcription factor binding sites derived from repetitive DNA. BMC Genomics 9:226

Rajput MK, Upadhyaya KC (2009) CARE1, a Ty3-gypsy like LTR-retrotransposon in the food legume chickpea (*Cicer arietinum* L.). Genetica 136:429–437

Rakocevic A, Mondy S, Tirichine L, Cosson V, Brocard L, Iantcheva A, Cayrel A, Devier B, Abu El-Heba GA, Ratet P (2009) MERE1, a low-copy-number copia-type retroelement in *Medicago truncatula* active during tissue culture. Plant Physiol 151:1250–1263

Ramallo E, Kalendar R, Schulman AH, Martínez-Izquierdo JA (2008) Reme1, a Copia retrotransposon in melon, is transcriptionally induced by UV light. Plant Mol Biol 66:137–150

Rico-Cabanas L, Martínez-Izquierdo JA (2007) CIRE1, a novel transcriptionally active Ty1-copia retrotransposon from *Citrus sinensis*. Mol Genet Genomics 277:365–377

Robins DM, Samuelson LC (1992) Retrotransposons and the evolution of mammalian gene expression. Genetica 86:191–201

Romanish MT, Lock WM, van de Lagemaat LN, Dunn CA, Mager DL (2007) Repeated recruitment of LTR retrotransposons as promoters by the anti-apoptotic locus NAIP during mammalian evolution. PLoS Genet 3:e10

Rotter D, Bharti AK, Li HM, Luo C, Bonos SA, Bughrara S, Jung G, Messing J, Meyer WA, Rudd S, Warnke SE, Belanger FC (2007) Analysis of EST sequences suggests recent origin of allotetraploid colonial and creeping bentgrasses. Mol Genet Genomics 278:197–209

Sabot F, Sourdille P, Chantret N, Bernard M (2006) Morgane, a new LTR retrotransposon group, and its subfamilies in wheats. Genetica 128:439–447

Sabot F, Picault N, El-Baidouri M, Llauro C, Chaparro C, Piegu B, Roulin A, Guiderdoni E, Delabastide M, McCombie R, Panaud O (2011) Transpositional landscape of the rice genome revealed by paired-end mapping of high-throughput re-sequencing data. Plant J 66:241–246

Salazar M, González E, Casaretto JA, Casacuberta JM, Ruiz-Lara S (2007) The promoter of the TLC1.1 retrotransposon from *Solanum chilense* is activated by multiple stress-related signaling molecules. Plant Cell Rep 26:1861–1868

Schmidt D, Schwalie PC, Wilson MD, Ballester B, Gonçalves A, Kutter C, Brown GD, Marshall A, Flicek P, Odom DT (2012) Waves of retrotransposon expansion remodel genome organization and CTCF binding in multiple mammalian lineages. Cell 148:335–348

Shapiro JA (2005) Retrotransposons and regulatory suites. Bioessays 27:122–125

Slotkin RK, Martienssen R (2007) Transposable elements and the epigenetic regulation of the genome. Nat Rev Genet 8:272–285

Steward N, Ito M, Yamaguchi Y, Koizumi N, Sano H (2002) Periodic DNA methylation in maize nucleosomes and demethylation by environmental stress. J Biol Chem 277:37741–37746

Studer A, Zhao Q, Ross-Ibarra J, Doebley J (2011) Identification of a functional transposon insertion in the maize domestication gene tb1. Nat Genet 43:1160–1163

Suoniemi A, Narvanto A, Schulman AH (1996) The BARE-1 retrotransposon is transcribed in barley from an LTR promoter active in transient assays. Plant Mol Biol 31:295–306

Tahara M, Aoki T, Suzuka S, Yamashita H, Tanaka M, Matsunaga S, Kokumai S (2004) Isolation of an active element from a high-copy-number family of retrotransposons in the sweetpotato genome. Mol Genet Genomics 272:116–127

Takeda S, Sugimoto K, Kakutani T, Hirochika H (2001) Linear DNA intermediates of the Tto1 retrotransposon in Gag particles accumulated in stressed tobacco and *Arabidopsis thaliana*. Plant J 28:307–317

Tanurdzic M, Vaughn MW, Jiang H, Lee TJ, Slotkin RK, Sosinski B, Thompson WF, Doerge RW, Martienssen RA (2008) Epigenomic consequences of immortalized plant cell suspension culture. PLoS Biol 6:2880–2895

Tapia G, Verdugo I, Yañez M, Ahumada I, Theoduloz C, Cordero C, Poblete F, González E, Ruiz-Lara S (2005) Involvement of ethylene in stress-induced expression of the TLC1.1 retrotransposon from *Lycopersicon chilense* Dun. Plant Physiol 138:2075–2086

Thomson SJ, Goh FG, Banks H, Krausgruber T, Kotenko SV, Foxwell BM, Udalova IA (2009) The role of transposable elements in the regulation of IFN-lambda1 gene expression. Proc Natl Acad Sci USA 106:11564–11569

Twell D, Yamaguchi J, Wing RA, Ushiba J, McCormick S (1991) Promoter analysis of genes that are coordinately expressed during pollen development reveals pollen-specific enhancer sequences and shared regulatory elements. Genes Dev 5:496–507

Ungerer MC, Strakosh SC, Zhen Y (2006) Genome expansion in three hybrid sunflower species is associated with retrotransposon proliferation. Curr Biol 16:R872–R873

Van de Lagemaat LN, Landry JR, Mager DL, Medstrand P (2003) Transposable elements in mammals promote regulatory variation and diversification of genes with specialized functions. Trends Genet 19:530–536

Verne S, Jaquish B, White R, Ritland C, Ritland K (2011) Global transcriptome analysis of constitutive resistance to the white pine weevil in spruce. Genome Biol Evol 3:851–867

Vicient CM (2010) Transcriptional activity of transposable elements in maize. BMC Genomics 11:601

Vicient CM, Jääskeläinen MJ, Kalendar R, Schulman AH (2001) Active retrotransposons are a common feature of grass genomes. Plant Physiol 125:1283–1292

Wang T, Zeng J, Lowe CB, Sellers RG, Salama SR, Yang M, Burgess SM, Brachmann RK, Haussler D (2007) Species-specific endogenous retroviruses shape the transcriptional network of the human tumor suppressor protein p53. Proc Natl Acad Sci USA 104:18613–18618

Wang J, Bowen NJ, Mariño-Ramírez L, Jordan IK (2009) A c-Myc regulatory subnetwork from human transposable element sequences. Mol Biosyst 5:1831–1839

Weber B, Wenke T, Frömmel U, Schmidt T, Heitkam T (2010) The Ty1-copia families SALIRE and Cotzilla populating the *Beta vulgaris* genome show remarkable differences in abundance, chromosomal distribution, and age. Chromosome Res 18:247–263

White SE, Habera LF, Wessler SR (1994) Retrotransposons in the flanking regions of normal plant genes: a role for copia-like elements in the evolution of gene structure and expression. Proc Natl Acad Sci USA 91:11792–11796

Wicker T, Sabot F, Hua-Van A, Bennetzen JL, Capy P, Chalhoub B, Flavell A, Leroy P, Morgante M, Panaud O, Paux E, SanMiguel P, Schulman AH (2007) A unified classification system for eukaryotic transposable elements. Nat Rev Genet 8:973–982

Woodrow P, Pontecorvo G, Ciarmiello LF, Fuggi A, Carillo P (2011) Ttd1a promoter is involved in DNA-protein binding by salt and light stresses. Mol Biol Rep 38:3787–3794

Zeller G, Henz SR, Widmer CK, Sachsenberg T, Rätsch G, Weigel D, Laubinger S (2009) Stress-induced changes in the *Arabidopsis thaliana* transcriptome analyzed using whole-genome tiling arrays. Plant J 58:1068–1082

Zhang JJ, Zhou ZS, Song JB, Liu ZP, Yang H (2012) Molecular dissection of atrazine-responsive transcriptome and gene networks in rice by high-throughput sequencing. J Hazard Mater 219–220:57–68

Chapter 15
Rider Transposon Insertion and Phenotypic Change in Tomato

Ning Jiang, Sofia Visa, Shan Wu, and Esther van der Knaap

Abstract The *Rider* retrotransposon is ubiquitous in the tomato genome and is likely an autonomous element that still transposes to date. The majority of approximately 2,000 copies of *Rider* are located near genes. Phenotypes associated with *Rider* insertion are diverse and often the result of knock out of the underlying genes. One unusual *Rider*-mediated phenotype resulted from a gene duplication event. By means of read-through transcription, *Rider* copied part of the surrounding sequence to another location in the genome, leading to high expression of one of the transposed genes, *SUN*, resulting in an elongated fruit shape. Transcription studies demonstrated that *Rider* is expressed to levels comparable to the expression of other tomato genes and that control of transposition may be regulated by antisense transcription. Taken together, *Rider* is a unique retrotransposon that may have played important roles in the evolution of tomato and its closest relatives.

Keywords LTR Copia • Phenotype • Rider • Tomato • Transcription

Abbreviations

ATP	Adenosine triphosphate
bHLH	Basic helix–loop–helix
BL	Blind
C	Cut leaf or potato leaf mutation

N. Jiang
Department of Horticulture, Michigan State University, East Lansing, MI 48824, USA

S. Visa
Department of Mathematics and Computer Science, College of Wooster, Wooster, OH 44691, USA

S. Wu • E. van der Knaap (✉)
Department of Horticulture and Crop Science, The Ohio State University, Wooster, OH 44691, USA
e-mail: Vanderknaap.1@osu.edu

M.-A. Grandbastien and J.M. Casacuberta (eds.), *Plant Transposable Elements*, Topics in Current Genetics 24, DOI 10.1007/978-3-642-31842-9_15,

CP	Coat protein
DNA	Deoxyribonucleic acid
EST	Expressed sequence tag
FER	Iron inefficient mutant
INT	Integrase
LTR	Long terminal repeat
Mb	Mega base pair
MITE	Miniature inverted repeat transposable element
mRNA	messenger RNA
MULE	Mutator-like element
MYA	Million years ago
MYB	Myeloblastosis transcription factor
PBS	Primer binding site
PPT	Polypurine tract
PR	Protease
PSY1	Phytoene synthase 1
R	Red or yellow flesh mutation
RAX1	Regulator of axillary meristem 1
RH	RNase H
RISC	RNA-induced silencing complex
RNA	Ribonucleic acid
TE	Transposable element
TIR	Terminal inverted repeat
TSD	Target site duplication

15.1 The Abundance of TEs in Genomes and the Phenotypic Consequences of their Insertions

Transposable elements (TEs), DNA fragments capable of replication and movement, are major components of eukaryotic genomes. Depending on the timing of their transposition activity, they may display different insertion sites among closely related genomes and hence contribute to genome diversity. TEs are divided into two classes. Class I elements or RNA elements (retrotransposons) use the element-encoded mRNA as the transposition intermediate. These transposons are either flanked by a long terminal repeat (LTR) or lack terminal repeat sequences (non-LTR transposons). Class II elements or DNA transposons are often characterized by the terminal inverted repeats (TIRs) and transposition through a DNA intermediate. Autonomous TEs encode a transposase and other proteins required for transposition, while nonautonomous elements lack functional transposition proteins and rely on the cognate autonomous TEs for their transposition. In plants, LTR retrotransposons are very abundant and are largely responsible for the genome size expansion in grass species (Bennetzen 1996). This is also the case for species in the *Solanaceae* family that includes tomato (*Solanum lycopersicum*), potato (*S. tuberosum*), pepper (*Capsicum* spp.), eggplant (*S. melongena*), petunia (*Petunia* spp.), and tobacco (*Nicotiana* spp.). The different genome size that ranges from 844 [potato, (Consortium 2011)] to 4,500 Mb (*Nicotiana tabacum*) is largely attributed to differences in the number of LTR elements, some of which are found in the euchromatic parts of the genome (Park et al. 2011a, b). Reduction in genome size also occurs by unequal recombination between the two LTRs of a single element (Ma et al. 2004). This often leads to the deletion of the internal region and one of the LTRs resulting in the formation of a "solo" LTR.

Transposons are mostly known for the disruption of genes when they insert into or very close to genes. However, they are also known to duplicate and mobilize gene sequences. Recent studies indicate most major types of TEs are capable of duplicating and amplifying genes or gene fragments (Kazazian 2004; Bennetzen 2005; Feschotte and Pritham 2007; Schnable et al. 2009). For example, the maize *Bs1* LTR retrotransposon carries part of a plasma membrane proton-translocating ATPase gene without its intron sequences (Bureau et al. 1994; Jin and Bennetzen 1994). Subsequently, it was shown that this chimeric element was transcribed and translated in early ear development and might have a function in the reproductive pathway (Elrouby and Bureau 2010). In rice, over a thousand genes that duplicated through retrotransposition (retrogenes) have been identified, and many recruited new exons from flanking regions, resulting in the formation of chimeric genes (Wang et al. 2006a). Similarly, there are thousands of *Mutator*-like elements (MULE) that carry genes or gene fragments in the rice genome (Jiang et al. 2004; Juretic et al. 2005). Due to the ability to duplicate genes or gene fragment, transposons themselves may represent the structural variation among species or individuals in the population. For example, there are thousands of *Helitrons* carrying genes in maize (Du et al. 2009; Yang and Bennetzen 2009), and they contribute significantly to many fragments that are not shared among different maize cultivars at the orthologous position (Fu and Dooner 2002; Morgante et al. 2005).

Despite the abundance of transposons in the tomato genome, few are known to result in an altered phenotype. In this chapter, we summarize the findings of what is known about *Rider*, a high copy *Copia* element found in tomato and its closest wild relatives. The element was first described as the cause of the elongated fruit shape at the locus *sun* and its ability to duplicate genes from one chromosome to another (Xiao et al. 2008). In addition, there are many unusual features of *Rider* that warrant further investigations as will be demonstrated below.

15.2 Features Associated with *Rider*

15.2.1 The Structure of the Rider *Element*

The structure of *Rider* element resembles that of a typical *Copia*-like element from many perspectives. The element is 4,867 in length with two identical LTRs on each terminus (Fig. 15.1). The LTR of the *Rider* element at *SUN* is 398 bp in length and

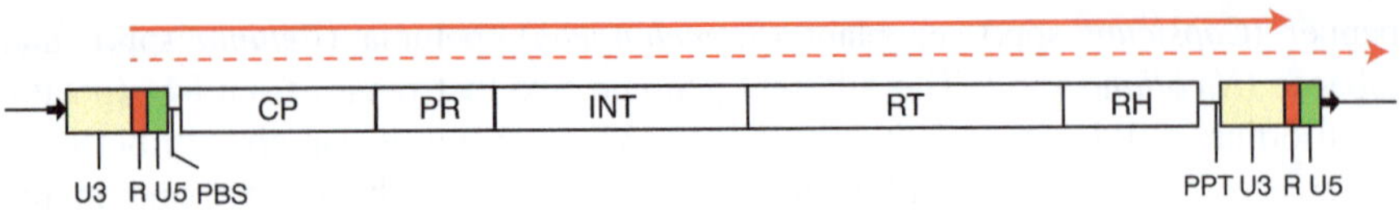

Fig. 15.1 The structure of *Rider*. *Color boxes* indicate distinct regions in LTR (U3, R and U5). Coding regions are indicated as *white boxes*. The genes within *Rider* are shown as *white boxes* and encode capsid-like proteins (CP), protease (PR), integrase (INT), reverse transcriptase (RT), and RNase-H (RH). Other sequence features are primer binding site (PBS), polypurine tract (PPT). *Black arrows* flanking the LTRs indicate target site duplication (TSD). For *Rider* elements, TSD are 5 bp. *Red solid arrow* represents a normal transcript from *Rider*, while the *dashed arrow* exemplifies a read-through transcript

includes the three classical LTR domains called U3, R, and U5. U3 region contains the promoter of the element, and its size is highly variable among individual *Rider* elements (Jiang et al. 2009). Sequences in R and U5 are responsible for the termination of transcription of the element, and they are well conserved among most individual elements (also see below). The internal region of *Rider* is 4,071 bp and encodes a single polyprotein of 1,307 amino acids, accounting for 96% of the internal region. The polyprotein contains all typical proteins or domains that a *Copia*-like element encodes, including capsid-like protein, protease, integrase, reverse transcriptase, and RNase H (Fig. 15.1) (Kumar and Bennetzen 1999). The internal region also contains the cis-elements required for transposition, such as the primer binding site and polypurine track (Lewin 2008). Thus, *Rider* is likely to be an autonomous *Copia*-like element.

15.2.2 The Timing of Rider *Amplification*

Database searches and DNA blots using the LTR as probe indicate that *Rider* element is present in all *Solanum* section *Lycopersicon* species tested and absent from related species such as potato, tobacco, and coffee (Cheng et al. 2009; Jiang et al. 2009). Therefore, it appears that the initial amplification of *Rider* occurred prior to the speciation of *Lycopersicon* section species and after the divergence of tomato and potato, which is estimated to be between 5.1 and 7.3 MYA (Wang et al. 2008). Among the section *Lycopersicon* species, variation of copy number was observed. For example, the copy number of *Rider* appears to be lower in the genomes of *S. habrochaites* and *S. chilense* compared to other species (Cheng et al. 2009; Jiang et al. 2009). The tomato genome harbors about 2,000 copies of *Rider* based on partial genome sequence surveys (Jiang et al. 2009). Two-thirds of the intact *Rider* elements inserted after the divergence of *S. lycopersicum* and *S. pimpinellifolium*, which occurred about 1.3 MYA. This finding suggests that the majority of *Rider* elements arose well after the speciation in the section *Lycopersicon* (Jiang et al. 2009). Moreover, insertion polymorphism of *Rider* and transcript accumulation were detected among different tomato cultivars

(Cheng et al. 2009; Jiang et al. 2009). Due to the high insertion polymorphism among tomato species, *Rider* would be useful as a tool for studying the phylogenic relationship in this important group.

15.2.3 The Origin of the Rider *Element*

The origin of *Rider* is mysterious. The presence of TEs in a certain genome can be either due to vertical transmission from ancestral genomes or horizontal transfer from an unrelated species. As mentioned above, *Rider* is absent from potato, tobacco, and coffee. Meanwhile, two individual LTR elements in *Arabidopsis*, named *Rider-like* 1 and *Rider-like* 2, have moderate nucleotide similarity (~75%) with *Rider* in the internal region and part of the LTR sequence (Cheng et al. 2009). For this reason, it was proposed that *Rider* was introduced into the tomato genome 1–6 MYA from *Arabidopsis* or a relative of *Arabidopsis* (Cheng et al. 2009). While the similarity between *Rider-like* elements and *Rider* is unusually high given the genetic distance between *Arabidopsis* and tomato, there is not sufficient evidence to support an unambiguous case of direct transfer between the two species in the proposed timeframe. Elements highly similar to *Rider-like* elements are not present in genomes of species related to *Arabidopsis*, such as *A. lyrata* and *B. oleracea* ((Cheng et al. 2009), Jiang, unpublished data). As a result, the ultimate donor or ancestor of *Rider* is unclear if it indeed resulted from horizontal transfer from one to the other species.

An equally plausible explanation for the occurrence of *Rider* and *Rider*-like elements in two distant genomes is that *Rider* is inherited from the ancestral genome of tomato and lost from related species. This is because most TE families experience a life cycle of "birth–burst–extinction" (Hartl et al. 1997). Once a TE family is no longer transpositionally active, mutations and deletions accumulate and the particular family will eventually disappear from the genome. According this scenario, loss of TEs from a genome is a common event and only a small subset of TEs can achieve long-term success. Due to the fact that *Rider* is a compact element without obviously nonessential sequences, the conservation between *Rider* and *Rider-like* elements could be due to functional constraints. Consequently, the origin of *Rider* is still an open question. The clarification of this issue awaits the availability of more genomic sequences in *Brassica* and *Solanaceae*, and other plant species.

15.3 Distribution and Targeting Preference of *Rider*

Plant genomes harbor numerous types of transposons, and different transposons have distinct niches. The distribution pattern of any transposons, including LTR elements, is the consequence of target specificity and selection against deleterious insertions or selection for favorable insertions (Pereira 2004; Peterson-Burch et al. 2004).

Many high copy number LTR elements are nested in the intergenic or heterochromatic regions (SanMiguel et al. 1996; Ananiev et al. 1998; Jiang et al. 2002). In contrast, low copy number LTR elements, such as *Tpv2* elements (40 copies) in common bean (*Phaseolus vulgaris*) and *Tos17* (a few copies in natural populations) in rice (*Oryza sativa*), are frequently found in genic regions (Garber et al. 1999; Miyao et al. 2003). Given the fact that *Tos17* can amplify rapidly under artificial conditions (Hirochika et al. 1996), its low copy number in natural populations suggests that the preference for genic regions may result in deleterious effect on the host organism, which prevents the element from further amplifications. The only known exception is the *Tnt1* element from tobacco, which has a relatively high copy number (a few hundred copies), yet is located in genic regions (Grandbastien et al. 1989; Le et al. 2007). Nevertheless, *Tnt1*-related elements are only present in a few dozens in tomato and those are mostly mapped to pericentromeric regions (Tam et al. 2007), suggesting host environment may have important influence on amplification and distribution of LTR elements.

Unlike any of the known LTR elements, *Rider* does not appear to be concentrated in certain regions of the genome (Cheng et al. 2009). Moreover, about half of the *Rider* elements are located within 1 kb of a gene. This ratio is much higher than that of another high copy number tomato LTR element *Jinling*, for which only 20% of the elements are within the same distance to a gene (Jiang et al. 2009). This can be explained by the difference in their chromosomal distribution patterns since most *Jinling* elements are located in heterochromation regions where the gene density is low (Wang et al. 2006b). In contrast, *Rider* elements are located in both heterochromatic and euchromatic regions so they are more likely surrounded by genes.

Despite its high copy number and frequent associations with genes, *Rider* does not seem to disrupt genes at a high level that would render the tomato genome unstable. This could be due to the regulation of its expression (see below) and to its insertion preference. *Rider* appears to insert into AT-rich sequence (Jiang et al. 2009). Since coding regions are usually more GC-rich than noncoding regions (Salinas et al. 1988; Mizuno and Kanehisa 1994), such a preference allows *Rider* elements to select noncoding regions as their targets and minimize possible deleterious effects. In this case, the amplification of *Rider* is largely silent despite the fact that many elements are close to genes. Meanwhile, being located in the genic regions may favor the element amplification since the element is more accessible to the transcription machinery. This might partially explain the success of *Rider* in the tomato genome.

15.4 Rider Expression, Read-Through Transcription and its Correlation with Mutations in LTR

Based on Northern blot and RT-PCR experiments as well as database searches, *Rider* is constitutively expressed in tomato (Cheng et al. 2009; Jiang et al. 2009). Transcript sizes suggest that most *Rider* RNA is intact and has the potential to transpose to new positions. Mining through mRNA seq data sets also showed that *Rider* is expressed in certain tissues at a level comparable to the tomato fruit shape gene *OVATE*. *SUN* and *R* (the latter corresponding a phytoene synthase gene, see below) are expressed higher than *Rider* while *DEFL2* (encoding a defensin protein, see below) is expressed the highest in the tissues examined (Table 15.1). Interestingly, while only sense expression of the genes *SUN*, *OVATE*, *R*, and *DEFL2* is found, *Rider* appears to be expressed equally in both sense and antisense direction (Table 15.1), raising the interesting question of whether the regulation of transposition is mediated in part by posttranscriptional silencing. Further examination of the position of the mRNA seq reads relative to *Rider* revealed that the reads are evenly distributed along the transposon in both directions (Fig. 15.2). Due to the finding that intact *Rider* elements outnumber truncated elements by 3.5 to 1, this suggests that transcription in the sense and antisense direction are derived from intact elements. However, spurious expression from exogenous promoters into truncated *Rider* elements cannot be excluded either. Regardless, double stranded RNAs are commonly resulting in rapid mRNA degradation via the RNA-induced silencing complex (RISC). Therefore, the potential gene silencing of *Rider* might explain the results from Northern blots that showed smears instead of one distinct band (Jiang et al. 2009). This finding is also consistent with the observation that the insertion polymorphism of *Rider* among tomato cultivars is relatively low compared to that among *Solanum* subsection *Lycopersicon* species (Jiang et al. 2009). In other words, the high copy number of *Rider* is likely due to its high transposition activity in recent past, which may have declined due to potential silencing arising from the abundance of elements.

A low number of ESTs were found to be chimeric between *Rider* LTR and an unrelated sequence (Cheng et al. 2009; Jiang et al. 2009). These chimeric elements can be explained by artifacts in the construction of the library for EST. Alternatively, these aberrant RNAs could also lead to gene silencing in cases where the chimeric part exhibits high sequence similarity to an endogenous gene. The finding of chimeric EST reads could also be the result of read-through transcription. Normally, *Rider* transcription starts in the R region of the 5′ LTR and ends in the R region of the 3′ LTR. Read-through transcription would extend past the R region into the U5 and neighboring genome region. Indeed, read-through transcription of *Rider* is found in all the tissues examined (Jiang et al. 2009).

Read-through transcription is at the heart of the *SUN* duplication as will be discussed in detail below. The *Rider* element that created the locus carried a mutation in one of the two “TTGT” sequences required for transcript termination (Jiang et al. 2009). The sequencing of read-through transcripts over the region that is required for termination indeed showed that the majority of the transcripts carried the mutation in the LTR found in the *Rider* element at the *sun* locus. These findings strongly suggest that read-through transcripts are indeed associated with *Rider* elements and are more prevalent when the LTR carries the “TTAT” mutation in one of the “TTGT” copies in the U5 region.

Table 15.1 Expression of *Rider* and four genes in tomato flower buds

Sample	Strand	RIDER	RIDER RPKM	SUN	SUN RPKM	OVATE	OVATE RPKM	DEFL2	DEFL2 RPKM	R	R RPKM	Number of illumina reads aligned to the genome (millions)
Replicate 1	Antisense	183	2.92	3	0.11	0	0	0	0	0	0	
	Sense	225	3.59	1,134	39.83	84	5.73	1,053	165.64	383	18.23	
	Total	408	6.52	1,137	39.94	84	5.73	1,053	165.64	383	18.23	13.82
Replicate 2	Antisense	268	3.83	1	0.03	0	0	0	0	0	0	
	Sense	342	4.89	1,209	38.01	87	5.32	1,196	168.39	959	40.86	
	Total	610	8.72	1,210	38.04	87	5.32	1,196	168.39	959	40.86	15.44
Replicate 3	Antisense	382	5	2	0.06	0	0	0	0	0	0	
	Sense	278	3.64	1,554	44.74	141	7.89	710	91.55	477	18.61	
	Total	660	8.64	1,556	44.8	141	7.89	710	91.55	477	18.61	16.86
Gene length (Kb)		4.53		2.06		1.06		0.46		1.52		

Small flower bud tissues from the tomato accession SA2 (LA1589 near isogenic line carrying the *SUN* gene duplication; Xiao et al. 2008) were harvested and RNA was extracted. Directional libraries were constructed according to Zhong et al. (2011). Reads were aligned to the five sequences using Tophat and allowing one mismatch. The first column for each sequence shows the number of raw reads. RPKM = reads per kb per million reads. Accession numbers: Rider (SGN-U569744), SUN (EU491503), OVATE (AY140893), DEFL2 (Solyc07g007750), R (Solyc03g031860). Due to the high similarity of *Rider* elements, the aligned reads cannot be assigned to specific elements in the tomato genome

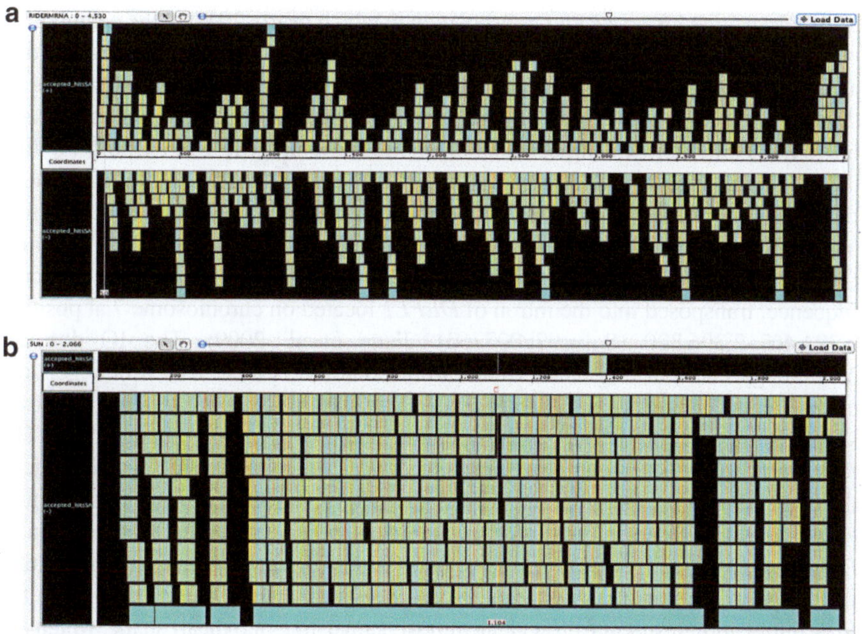

Fig. 15.2 Alignment of the mRNA seq reads to (**a**) *Rider* and (**b**) *SUN*. The libraries were constructed strand-specifically such that only the first-strand cDNA will yield reads (Zhong et al. 2011). The SAM files generated by Tophat alignment were visualized using the IGB viewer (http://bioviz.org/igb/faq.shtml). *Rider* reads are found in both directions along the transposon while *SUN* reads are only found for the sense strand (– strand in the viewer), with the exception of one read. Results are from replicate 2 in Table 15.1. The *solid green bar* on the bottom indicates that there are more reads corresponding to *SUN* that are not displayed in the viewer due to space constraints

15.5 Case Studies of Phenotypic Changes Caused by *Rider* and Genomic Landscape in which the Element Inserts

As demonstrated in the previous sections, the insertion preference of *Rider* is found near genes, *Rider* is constitutively expressed in tomato tissues albeit in both directions and *Rider* read-through transcription is occurring. Also, *Rider* elements are only found in the species of the *Solanum* section *Lycopersicon* and not in other Solanaceous relatives such as potato and tobacco. In addition, *Rider* has been shown to be involved in phenotypic changes that are found in the *Lycopersicon* section of the Solanum genus, including those that impact domestication-related phenotypes as well as spontaneously arising mutations.

15.5.1 Rider *and Fruit Shape*

One of the most striking examples of phenotypic change mediated by *Rider* transposition is found at the fruit shape locus *sun* located on chromosome 7 (Xiao et al. 2008).

The locus resulted from a *Rider* transposition in which nearly 20 kb of the neighboring genome was included in the event. Based on sequence comparisons, the transposition and resulting genomic duplication was deduced to have happened as follows. Read-through transcription of the *Rider* element on chromosome 10 found at position 60,134,479–60,139,738 (http://www.solgenomics.net unigene SGN-U569744) into the neighboring genes, followed by a template switch in the first intron of a SDL1-like gene to downstream of an IQ domain-containing gene found at position 60,140,568–60,142,797. Transcription continued until the first LTR of *Rider* (Xiao et al. 2008). This giant retroelement, that includes *Rider* and nearby genome sequence, transposed into the intron of *DEFL1* located on chromosome 7 at position 2,394,467–2,396,320 (Solyc07g007760) (Jiang et al. 2009). The IQ domain-containing gene that originated from chromosome 10 is located in a new genome environment leading to high expression in the fruit resulting in an elongated fruit shape (Xiao et al. 2008). Thus the IQ domain containing gene was renamed *SUN*. The *Rider* insertion knocked out the expression of *DEFL1* (Solyc07g007760) and reduced the expression of the neighboring *DEFL2* gene (Solyc07g007750) by at least fivefold (unpublished mRNA seq data). Further studies have shown that the transposition of *Rider* and duplication of *SUN* was most likely a post-domestication event originating in Europe in the last 200–500 years (Rodriguez et al. 2011). Varieties carrying the *SUN* duplication result in fruit with an almost pepper-like or oxheart shape, which are typically found in heirloom tomatoes (Fig. 15.3). The genome environment of the ancestral locus on chromosome 10 showed no class I transposons except for *Rider*, but instead a high number of class II DNA transposons. At the *sun* locus, the number of class II transposons was higher than found on the ancestral locus (Jiang et al. 2009).

15.5.2 **Rider** *and Iron Deficiency*

The chlorotic tomato mutant *fer* was a spontaneous mutant identified in the 1960s (Brown et al. 1971). The mutant plant exhibits defects in all the typical responses to iron deficiency and uptake of Fe^{3+} (Brown et al. 1971; Ling et al. 1996). Although located in the pericentromeric region of chromosome 6, which might exhibit reduced recombination rates, the *FER* gene was identified by positional cloning and found to encode a bHLH protein involved in the transcriptional regulation of plant iron nutrition (Ling et al. 2002; Brumbarova and Bauer 2005; Guyot et al. 2005). The gene is located on chromosome 6 at position 31,549,026–31,547,113 (Solyc06g051550). The mutation in tomato *FER* was due to a spontaneous insertion of *Rider* in the first exon resulting in disruption of the gene (Ling et al. 2002; Cheng et al. 2009). The *fer Rider* element is 100% identical, including the LTRs, to the *Rider* element found at the *sun* locus and the ancestral locus on chromosome 10 (Cheng et al. 2009). A high level of transposable elements, including class I, class II and unclassified repeats, are found at the *fer* locus demonstrating a highly diverse TE landscape in the pericentromeric region of chromosome 6. The *fer* locus also features a relatively low density of genes of 19.8 kb per gene (Guyot et al. 2005).

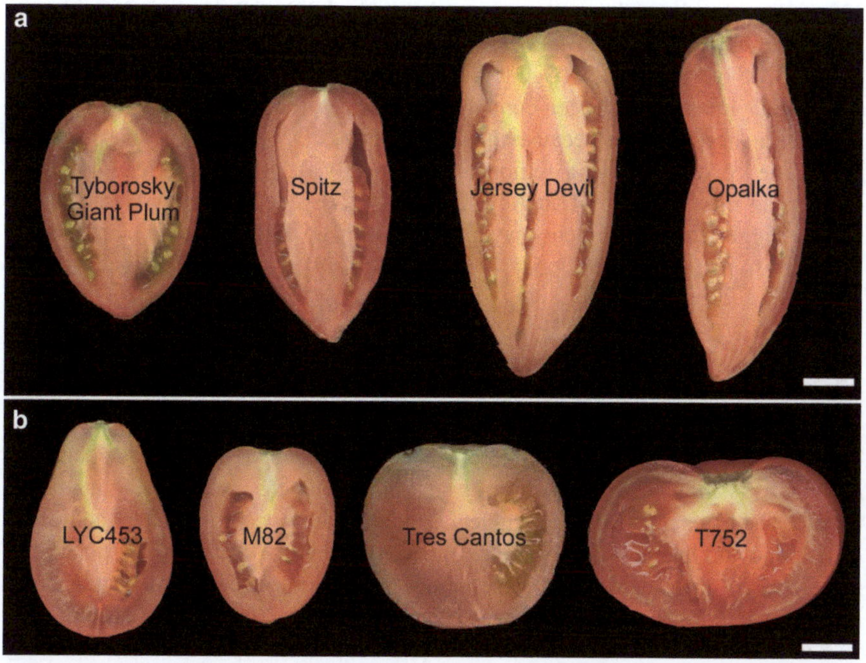

Fig. 15.3 Tomato fruit shape affected by *Rider*. (**a**) Varieties with the *SUN* gene resulting from *Rider* transposition and gene duplication. (**b**) Varieties without the *SUN* gene duplication. The variety names are written in each fruit (Rodriguez et al. 2011). Note the characteristically long fruit and pointed shape as a result of *SUN*. Pear-shaped fruit (LYC453 in B) is controlled by *OVATE*. *Bar* corresponds to 2 cm

This is in contrast to the *sun* locus and the ancestral locus on chromosome 10, where gene density approached that of what is typically found in euchromatin in the range of 5–7 kb per gene (Jiang et al. 2009). Other than *Rider*, the LTR transposons found at the *fer* locus are neither active nor autonomous as they have accumulated numerous mutations (Guyot et al. 2005).

15.5.3 Rider *and Fruit Color*

The yellow flesh mutation in tomato confers a yellow instead of the wild type red fruit and the locus is named "r" (Price and Drinkard 1908). The underlying gene is *phytoene synthase 1* (*PSY1*) that encodes the first enzyme in the carotenoid biosynthesis pathway. Initially, the gene was identified in a screen for ripening-induced genes (Bartley et al. 1992; Fray and Grierson 1993). The cDNA cloning and sequencing of the two allelic versions of the yellow flesh mutant alleles, *r* and r^y, showed that the older allele, *r*, was due to an insertion of a repetitive element (Fray and Grierson 1993). Sequence comparisons of the inserted fragment of

328 nucleotides showed that it corresponded to the LTR of *Rider* with 96% identity to the element found at the *sun* locus. *PSY1* is found on chromosome 3 at position 8,606,368–8,610,361 (Solyc03g031860). A detailed analysis of the genome structure at the *r* locus has not been conducted. However, the r^y allele appears to be the result of a short deletion because the 3′ end of the cDNA sequence of the mutant *psy1* gene corresponds to a region approximately 4.5 kb downstream of *PSY1* comprising the first exon of an Acyl-CoA synthase gene (Solyc03g031870). This finding suggests that the *r* locus may have experienced other types of rearrangement unrelated to *Rider* transposition.

15.5.4 Rider *and Leaf Complexity*

The last and most recently reported example of a phenotypic change mediated by *Rider* transposition is exemplified by the gene underlying the "potato leaf" mutation in tomato. The locus is called *C*, for cut leaf. Tomato features complex leaves comprised of terminal and lateral leaflets that are often serrated at the margins. The potato leaf represents an old tomato mutation resulting in reduced leaf complexity and smooth leaf blade margins (Price and Drinkard 1908; Busch et al. 2011). The underlying gene is a member of the R2R3 MYB transcription factor family that is evolutionarily very closely related to the tomato *BLIND* (*BL*) gene regulating shoot branching. *C* (Solyc06g074910) maps to chromosome 6 at position 42,804,036–42,806,196. *C* has acquired a new but related function compared to *BL* and both correspond to *RAX1* in Arabidopsis regulating shoot branching (Busch et al. 2011). *Rider* inserted near the 3′ end of *C* disrupting the coding region resulting in a null mutation. The *Rider* element found at *c* is identical in sequence to the element found at *sun* (Busch et al. 2011). Except for the *Rider* insertion allele which is spontaneous, most of the other reported *c* alleles were derived from mutagenesis screens (Busch et al. 2011). Of these induced mutations, two resulted from a deletion event of 286 bp and 40.6 kb, respectively. Although a detailed genome analysis of the locus has not been conducted, the *c* locus also appears prone to genome rearrangements in addition to transposon insertions.

15.6 Concluding Remarks

Transposable elements achieve their success through different strategies. Some elements, such as *Jinling* in tomato, are preferentially located in the pericentromeric heterochromatin, which is the "safe haven" for insertion. Other elements, such as the miniature inverted repeat transposable element (MITE) *mPing* in rice, are preferentially located in genic regions. Nevertheless, the impact of MITE insertion is often subtle due to their small size (usually less than 500 bp) as well as avoidance of insertion into coding region (Naito et al. 2009). Moreover, *mPing* harbors regulatory motifs that enable the adjacent genes to become stress inducible (Naito et al. 2009). In other words, a successful transposable element must either have minimal detrimental impact or bring about favorable mutations for the host genome, especially when the element is capable of transposition. From this point of view, *Rider* has developed many features for its success despite its relatively large size. First of all, *Rider* elements have been active in transposition since it amplified to thousands of copies in just a few million years. The most recent known transposition occurred in the 1960s with the creation of the *fer* locus (Cheng et al. 2009). Second, it targets all chromosomal regions but appears to avoid inserting into coding regions by selective insertions into AT-rich regions. Third, the transposition activity of *Rider* is likely regulated by antisense transcription of the element, thereby limiting the extent of transposition per generation. Finally, *Rider* creates read-through transcripts which may allow the duplication of flanking sequences including genes. The duplication of genes may create novel phenotypes that are favored by selection. Taken together, *Rider* is a unique retrotransposon that has been successfully amplified in the genome of tomato and may have played important roles in the evolution of tomato and its closest relatives.

Acknowledgments Funding in the Jiang laboratory is provided by National Science Foundation Molecular and Cellular Biosciences grant number 1121650. Funding in the van der Knaap laboratory is provided by National Science Foundation Integrative Organismal Systems grant number 0922661.

References

Ananiev EV, Phillips RL, Rines HW (1998) Chromosome-specific molecular organization of maize (*Zea mays* L.) centromeric regions. Proc Natl Acad Sci USA 95:13073–13078

Bartley GE, Viitanen PV, Bacot KO, Scolnik PA (1992) A tomato gene expressed during fruit ripening encodes an enzyme of the carotenoid biosynthesis pathway. J Biol Chem 267:5036–5039

Bennetzen JL (1996) The contributions of retroelements to plant genome organization, function and evolution. Trends Microbiol 4:347–353

Bennetzen JL (2005) Transposable elements, gene creation and genome rearrangement in flowering plants. Curr Opin Genet Dev 15:621–627

Brown JC, Chaney RL, Ambler JE (1971) A new tomato mutant inefficient in the transport of iron. Physiol Plant 25:48–53

Brumbarova T, Bauer P (2005) Iron-mediated control of the basic helix-loop-helix protein FER, a regulator of iron uptake in tomato. Plant Physiol 137:1018–1026

Bureau TE, White SE, Wessler SR (1994) Transduction of a cellular gene by a plant retroelement. Cell 77:479–480

Busch BL, Schmitz G, Rossmann S, Piron F, Ding J, Bendahmane A, Theres K (2011) Shoot branching and leaf dissection in tomato are regulated by homologous gene modules. Plant Cell 23:3595–3609

Cheng X, Zhang D, Cheng Z, Keller B, Ling H-Q (2009) A new family of Ty1-*copia*-like retrotransposon originated in the tomato genomes by a recent horizontal transfer event. Genetics 181:1183–1193

Consortium PGS (2011) Genome sequence and analysis of the tuber crop potato. Nature 475:189–195

Du C, Fefelova N, Caronna J, He L, Dooner HK (2009) The polychromatic Helitron landscape of the maize genome. Proc Natl Acad Sci USA 106:19916–19921

Elrouby N, Bureau TE (2010) *Bs1*, a new chimeric gene formed by retrotransposon-mediated exon shuffling in maize. Plant Physiol 153:1413–1424

Feschotte C, Pritham E (2007) DNA transposons and the evolution of eukaryotic genomes. Annu Rev Genet 41:331–368

Fray RG, Grierson D (1993) Identification and genetic analysis of normal and mutant phytoene synthase genes of tomato by sequencing, complementation and co-suppression. Plant Mol Biol 22:589–602

Fu H, Dooner HK (2002) Intraspecific violation of genetic colinearity and its implications in maize. Proc Natl Acad Sci USA 99:9573–9578

Garber K, Bilic I, Pusch O, Tohme J, Bachmair A, Schweizer D, Jantsch V (1999) The *Tpv2* family of retrotransposons of *Phaseolus vulgaris*: structure, integration characteristics, and use for genotype classification. Plant Mol Biol 39:797–807

Grandbastien MA, Spielmann A, Caboche M (1989) Tnt1, a mobile retroviral-like transposable element of tobacco isolated by plant cell genetics. Nature 337:376–380

Guyot R, Cheng X, Su Y, Cheng Z, Schlagenhauf E, Keller B, Ling H-Q (2005) Complex organization and evolution of the tomato pericentromeric region at the *FER* gene locus. Plant Physiol 138:1205–1215

Hartl DL, Lozovskaya ER, Nurminsky DI, Lohe AR (1997) What restricts the activity of mariner-like transposable elements? Trends Genet 13:197–201

Hirochika H, Sugimoto K, Otsuki Y, Tsugawa H, Kanda M (1996) Retrotransposons of rice involved in mutations induced by tissue culture. Proc Natl Acad Sci USA 93:7783–7788

Jiang N, Bao Z, Temnykh S, Cheng Z, Jiang J, Wing RA, McCouch SR, Wessler SR (2002) *Dasheng*: a recently amplified nonautonomous long terminal repeat element that is a major component of pericentromeric regions in rice. Genetics 161:1293–1305

Jiang N, Bao Z, Zhang X, Eddy SR, Wessler SR (2004) Pack-MULE transposable elements mediate gene evolution in plants. Nature 431:569–573

Jiang N, Gao D, Xiao H, van der Knaap E (2009) Genome organization of the tomato *sun* locus and characterization of the unusual retrotransposon *Rider*. Plant J 60:181–193

Jin YK, Bennetzen JL (1994) Integration and nonrandom mutation of a plasma membrane proton ATPase gene fragment within the *Bs1* retroelement of maize. Plant Cell 6:1177–1186

Juretic N, Hoen DR, Huynh ML, Harrison PM, Bureau TE (2005) The evolutionary fate of MULE-mediated duplications of host gene fragments in rice. Genome Res 15:1292–1297

Kazazian HH Jr (2004) Mobile elements: drivers of genome evolution. Science 303:1626–1632

Kumar A, Bennetzen JL (1999) Plant retrotransposons. Annu Rev Genet 33:479–532

Le QH, Melayah D, Bonnivard E, Petit M, Grandbastien MA (2007) Distribution dynamics of the Tnt1 retrotransposon in tobacco. Mol Gen Genet 278:639–651

Lewin B (2008) Genes IX. Jones and Bartlett Publishers, Sudbury, MA

Ling H-Q, Pitch A, Scholz G, Ganal MW (1996) Genetic analysis of two tomato mutants affected in the regulation of iron metabolism. Mol Gen Genet 252:87–92

Ling HQ, Bauer P, Bereczky Z, Keller B, Ganal M (2002) The tomato *fer* gene encoding a bHLH protein controls iron-uptake responses in roots. Proc Natl Acad Sci USA 99:13938–13943

Ma J, Devos KM, Bennetzen JL (2004) Analyses of LTR-retrotransposon structures reveal recent and rapid genomic DNA loss in rice. Genome Res 14:860–869

Miyao A, Tanaka K, Murata K, Sawaki H, Takeda S, Abe K, Shinozuka Y, Onosato K, Hirochika H (2003) Target site specificity of the *Tos17* retrotransposon shows a preference for insertion within genes and against insertion in retrotransposon-rich regions of the genome. Plant Cell 15:1771–1780

Mizuno M, Kanehisa M (1994) Distribution profiles of GC content around the translation initiation site in different species. FEBS Lett 352:7–10

Morgante M, Brunner S, Pea G, Fengler K, Zuccolo A, Rafalski A (2005) Gene duplication and exon shuffling by helitron-like transposons generate intraspecies diversity in maize. Nat Genet 37:997–1002

Naito K, Zhang F, Tsukiyama T, Saito H, Hancock CN, Richardson AO, Okumoto Y, Tanisaka T, Wessler SR (2009) Unexpected consequences of a sudden and massive transposon amplification on rice gene expression. Nature 461:1130–1134

Park M, Jo S-H, Kwon J-K, Park J, Ahn J-H, Kim S, Lee Y-H, Yang T-J, Hur C-G, Kang B-C, Kim B-D, Choi D (2011a) Comparative analysis of pepper and tomato reveals euchromatin expansion of pepper genome caused by differential accumulation of Ty3/Gypsy-like elements. BMC Genomics 12:85

Park M, Park J, Kim S, Kwon J-K, Park HM, Bae IH, Yang T-J, Lee Y-H, Kang B-C, Choi D (2011b) Evolution of the large genome in *Capsicum annuum* occurred through accumulation of single-type long terminal repeat retrotransposons and their derivatives. Plant J 69:1018–1029

Pereira V (2004) Insertion bias and purifying selection of retrotransposons in the *Arabidopsis thaliana* genome. Genome Biol 5:R79

Peterson-Burch BD, Nettleton D, Voytas DF (2004) Genomic neighborhoods for Arabidopsis retrotransposons: a role for targeted integration in the distribution of the Metaviridae. Genome Biol 5:R78

Price HL, Drinkard AWJ (1908) Inheritance in tomato hybrids. Virginia Agr Exp Sta Bull 177:18–53

Rodriguez GR, Munos S, Anderson C, Sim SC, Michel A, Causse M, Gardener BB, Francis D, van der Knaap E (2011) Distribution of SUN, OVATE, LC, and FAS in the tomato germplasm and the relationship to fruit shape diversity. Plant Physiol 156:275–285

Salinas J, Matassi G, Montero LM, Bernardi G (1988) Compositional compartmentalization and compositional patterns in the nuclear genomes of plants. Nucl Acids Res 16:4269–4285

SanMiguel P, Tikhonov A, Jin Y-K, Motchoulskaia N, Zakharov D, Melake-Berhan A, Springer PS, Edwards KJ, Lee M, Avramova Z, Bennetzen JL (1996) Nested retrotransposons in the intergenic regions of the maize genome. Science 274:765–768

Schnable PS, Ware D, Fulton RS, Stein JC, Wei F, Pasternak S, Liang C, Zhang J, Fulton L, Graves TA, Minx P, Reily AD, Courtney L, Kruchowski SS, Tomlinson C, Strong C, Delehaunty K, Fronick C, Courtney B, Rock SM, Belter E, Du F, Kim K, Abbott RM, Cotton M, Levy A, Marchetto P, Ochoa K, Jackson SM, Gillam B, Chen W, Yan L, Higginbotham J, Cardenas M, Waligorski J, Applebaum E, Phelps L, Falcone J, Kanchi K, Thane T, Scimone A, Thane N, Henke J, Wang T, Ruppert J, Shah N, Rotter K, Hodges J, Ingenthron E, Cordes M, Kohlberg S, Sgro J, Delgado B, Mead K, Chinwalla A, Leonard S, Crouse K, Collura K, Kudrna D, Currie J, He R, Angelova A, Rajasekar S, Mueller T, Lomeli R, Scara G, Ko A, Delaney K, Wissotski M, Lopez G, Campos D, Braidotti M, Ashley E, Golser W, Kim H, Lee S, Lin J, Dujmic Z, Kim W, Talag J, Zuccolo A, Fan C, Sebastian A, Kramer M, Spiegel L, Nascimento L, Zutavern T, Miller B, Ambroise C, Muller S, Spooner W, Narechania A, Ren L, Wei S, Kumari S, Faga B, Levy MJ, McMahan L, Van Buren P, Vaughn MW, Ying K, Yeh C-T, Emrich SJ, Jia Y, Kalyanaraman A, Hsia A-P, Barbazuk WB, Baucom RS, Brutnell TP, Carpita NC, Chaparro C, Chia J-M, Deragon J-M, Estill JC, Fu Y, Jeddeloh JA, Han Y, Lee H, Li P, Lisch DR, Liu S, Liu Z, Nagel DH, McCann MC, SanMiguel P, Myers AM, Nettleton D, Nguyen J, Penning BW, Ponnala L, Schneider KL, Schwartz DC, Sharma A, Soderlund C, Springer NM, Sun Q, Wang H, Waterman M, Westerman R, Wolfgruber TK, Yang L, Yu Y, Zhang L, Zhou S, Zhu Q, Bennetzen JL, Dawe RK, Jiang J, Jiang N, Presting GG, Wessler SR, Aluru S, Martienssen RA, Clifton SW, McCombie WR, Wing RA, Wilson RK (2009) The B73 maize genome: complexity, diversity, and dynamics. Science 326:1112–1115

Tam SM, Causse M, Garchery C, Burck H, Mhiri C, Grandbastien MA (2007) The distribution of copia-type retrotransposons and the evolutionary history of tomato and related wild species. J Evol Biol 20:1056–1072

Wang W, Zheng H, Fan C, Li J, Shi J, Cai Z, Zhang G, Liu D, Zhang J, Vang S, Lu Z, Wong GK, Long M, Wang J (2006a) High rate of chimeric gene origination by retroposition in plant genomes. Plant Cell 18:1791–1802

Wang Y, Tang X, Cheng Z, Mueller L, Giovannoni J, Tanksley SD (2006b) Euchromatin and pericentromeric heterochromatin: comparative composition in the tomato genome. Genetics 172:2529–2540

Wang Y, Diehl A, Wu F, Vrebalov J, Giovannoni J, Siepel A, Tanksley SD (2008) Sequencing and comparative analysis of a conserved syntenic segment in the *Solanaceae*. Genetics 180:391–408

Xiao H, Jiang N, Schaffner EK, Stockinger EJ, van der Knaap E (2008) A retrotransposon-mediated gene duplication underlies morphological variation of tomato fruit. Science 319:1527–1530

Yang L, Bennetzen JL (2009) Distribution, diversity, evolution, and survival of Helitrons in the maize genome. Proc Natl Acad Sci USA 106:19922–19927

Zhong S, Joung J-G, Zheng Y, Chen Y-R, Liu B, Shao Y, Xiang JZ, Fei Z, Giovannoni JJ (2011) High-throughput Illumina strand-specific RNA sequencing library preparation. Cold Spring Harb Protoc 2011(8):940–949, pdb.prot5652

Chapter 16
Retrotransposons and the Eternal Leaves

Antonella Furini

Abstract The resurrection plant *Craterostigma plantagineum* can tolerate up to 96% loss of its relative water content and recover within hours after rehydration. In callus tissue desiccation tolerance is induced by pre-incubation with Abscisic acid (ABA). In callus and plant ABA treatment and dehydration induce a set of dehydration-responsive genes. T-DNA activation tagging led to the identification of *CDT-1*, a dehydration- and ABA-responsive gene, which renders calli tolerant without ABA pre-incubation. Molecular analysis indicated that *CDT-1* is a retroelement, present in multiple copy in the genome, able to direct the synthesis of small RNAs responsible for desiccation tolerance. Transposition of *CDT-1* retroelements have progressively increased the capacity of the species to synthesize small RNAs and thus recover after desiccation. This may be a case of evolution towards the acquisition of a new trait, stimulated by the environment acting directly on intra-genomic DNA replication.

Keywords CDT-1 • Craterostigma plantagineum • Desiccation tolerance • Retrotransposon • Small RNA

16.1 Introduction

Water is essential to all physiological processes, and at cellular level, it is the major medium for transporting metabolites and nutrients. Water availability has determined the distribution of plants on Earth. In their natural environment plants often experience water stress episodes that affect normal growth. Many plants are able to withstand this challenge either by decreasing water flux through the plant or by increasing their water uptake. Water loss can be reduced by various mechanisms such as stomatal closure,

A. Furini (✉)
Department of Biotechnology, University of Verona, Strada Le Grazie, 15, 37134 Verona, Italy
e-mail: antonella.furini@univr.it

M.-A. Grandbastien and J.M. Casacuberta (eds.), *Plant Transposable Elements*,
Topics in Current Genetics 24, DOI 10.1007/978-3-642-31842-9_16,

reduction of leaf growth, or production of specialized leaf surfaces to reduce transpiration, whereas water uptake can be increased by the growth of specialized root structures (Phillips et al. 2002). Tolerance to desiccation—the ability to recover when most of the protoplasmic water is lost and only a very small amount of tightly bound water remains in the cell—is common in mosses, lichens, and ferns and in the reproductive structures of vascular plants, pollen, spores, and seeds but rare in vegetative organs (e.g., leaves) of tracheophytes (Bewley and Krochko 1982; Oliver and Bewley 1997; Kranner et al. 2005). However, a small group of angiosperms, termed resurrection plants, possesses desiccation-tolerant vegetative tissues with the unique ability to revive from an air-dried state (Gaff 1971), and the process of drying and rehydration causes only limited damage to the plant tissues. These plants have the advantage over other species in arid environments; they can remain quiescent during long period of drought. Upon watering they can resurrect, restore their photosynthetic activity within 24 h, grow, and reproduce long before non-resurrecting plants.

16.2 Resurrection Plants

It was postulated that initial evolution of vegetative desiccation tolerance has been a crucial step for primitive plants to colonize the land. It is thought that, during evolution, tolerance was lost in vegetative tissues with the acquisition of water transport in tracheophytes, but this trait has reevolved independently in plant species that are nowadays defined resurrection plants (Oliver et al. 2000). These plants are often small and low growing; they are found in all continents, except Antarctica, in places where substantial rains are seasonal and extremely sporadic. They are mainly concentrated in southern Africa, eastern South America, and western Australia (Gaff 1987), while only a few species have been found in Europe in the Balkan mountains (Stefanov et al. 1992). These areas show great variation in moisture availability, as a consequence the ability to survive dehydration becomes a necessity. Surprisingly, vegetative desiccation tolerance was recently discovered in *Linderia brevidens*, a species endemic to montane rainforest of coastal Africa, a niche that does not experience drought (Phillips et al. 2008).

About 330 species of angiosperms have been found to survive desiccation but no resurrection gymnosperms are known (Hartung et al. 1998). There are both monocotyledonous plants such as *Xerophyta viscosa* and *Sporobolus stapfianus* and dicotyledonous species such as *Myrothamnus flabellifolia*, *Craterostigma plantagineum*, and *Chamaegigas intrepidus*. The latter is the unique known example of aquatic resurrection plants (Hartung et al. 1998).

Acquisition of tolerance may depend mainly on changes in gene expression since genes necessary for tolerance in seed and pollen grain are already present but not expressed in vegetative tissues (Bartels and Salamini 2001). Studies aimed at understanding the molecular basis of desiccation tolerance have mainly focused on the dicotyledonous South African *Craterostigma plantagineum* (Bartels et al. 1990; Bartels and Salamini 2001), the monocotyledonous species *Sporobolus stapfianus*

(Neale et al. 2000), and the moss *Tortula ruralis* (Oliver and Bewley 1997). The molecular basis of desiccation tolerance is complex, and it is not clear yet how and whether mechanisms may vary between different species (Bartels 2005). For instance some species retain chlorophyll during dehydration, whereas others lose chlorophyll. Many proteins accumulate during drying in resurrection plants and some have been cloned and sequenced. *Late Embryogenesis Abundant* (LEA) proteins represent one major group of expressed proteins in vegetative tissues during desiccation. LEA proteins comprise a large number of plant proteins that accumulate in mature embryo during late stages of embryo development (Galau et al. 1986) and in vegetative tissues in response to water deficit. Their generally high expression is found in osmotically stressed or ABA-treated tissues in many cell types and predominantly in the cytosol. LEA proteins are characterized by being small, with a biased amino acid composition, which results in highly hydrophilic polypeptides, with just a few residues providing 20–30% of their total complement (Ingram and Bartels 1996). To date several molecular mechanisms have been proposed to describe functional aspects of LEA proteins, and they are thought to function as molecular chaperons protecting against aggregation of proteins under water stress (Goyal et al. 2005).

In addition to the synthesis of proteins, an increased concentration of soluble sugars in seeds and in vegetative tissues of resurrection plants at the onset of desiccation is an important factor for the acquisition of tolerance. In animals, fungi, yeast, and bacteria high level of trehalose ensures membrane osmoprotection during desiccation (Crowe et al. 1992). This sugar is extremely rare in plants where sucrose and other sugars may play a similar role in resurrection plants. Sugars may be effective in osmotic adjustment during water loss, but they may protect the cells by causing, during severe desiccation, glass formation with the mechanical properties of a solid (Williams and Leopold 1989). The relevance of ABA in desiccation tolerance of resurrection plants is also well documented. In general ABA content in leaves increased upon dehydration; in addition when leaves of the resurrection plants *Myrothamnus flabellifolia* and *Borya nitida* were too rapidly dehydrated, the increase in ABA content in leaves was not observed and plants did not resurrect (Gaff and Loveys 1984). Most of the proteins highly expressed during desiccation (i.e., LEA proteins) showed an induction upon ABA treatment. Genes induced at very low level during the initial stage of desiccation process have also been identified. The *SDG134c* isolated in *Sporobolus stafianus* encodes a protein translation initiation factor 1 and its transcript is present at very low level in fully hydrated tissues and increased in dehydrated tissues. It was suggested that SDG134c is necessary for the process of rehydration that fully restore the metabolic activity within several hours (Neale et al. 2000).

Resurrection plants, as other desiccation tolerant systems, attract particularly the research interests, since one major factor that limits the productive potential of higher plants is the availability of water. To better know the molecular mechanisms of drought resistance may have potential implications in the future development of drought tolerant crops and therefore increasing crop productivity in arid lands. In this respect most information is available for the resurrection plant *Craterostigma*

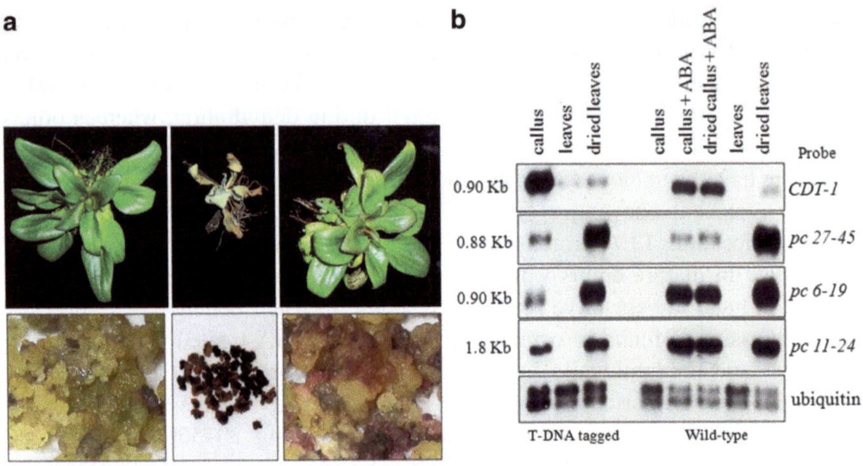

Fig. 16.1 (**a**) Effect of desiccation treatment on the resurrection plant *C. plantagineum* (*top*) and T-DNA tagged callus line (*bottom*). From *left* to *right*: fully turgid, desiccated, and rehydrated. (**b**) Northern analyses showing the expression patterns for the *CDT-1* gene and the ABA- and desiccation-induced *C. plantagineum Lea* genes *pc-27-45*, *pc-6-19*, and *pc-11-24* (Bartels et al. 1990). (**b** is reproduced from Furini et al. 1997, with permission)

plantagineum that has been extensively investigated at molecular level and has significantly contributed to our knowledge of molecular regulation of dehydration tolerance in vegetative tissues (Bartels et al. 1990; Bartels and Salamini 2001; Bernacchia and Furini 2004; Bartels 2005).

16.3 *Craterostigma plantagineum* as a Model System

C. plantagineum is a member of the *Schrophulariaceae* family of African origin and distributed in various ecological niches that must have been associated with long period of drought (Fischer 2004). This species can tolerate up to 96% loss of its relative water content and recovers within several hours from such extreme dehydration (Bernacchia et al. 1996). This resurrection response is expressed in differentiated tissues (Fig. 16.1a top). In vitro maintained callus is not desiccation tolerant and requires exposure to exogenous ABA in order to survive severe dehydration. This feature allows to compare gene expression in two systems with the same genetic background without developmental constrains. Dehydration of *C. plantagineum* plants as well as ABA treatment of leaves or callus induce the expression of similar sets of dehydration- and/or ABA-responsive genes (Bartels et al. 1990). Furthermore, in leaves of this species the desiccation phase is characterized by a massive conversion of the main C8 sugar in fully hydrated leaves, the 2-octulose, into sucrose. During the rehydration phase the sucrose level drops and octulose accumulates again (Bianchi et al. 1991). The synthesis of sucrose in water stressed *C. plantagineum* leaves is similar to that observed in seeds of higher plants and in lower eukaryotes, in which a specific sugar increases with tolerance acquisition. Desiccation-induced transcripts from *C. plantagineum* can be assigned to different type of *Lea* genes. The degree of homology varies: conservation may be restricted to particular sequence motifs and sequence structures or *C. plantagineum* genes can share high identity with *Lea* genes expressed in seeds at early stages of desiccation. The high expression of these different *Lea* type genes in ABA-treated fully hydrated and dehydrated leaves of *C. plantagineum* suggests that similar metabolic processes are occurring during seed maturation (when the ABA level naturally increases) and that in vegetative tissues of *C. plantagineum* the signal transduction pathway from water stress to gene expression requires the activation of specific genes that in desiccation-sensitive species are relevant to seed dehydration. This means that, at least with respect to *Lea* genes, the differences between desiccation tolerant and sensitive species are due to differences in expression patterns (Bartels and Salamini 2001). Furthermore, promoter studies of several genes isolated from dehydrated tissues of *C. plantagineum* revealed that in transgenic tobacco, these gene promoters were active only in naturally desiccation tolerant tissues (mature embryo and pollen), and the responsiveness to ABA in vegetative tissues decreases during plant development (Michel et al. 1993, 1994). It was hypothesized that the ABI3 protein contributes to the ABA-regulated gene expression in the *Arabidopsis* seed development (Giraudat et al. 1992) and the ectopic expression of the ABI3 protein induces, in response to ABA, the expression of seed-specific transcripts in leaves of transgenic *Arabidopsis* (Parcy et al. 1994). ABI3 proteins was effective also in the activation of *C. plantagineum Lea* gene promoters upon ABA treatments in vegetative tissues of transgenic *Arabidopsis* (Furini et al. 1996; Velasco et al. 1998), reinforcing the hypothesis that desiccation tolerance in *C. plantagineum* requires the induction of ABA and/or desiccation-inducible proteins that in desiccation sensitive plants are expressed only in seeds. However, the ABI3 homolog was identified in *C. plantagineum*, but its expression was not observed in fully developed leaves (Chandler and Bartels 1997), suggesting that other factors may be involved in the activation of *Lea* genes in *C. plantagineum*.

16.3.1 Isolation of ABA-Independent Desiccation Tolerant Callus

As a model system the polyploid *C. plantagineum* is a poor target for mutation approaches using chemical or insertional mutagens such as transposons or T-DNA insertions. However, an efficient transformation system (Furini et al. 1994) and a T-DNA activation tagging approach allowed the isolation of elements relevant to desiccation tolerance in *C. plantagineum* (Furini et al. 1997; Smith-Espinoza et al. 2005). The fact that wild-type dedifferentiated callus tissues do not survive desiccation unless pretreated with ABA suggests that a number of ABA-mediated pathways that lead to the acquisition of desiccation tolerance are silent during callus dehydration.

This information offered the opportunity to search for dominant mutations that activate the ABA and/or the dehydration signaling pathway and allows to select desiccation tolerant calli even in the absence of ABA.

T-DNA activation tagging carrying an enhancer domain from the gene 5 promoter (pg5) of *Agrobacterium tumefaciens* and capable to induce transcription in dedifferentiated proliferating tissues, such as calli growing in auxin rich medium, but not in differentiated leaves, was used for *C. plantagineum* leaf disc transformation (Furini et al. 1997). Transformed calli were selected for viable dominant mutants by severe cycles of dehydration–rehydration without exogenous ABA pretreatment. One callus line over 25,000 transformants passed the selection (Fig. 16.1a bottom). This callus showed a reddish color similar to that observed in ABA treated calli, and when cultured in differentiation medium it developed shoots and eventually fully developed plants. Callus was again dedifferentiated from these shoots, and it retained the ability to withstand desiccation. Furthermore, *Lea* genes previously identified in *C. plantagineum* (Bartels et al. 1990) and normally expressed in dried leaves and ABA-treated calli were expressed in the T-DNA tagged line without exogenous ABA application (Fig. 16.1b), suggesting that the pathway that leads to desiccation tolerance was switched on.

16.3.2 Identification of the Retrotransposon CDT-1

Molecular analysis of the mutant callus allowed the isolation of DNA sequences flanking the T-DNA insertion and the identification of a DNA fragment highly transcribed in the desiccation tolerant mutant callus line and wild-type ABA-treated callus or dried leaves (Fig. 16.1b). To prove that this identified DNA fragment was responsible for desiccation tolerance of ABA-untreated callus, it was cloned under the control of pg5, inserted into a plant transformation vector, and used for leaf disc transformation. Newly transformed calli were able to withstand dehydration in the absence of ABA, and these results confirmed the assumption that the fragment identified by T-DNA tagging approach was responsible for the gain-of-function phenotype observed in the desiccation tolerant mutant callus (Furini et al. 1997).

Screening of a cDNA library with the isolated fragment brought to the identification of many identical clones indicating that the identified gene, named *CDT-1* (*Craterostigma* Desiccation *T*olerant-1, NCBI accession n. Y11822), is part of a large gene family in the *C. plantagineum* genome. The characterization of *CDT-1* revealed that (1) it is flanked by direct repeats and it is present in multiple copies, suggesting that it is a transposable element; (2) it has a poly(A) tail and lack LTRs indicating that it is a non-LTR retrotransposon; (3) it is intronless since cDNA structure is similar to genomic clones; and (4) it does not possess large coding domain with similarities to LINEs coding sequences. In addition, no sequence homology to *CDT-1* was detected in current databases, and translation product was not observed in in vitro assay. An oligo(A) tract of 17–22 nucleotides was also found in the 5′ region of all cDNA and genomic clones. Most importantly, *CDT-1* transcription was never detected in hydrated leaves, but induced by dehydration and repressed by rehydration, whereas in callus is upregulated by ABA (Furini et al. 1997).

Mutated versions of *CDT-1* cDNA were tested in transgenic plants to verify whether the only translational region present in *CDT-1* sequence could be responsible for the activation of desiccation tolerance pathway in callus. It was found that the 3′ sequence of *CDT-1*—or part of it—is required for desiccation tolerance, whereas a translation product is not necessary (Hilbricht et al. 2008). Furthermore, a T-DNA activation tagging approach, similar to that previously used for the identification of *CDT-1*, led to the finding of other desiccation tolerant mutant callus lines. One of the characterized mutant, named *CDT-2*, as *CDT-1*, constitutively expresses known osmoprotective *Lea* genes in callus and leaves. Further analysis of this mutant revealed that the tagged locus is similar to the previously characterized *CDT-1*. The fact that two independently identified mutant loci are homologous was unexpected but offer strong proof that *CDT-1/2* retroelements are crucial for the acquisition of desiccation tolerance in callus tissue. Surprisingly, *CDT-1* and *CDT-2* not only showed high sequence similarity, but they also share sequence motifs within the 3′ region (Smith-Espinoza et al. 2005). Other desiccation tolerant species of the genus *Craterostigma*, such as *C. hirsutum*, *C. pumilum*, and *C. lanceolatum*, were analyzed for the presence of *CDT-1* homologs. This investigation brought to the identification of CDT-genes in the three species. In all cases sequence similarities were identified within the 3′ part of *CDT-1* sequence (Furini 2008). All these observations lent strong support that the *CDT* non-LTR retrotransposons function as regulatory noncoding RNA. The sequence similarity among the *CDT* retroelements strongly indicated that the functionally important elements, that have been maintained during evolution, are likely to be located in the conserved 3′ region of these non-LTR retrotransposons. Furthermore, the lack of homology with sequences present in databases suggest the specificity of this transposon family for the unique ability of resurrection plants—at least in the genus *Craterostigma*—to revive after long periods of drought.

16.3.3 CDT-1 *Role in Desiccation Tolerance is Mediated Through Small RNA*

Transcription analysis showed detection of both sense and antisense *CDT-1* RNA (Fig. 16.2a) and suggest that the role of *CDT-1* in desiccation tolerance could be mediated by small RNA. Low-molecular weight RNA from desiccation tolerant calli hybridized with sense and antisense 21mers (from nt 634 to 654) identified in the 3′ end of *CDT-1* cDNA, whereas accumulation of small transcripts was not detected when desiccation sensitive calli were examined (Fig. 16.2b). This oligonucleotide had some similarity to microRNA 159 (Achard et al. 2004), which is highly conserved in evolution. In addition, *C. plantagineum* callus-derived protoplast transfection was used to show that this small RNA alone was able to induce

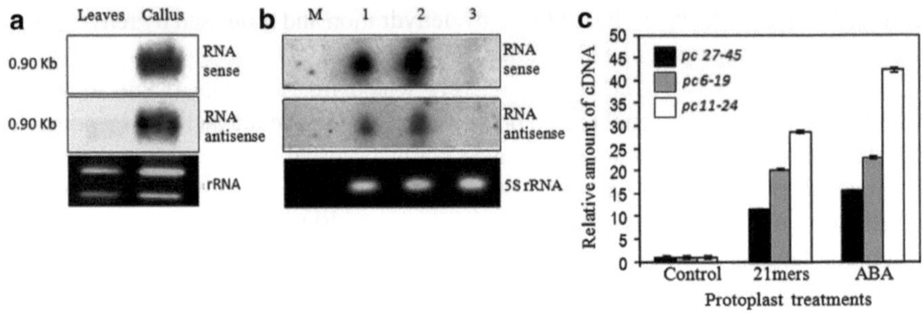

Fig. 16.2 (**a**) Northern analysis showing the transcription of sense and antisense *CDT-1* strands. Total RNA was extracted from leaves of wild-type plants and from transformed callus expressing *CDT-1*. (**b**) Northern analysis of low-molecular weight RNA hybridized with sense and antisense 21mers identified in the 3′ region of *CDT-1* sequence. RNA was isolated from ABA-treated wild-type callus (1), from transformed callus expressing *CDT-1* (2), and from untreated wild-type callus (3). (**c**) Real-time PCR measuring the level of transcription of the desiccation- and ABA-induced *C. plantagineum Lea* genes *pc-27-45*, *pc-6-19*, and *pc-11-24* (Bartels et al. 1990) in untransfected callus-derived protoplasts (control), in protoplasts transfected with the 21mers identified in the 3′ end of *CDT-1* sequence, and in protoplasts incubated with 10 μM ABA for 36 h. *Error bars* denote SE (reproduced from Hilbricht et al. 2008, with permission)

dehydration-responsive genes to the same extend as exogenous application of ABA (Fig. 16.2c) (Hilbricht et al. 2008).

16.3.4 CDT-1 *Retrotransposition and the Acquisition of Desiccation Tolerance*

The structure of several *CDT-1* genomic clones (schematized in Fig. 16.3a, b) shows that *CDT-1* elements are flanked by direct repeats of 5 to 22 bp (Fig. 16.3b regions a and d). In these clones the length of the poly(A) tail vary (from 10 to >60 bp; Fig. 16.3b, region c), and the presence of the same direct repeat core sequences (colored in Fig. 16.3b, c) in more clones made possible to reconstruct, at least in part, the temporal series of *CDT-1* transpositions (Fig. 16.3c). Sequence analysis of *CDT-1* cDNA reveals that transcription occurs from different loci and gave rise to almost identical *CDT-1* mRNAs. The only variant is the length of the 5′ oligo sequence (17, 18, 20, or 21) which was of 19, 21, and 22 bp in three sequenced genomic clones (Hilbricht et al. 2008).

The abundance of *CDT-1* transcripts induced by dehydration and/or by ABA treatment may be recognized by the cell as signal of stress and, with the formation of double stranded RNA, these transcripts may be converted to small RNA which in turn may control the expression of gene(s) responsible for desiccation tolerance in *C. plantagineum* but thus far unknown. Interestingly, there is a functional link between retrotransposition and increased level of small RNA transcription and thus of desiccation tolerance: *CDT-1* mRNA accumulates in wild-type plants only

a

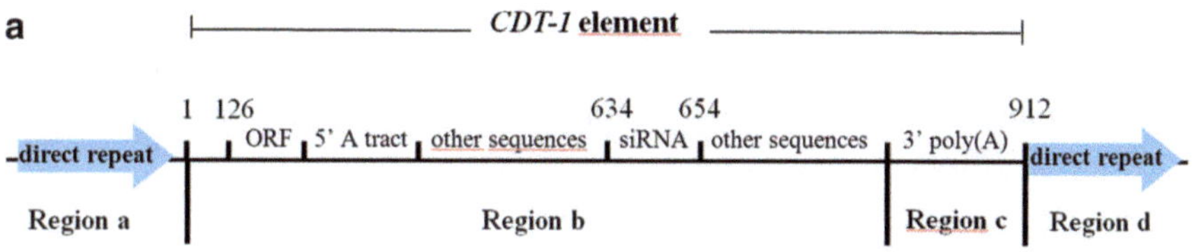

b

Clone	Region a	Start of Region b	Region c	Direct repeat
1:	TTGCATCGTTGAACAATGGCTTCGAAAAAA	AAAAAACCCTT	25A	GAACAATGGCTTCGAAATT
2:	AAGAGAGAGAGAGAATCTGATATACTCCACGCTTGTCTC	AAAAAACCCTT	17A	GAGAGAGAGAATCTGATGC
3:	ACTCAACTTTTTTTCTCTACTCACGGTCTC	AAAAAACCCTT	15A	TTTTTTTCTCTATTCACA
4:	AAATGACACGTTTGAAATTACGCTTGTCTC	AAAAAACCCTT	10A	GAAATTACGCTTGTCTCGT
5:	ACTCAACTTTTTTTCTCTACTCACGGTCTC	AAAAAACCCTT	15A	CTTTTTTTCTCTACTCACAAAG
6:	GGTATTGAATATTGTTGTATTTCTTTCACGGTCTC	AAAAAACCCTT	10A	GAATATTGTTGTATTTCTTTC
7:	TTCAACTAGGAATGGACTCGAAATTACGCTTGTCTC	AAAAAACCCTT	15A	CTAGGAATGGACTCAAGG
8:	GCAAGTATT	AAAAAACCCTT	10A	GTATTAAGA
9:	TATCAAGTAATAACCCCGAGAA	CCCTCCCATCT	>60A	TAATAACCCCGAGAAGAA
10:	TGACACGTTTGAAATTACGCTTGTTCT	AAAAAACCCTT	10A	GAAATTACGCTTGTTCTCGTG

c

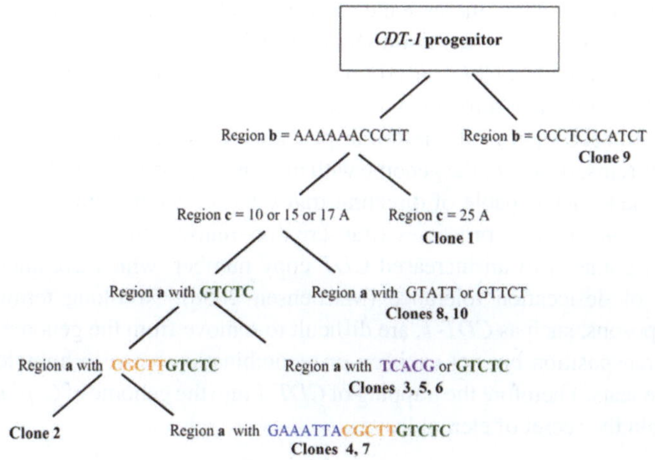

Fig. 16.3 (**a**) Schematic representation of *CDT-1* structure. (**b**) Ten sequenced *CDT-1* genomic clones. Regions **a** and **d** (as in **a**) represent the direct repeats flanking the *CDT-1* element. Motifs present in both regions are underlined. Region **b** represent the starting sequence of *CDT-1* (only clone 9 showed a shorter sequence). Region **c** represent the number of bp in the 3′ poly(A) tail. *Colored bases* in region a represent core motifs present in more than one clone. They made it possible to reconstruct, at least in part, the temporal series of transposition as shown in **c**. (**c**) Putative succession of transpositions starting from a common *CDT*-1 progenitor present in 10 genomic clones. *Color letters* indicate core motifs that are or have been parts of direct repeats. The sequence at the start of region **b**, the number of base pairs in the 3′ poly(A) (region **c**), and the type of core motifs establish the succession of transposition events. (reproduced from Hilbricht et al. 2008, with permission)

during dehydration (Furini et al. 1997) this implies that both the level of transcription of retroelements—with potential small RNA activity—and their reinsertion into the genome are environmentally controlled. Since plants do not have a sequestered germ line, new insertions of *CDT-1* retroelement in the genome of meristematic cell may contribute to increase desiccation tolerance in the progeny providing that new copies of *CDT-1* element can be transcribed under stress.

16.4 Conclusions and Implications

The almost complete invariance of the *CDT-1* genomic clones is an unusual finding for plant transposons (Kumar and Bennetzen 1999), as if selection acted to preserve this retroelement and highlights the importance of maintaining the integrity of DNA information necessary for *CDT-1* transcription, double stranded and small RNA synthesis and hence reinforcing the phenomenon of desiccation tolerance of vegetative tissues during period of drought stress. This mechanism offers an evolutionary explanation of the interaction between environment and genome. In fact, it is well known the expression of transposons under environmental stress, and the resulting transposition is thought to increase the chances of inheritance by the next generation, ensuring survival of the transposon (Slotkin and Martienssen 2007). But, what is singular in the case of *CDT-1* transposon is that its transcription during environmental stress has been selected, through evolution, to ensure plant desiccation tolerance: the higher the transcription of the retrotransposon under severe water stress, the more frequent its reinsertion into the genome with increasing probability of being reinserted in a DNA sequence capable of directing transcription under water stress condition. The reiteration of these processes (transcription–reinsertion) over generations has resulted in plants with an increased *CDT* copy number, which eventually triggers the onset of desiccation tolerance (Martiensen 2008). Non-long terminal repeat retrotransposons, such as *CDT-1*, are difficult to remove from the genome since they undergo transposition but not excision or recombination between homologous long terminal repeats. Therefore the trapping of *CDT-1* into the genome of *C. plantagineum* may explain the secret of eternal leaves.

Acknowledgments This work was mainly carried out at the Max-Planck-Institute (Cologne, Germany), supported in part by a grant of the EC BRIDGE programme. The supervision of Profs. D. Bartels and F. Salamini is fully acknowledged.

References

Achard P, Herr A, Baulcombe DC, Harberd NP (2004) Modulation of floral development by a gibberellin-regulated microRNA. Development 13:3357–3365

Bartels D (2005) Desiccation tolerance studied in the resurrection plant *Craterostigma plantagineum*. Integr Comp Biol 45:696–701

Bartels D, Salamini F (2001) Desiccation tolerance in the resurrection plant *Craterostigma plantagineum*. A contribution to the study of drought tolerance at the molecular level. Plant Physiol 127:1346–1353

Bartels D, Schneider K, Terstappen G, Piatkowski D, Salamini F (1990) Molecular cloning of abscisic acid modulated genes which are induced during desiccation of the resurrection plant *Craterostigma plantagineum*. Planta 181:27–34

Bernacchia G, Furini A (2004) Biochemical and molecular responses to water stress in resurrection plants. Physiol Plant 121:175–181

Bernacchia G, Salamini F, Bartels D (1996) Molecular characterization of the rehydration process in the resurrection plant *Craterostigma plantagineum*. Plant Physiol 111:1043–1050

Bewley JD, Krochko JE (1982) Desiccation tolerance. In: Lange OL, Nobel PS, Osmond CB, Ziegler H (eds) Encyclopedia of plant physiology, vol 12B, Physiological ecology II. Springer, Berlin

Bianchi G, Gamba A, Morelli C, Salamini F, Bartels D (1991) Novel carbohydrate metabolism in the resurrection plant *Craterostigma plantagineum*. Plant J 1:355–359

Chandler J, Bartels D (1997) Structure and function of the vp1 gene homologue from the resurrection plant *Craterostigma plantagineum* Hochst. Mol Gen Genet 256:539–546

Crowe JH, Hoekstra FA, Crowe LM (1992) Anhydrobiosis. Annu Rev Physiol 54:579–599

Fischer E (2004) *Scrophulariaceae*. In: Kubitzki K (ed) The families and genera of vascular plants. Springer, Berlin, pp 333–432

Furini A (2008) CDT retroelement: the stratagem to survive extreme vegetative dehydration. Plant Signal Behav 3:1–3

Furini A, Koncz C, Salamini F, Bartels D (1994) *Agrobacterium*-mediated transformation of the desiccation-tolerant plant *Craterostigma plantagineum*. Plant Cell Rep 14:102–106

Furini A, Parcy F, Salamini F, Bartels D (1996) Differential regulation of two ABA-inducible genes from *Craterostigma plantagineum* in transgenic *Arabidopsis* plants. Plant Mol Biol 30:343–349

Furini A, Koncz C, Salamini F, Bartels D (1997) High level transcription of a member of a repeated gene family confers dehydration tolerance to callus tissue of *Craterostigma plantagineum*. EMBO J 16:3599–3608

Gaff DF (1971) Desiccation-tolerant flowering plants in Southern Africa. Science 174:1033–1034

Gaff DF (1987) Desiccation tolerant plants in South America. Oecologia 74:133–136

Gaff DF, Loveys BR (1984) Abscisic acid content and effects during dehydration of detached leaves of desiccation tolerant plants. J Exp Bot 35:1350–1358

Galau GW, Hugles DW, Dure L III (1986) Abscisic acid induction of cloned cotton late embryogenesis abundant (LEA) messenger RNAs. Plant Mol Biol 7:155–170

Giraudat J, Hauge BM, Valon C, Smalle J, Parcy F, Goodman HM (1992) Isolation of the *Arabidopsis ABI3* gene by positional cloning. Plant Cell 4:1251–1261

Goyal K, Walton LJ, Tunnacliffe A (2005) LEA proteins prevent protein aggregation due to water stress. Biochem J 388:151–157

Hartung W, Schiller P, Dietz KJ (1998) Physiology of poikilohydric plants. Cell Biol Physiol Prog Bot 59:299–327

Hilbricht T, Varotto S, Sgaramella V, Bartels D, Salamini F, Furini A (2008) Retrotransposons and siRNA have a role in the evolution of desiccation tolerance leading to resurrection of the plant *Craterostigma plantagineum*. New Phytol 179:877–887

Ingram J, Bartels D (1996) The molecular basis of dehydration tolerance in plants. Annu Rev Plant Physiol Plant Mol Biol 47:377–403

Kranner I, Cram WJ, Zorn M, Wornik S, Yoshimura I, Stabentheiner E, Pfeifhofer HW (2005) Antioxidants and photoprotection in a lichen as compared with its isolated symbiotic partners. Proc Natl Acad Sci USA 102:3141–3146

Kumar A, Bennetzen JL (1999) Plant retrotransposons. Annu Rev Genet 33:479–532

Martiensen R (2008) Great leap forward? Transposable elements, small interfering RNA and adaptive Lamarckian evolution. New Phytol 179:570–572

Michel D, Salamini F, Bartels D, Dale P, Baga M, Szalay A (1993) Analysis of a desiccation and ABA-responsive promoter isolated from the resurrection plants *Craterostigma plantagineum*. Plant J 4:29–40

Michel D, Furini A, Salamini F, Bartels D (1994) Structure and regulation of an ABA- and desiccation-responsive gene. Plant Mol Biol 24:549–560

Neale AD, Blomstedt CK, Bronson P, Le T-N, Guthridge K, Evans J, Gaff DF, Hamill JD (2000) The isolation of genes from the resurrection grass *Sporobolus stapfianus* which are induced during severe drought stress. Plant Cell Environ 23:265–277

Oliver MJ, Bewley JD (1997) Desiccation tolerance of plant tissues: a mechanistic overview. Hort Rev 18:171–214

Oliver MJ, Tuba Z, Mishler BD (2000) The evolution of vegetative desiccation tolerance in land plants. Plant Ecol 151:85–100

Parcy F, Valon C, Raynal M, Gaubier-Comella P, Delseny M, Giraudat J (1994) Regulation of gene expression programs during *Arabidopsis* seed development: roles of *ABI3* locus and endogenous abscisic acid. Plant Cell 6:1567–1582

Phillips JR, Oliver MJ, Bartels D (2002) Molecular genetics of desiccation and tolerant systems. In: Black M, Pritchard HW (eds) Desiccation and survival in plants: drying without dying. CABI Publishing, Wallingford, UK

Phillips JR, Fischer E, Baron M, van den Dries N, Facchinelli F, Kutzer M, Rahmanzadeh R, Remus D, Bartels D (2008) *Lindernia brevidens*: a novel desiccation-tolerant vascular plant, endemic to ancient tropical rainforests. Plant J 54:938–948

Slotkin RK, Martienssen R (2007) Transposable elements and the epigenetic regulation of the genome. Nat Rev Genet 8:272–285

Smith-Espinoza CJ, Phillips JR, Salamini F, Bartels D (2005) Identification of further *Craterostigma plantagineum* cdt mutants affected in abscisic acid mediated desiccation tolerance. Mol Gen Genet 274:364–372

Stefanov K, Markovska Y, Kimenov G, Popov S (1992) Lipid and sterol changes in leaves of *Haberlea rhodopensis* and *Ramonda serbica* at transition from biosis into anabiosis and vice versa caused by water stress. Phytochemistry 31:2309–2314

Velasco R, Salamini F, Bartels D (1998) Gene structure and expression analysis of the drought- and abscisic acid-responsive CDeT11-24 gene family from the resurrection plant *Craterostigma plantagineum* Hochst. Planta 204:459–471

Williams RJ, Leopold AC (1989) The glassy state in corn embryos. Plant Physiol 89:977–981

Index

A

ABA. *See* Abscisic acid (ABA)
ABA-responsive gene, 316
Abscisic acid (ABA), 121, 277, 281, 315–320
Aegilops, 158
AGO9, 135, 136, 138
Alu, 63, 82, 236–238, 258–261
AP. *See* Aspartic proteinase (AP)
Arabidopsis (*Arabidopsis thaliana*), 4, 41, 47, 59, 72, 120, 128–132, 156, 178, 196, 202, 208, 210, 233, 259–261, 287, 301, 317
  A. lyrata, 41, 47, 50–52, 156, 196, 209, 287
ARGONAUTE, 51, 132, 134, 135
Aspartic proteinase (AP), 73, 75, 76
Athila, 27, 94, 101–103, 105, 134, 137, 161, 184, 287
Au-SINE, 264, 265

B

B1, 258, 260, 286
B2, 256, 258, 260, 266
BARE, 9, 78, 81, 276, 277, 281
BC1, 256, 258, 259, 261
BC200, 256, 259
BOT1, 161
Brachypodium distachyon, 2, 3, 10, 96, 186, 196
  Brachypodium repeat annotation consortium (BRAC), 10
Brassica, 4, 161, 301
BraSto, 161

C

CACTA, 6,10, 116, 128, 129, 156, 157, 161, 201, 204, 205
Cassandra, 81, 156, 158
Ceph-SINE, 264, 265
CG methylation, 127–131, 138, 155, 159
CHG methylation, 128, 130, 131, 152
CHH methylation, 129
Chimeric transcripts (cotranscripts), 208, 288
Chromomethylase3 (CMT3), 127, 128, 130, 131, 133, 134
Citrus sinensis (orange), 277
Class I. *See* Retrotransposons (Class I)
Class II. *See* DNA transposons (Class II)
Classification, 1–15, 24, 25, 73, 82, 83, 98
  biological meaning *vs*. pragmatism, 9
  class, 6–7
  evolutionary lineage, 7
  family, 7–8
  Repbase update, 57
  REPCLASS, 25
  "80-80-80" rule, 8–9
  subclass, 6–7
  superfamily, 7
  TEclass, 10–11
  TEclassifier, 6–7, 25
Cluster
  PILER-DF, 20
  RECON, 20
CMT3. *See* Chromomethylase3 (CMT3)
Coffea, 278
Copy-and-paste (replication), 6, 72, 210
CORE-SINE, 256, 262, 263, 265
Cornucopious, 205
Craterostigma desiccation tolerant-1 (CDT-1), 316, 318–322
Craterostigma plantagineum, 314, 316–317
CRM1, 157, 163
Curation (manual), 18, 21, 25–31
Cut-and-paste (excision), 7, 60, 72, 118, 206–207, 210

M.-A. Grandbastien and J.M. Casacuberta (eds.), *Plant Transposable Elements*, Topics in Current Genetics 24, DOI 10.1007/978-3-642-31842-9,

D

DDM1. *See* Decrease in DNA methylation1 (DDM1)
Decrease in DNA methylation1(DDM1), 127–130, 134, 135, 137, 138, 209
Dehydration, 314–321
Desiccation tolerance, 314, 315, 317–322
Dimerization signal (DIS), 76, 83
DIS. *See* Dimerization signal (DIS)
DNA methylation, 42, 52, 61, 126–138, 152, 155, 209, 232, 240, 284
DNA transposons (Class II), 6, 7, 9, 25, 83, 121, 128, 158, 174, 176, 178, 194, 195, 197, 201, 203–205, 207, 209, 222, 224, 226, 234–239, 242, 283, 298, 306
 classification, 25, 128, 174, 194, 195, 203, 298, 306
Domains rearranged methyltransferase2 (DRM2), 127, 132–134
DRM2. *See* Domains rearranged methyltransferase2 (DRM2)
Drosophila, 30, 66, 76, 104, 132, 153, 221, 222, 254, 282
 hybrid dysgenesis, 153, 222
 P elements, 222, 223

E

Endogenous retrovirus (ERV), 90–94, 104–107
 evolution of regulatory networks, 285
 LTR-derived gene promoters, 283–284
 transcription factor binding sites, 284
Endosperm, 136, 137, 152, 153, 209, 210, 240
Endovir1, 95–97, 99, 103, 104
Envelope *(Env)*, 73, 76, 89–112
 capture, 92
 coiled coil, 94, 107
 enveloped viruses, 91, 106
 envelope-like, 90, 93, 94
 Errantiviruses, 93
 membrane fusion, 91, 94, 105, 107
 proteins, 73, 76, 90–94, 104, 107, 226
 transmembrane, 76, 91, 94, 107
Epigenetic control, 51–52, 61, 125–139, 209, 210, 239–241, 279
 gametes, 136, 138
 impact on host, 289
 methylation, 42, 52, 128
 reprogramming of TE silencing, 136–137
 siRNA (smRNA) (*see* Small interfering RNA (siRNA))
ERV. *See* Endogenous retrovirus (ERV)
Evade (EVD), 63, 65, 128, 129, 131, 134, 135, 255–257
Exaptation, 176–178, 219–243, 253–277
 adaptation, 53, 220, 221, 227, 242, 266
 DAYSLEEPER, 229, 232–235, 239
 domesticated transposable elements (DTEs), 223, 228–236
 evolution, 221, 222
 exonization, 222, 236–238, 242
 FHY3/FAR1, 229–231
 frequent birth model, 223–228, 240, 241
 function, 220, 221, 223–232, 234–239, 243
 level of selection, 221, 223, 224, 227, 242
 molecular domestication, 222–227, 231, 233–239, 241–243
 MUSTANG, 233–234
 phenotypic selection, 221, 223–227, 238, 241–243
 protein-coding, 221, 236–238, 241
 regulatory functions, 226
 self-repression, 225, 227
Excision. *See* Cut-and-paste

F

FISH. *See* Fluorescent in situ hybridisation (FISH)
Fluorescent in situ hybridisation (FISH), 181–184, 186, 187

G

Gag, 7, 73, 75, 76, 79, 81, 89, 94, 98, 101–103
Genome size, 1–4, 41–43
 C-value paradox, 2–4
 gene space, 2
Genome size variation, 41–58, 72, 187
 ecological variables, 42
 effective population size, 44, 47, 51
 genome shrinkage, 51
 intron loss, 51
 mechanistic model, 44
 plant development, 42, 138, 152
 positive selection, 51, 210
 removal, 25, 42, 44, 104, 129, 138, 184, 205, 207, 254
Glycine max (soybean), 94, 96, 97, 196, 198
Gossypium (cotton), 49, 51, 97, 157, 161

H

hAT, 6, 116, 158, 174, 175, 177–179, 187, 229, 232, 234, 235, 237, 283
 sugarCane hAT sequences (SChAT), 179
Helitron, 6, 7, 30, 72, 193–217
 annotation, 10, 23, 201–203, 205

autonomous, 195–199, 203, 206, 210, 211
colinearity disruptors, 201
computational discovery, 197
DNA helicase, 195, 206
excision, 207, 210
gene-fragment capture, 193–211
haplotypic diversity, 199
HelitronFinder, 201
HelSearch, 201
nonautonomous, 195–199, 201, 203, 204, 210
plus-minus variation, 201
proliferation, 193–211
RepHel proteins, 197, 198, 211
rolling circle (RC) replication, 194, 195, 206
Histone methyltransferases, 131, 237
H3K9me2, 128, 130–134, 138

I
ID-SINE, 258, 261
Integrase, 7, 73–76, 78, 79, 84, 85, 89, 300

J
Joining TE fragments, 29
MATCHER, 29

K
k-mers
P-CLOUDS, 18
ReAS, 19
REPEATSCOUT, 19
TALLYMER, 18
KRYPTONITE/SUVH (KYP), 130

L
LARD. *See* Large retrotransposon derivative (LARD)
Large retrotransposon derivative (LARD), 73, 81
Late embryogenesis abundant (LEA), 315
LF-SINE, 263–265
Long interspersed nuclear element (LINE), 22, 73, 82, 161, 254, 256, 282, 318
Long terminal repeat (LTR), 8, 74, 75, 78, 84, 275, 277, 278
antisense promoters, 279, 282
binding sites for transcription factors, 279, 282
LTR-derived promoters, 288–290
LTR-driven readout transcripts, 283, 288, 290
promoter/regulatory sequences, 274, 282, 283, 285
U3 region, 180, 181, 276, 280, 289, 300
LTR retrotransposon identification
LTRdigest, 22, 23
LTR_FINDER, 22, 23
LTRharvest, 22, 23
LTR_MINER, 22, 23
LTR_par, 22, 23
LTR_STRUC, 22, 23
MGEscanLTR (find_LTR), 22, 23
SmaRTFinder, 22, 23
LTR retrotransposon, 89, 90, 235, 243, 273
autonomous, 53, 72–79, 83–85, 92, 94, 243, 276, 298
functional impact, 61, 281–290
host gene expression, 274, 281, 282, 287
impact on adjacent gene, 279, 282, 286, 288, 309
lifecycle, 74–76, 79, 81, 85
nonautonomous, 71–86, 260
packaging, 74–76, 81–83, 85, 86
packaging signal (PSI), 76
replication, 72–79, 86
translation, 74–76, 80, 85, 90, 95, 102, 318, 319

M
Mammalian-wide interspersed repeats (MIRs), 263
Mariner, 6, 10, 114, 118, 119, 205, 210, 237, 238, 258
Maverick/*Polinton*, 72
Maximus, 9, 95, 96, 98, 109, 181, 182, 187
McClintock, B., 60, 127, 281
MET1. *See* Methyltransferase1 (MET1)
Methyl-sensitive transposon display (MSTD), 155
Methyltransferase1 (MET1), 127
Miniature inverted repeat transposable element (MITE), 9, 30, 114, 229, 308
autonomous, 115, 116, 118, 134
FindMITE, 22, 23, 116
MAK, 22, 23, 116
MITE-hunter, 22, 23, 116
MUST, 23, 116
SPAT, 116
TRANSPO, 116
transposition, 60, 65, 114, 116, 118, 119, 137, 161, 205
TS clustering, 23, 24
MIRs. *See* Mammalian-wide interspersed repeats (MIRs)

MITE. *See* Miniature inverted repeat transposable element (MITE)
Molecular domestication. *See* Exaptation
Molecular fingerprint techniques, 155
Morgane, 73, 81, 277
mPing, 61, 62, 65, 114, 115, 118, 120, 121, 205, 308
MSTD. *See* Methyl-sensitive transposon display (MSTD)
Multiple alignment
  MAFFT, 20, 26
  MAP, 20, 26
Mutator, 6, 116, 175–177, 228, 233, 234
  Mu killer (Muk), 134, 224, 241
  Pack-MULE (*see* Transduplication)
Mutator-like element (MULE). *See* Mutator

N
Next generation sequencing (NGS), 41–58, 59–70
  AAARF, 53
  de novo assembly, 44
  depth of coverage (DOC), 67
  Illumina, 48, 49, 53, 67
  Nanopore, 67
  Pacific Biosciences, 67
  paired-end mapping (PEM), 63–67
  resequencing, 44
  retrotransposon capture (RC-seq), 63
  split read mapping (SRM), 67
  structural variations, 63
  TEASV detection, 66
  Tos17 mutants, 65
  transpositional landscape, 62
  whole genome shotgun, 66
NGS. *See* Next generation sequencing (NGS)
Nicotiana, 96, 156, 162, 277, 278, 280, 299
  tobacco, 103, 302
Nin-DC-SINE
  AmnSINE, 264
  Deu-SINE, 264
Non-LTR retrotransposon identification
  MGEScannon-LTR, 23
  RTAnalyzer, 22, 23
  SINEDR, 22, 23
  TSDfinder, 22, 23

O
ONSEN, 138, 277, 286
Oryza (rice), 49, 51, 161, 170, 196, 202, 203, 277–279, 302

P
Paired-end mapping. *See* Next generation sequencing (NGS)
Pairwise alignments
  BLASTER, 19, 28
  CENSOR, 28
  MATCHER, 28, 29
  REPEATMASKER, 28
Pararetrovirus, 85, 94
PBS. *See* Primer binding site (PBS)
Pearly-s, 199
Penelope-like element (PLE), 73
Phenotype, 42, 62, 63, 115–118, 220, 224–229, 232, 234, 243, 260, 286, 287, 289, 290, 299, 305, 309, 318
PIF, 114, 116
Pipeline, 21, 24, 28, 29, 31, 32
PLE. *See* Penelope-like element (PLE)
Pol IV, 129, 132, 133, 135, 136
Pol V, 132, 134–136
Polyploidy, 147–168
  allopolyploid, 149, 150, 160–162, 275, 288
  autopolyploid, 151, 164
  bottleneck, 151, 164
  epigenetic change, 149, 163
  genome reorganization, 148–151
  genome shock, 150, 151
  hybridization, 148, 150, 152, 155, 159, 160, 163
  polyploidization, 72, 149, 151, 153, 159, 162, 184
  redundancy, 147, 150, 151
  whole genome duplication (WGD), 45, 147, 148, 171
Polypurine tract (PPT), 73, 76, 77, 300
Population processes
  genetic drift, 44
  natural selection, 44
Post-transcriptional gene silencing (PTGS), 61, 132–135
Post-translational modification of histone, 129–131
PPT. *See* Polypurine tract (PPT)
Primer binding site (PBS), 22, 76, 77, 300
PTGS. *See* Post-transcriptional gene silencing (PTGS)

R
RdDM. *See* RNA-directed DNA methylation (RdDM)
Recombination, 42, 45, 90, 93, 150, 161, 163, 184, 205, 235, 238, 263, 286, 299, 306, 322

Regulatory networks, 121, 221, 226, 230, 238–240, 261, 274, 282–285, 290
Related empty sites, 116
Repbase, 5, 11, 25, 27
Repetitive DNA, 42, 44, 45, 49
REPET package
  TEannot, 8, 9, 18, 19, 28–30, 32, 53
  TEdenovo, 21, 23, 31
Resurrection plant, 314–316
Retrotransposons (Class I), 6, 7, 73, 90, 235, 243
  autonomous, 72–79, 82–85
  classification, 82–83
  nonautonomous, 71–86
Retrovirus, 76, 84, 85, 89–107
Reverse transcriptase, 7, 22, 62, 72–76, 79–81, 84, 85, 89, 95, 161, 180, 182, 276, 300
Riboregulators, 259–260
Rider, 297–309
RNA-directed DNA methylation (RdDM), 61, 65, 127, 129, 132–138, 209, 240
RNA polymerases IV and V, 127
RNA template, 76, 81, 84, 133, 275, 290

S

S_{36}, 199, 201
Saccharum (sugarcane), 162, 171–173, 184
Satellites filtering, 28–29
  PILER-TA, 29
SB1-SINE, 256, 258, 260
Self-alignment
  BLASTER, 19, 28
  PALS, 19
Selfish (junk) DNA, 5
Sequence-specific amplified polymorphism (SSAP), 155, 158, 162
sh2, 199, 200, 208
Short interspersed nuclear element (SINE), 30, 72, 82, 83, 254–271
  Alu/B2, 256
  Au-SINE, 265
  CORE-SINE, 262–265
  distribution, 254
  exaptation, 253–267
  LF-SINE, 256, 263–265
  Nin-DC-SINE, 256, 263–265
  organization, 254
  parasitic mode, 255
  RNA structure, 255–257
  SB1, 256, 258–260
  survival, 255, 265, 267
Short simple repeat (SSR) filtering
  MREPS, 28
  REPEATMASKER, 28
  TRF, 28
Silencing. *See* Epigenetic control
SINE. *See* Short interspersed nuclear element (SINE)
SIRE1, 94–99, 101, 106
Sirevirus, 95–101, 103, 105
7SL RNA, 236, 254, 255, 260, 267
Small interfering RNA (siRNA, sRNA), 2, 42, 51, 61, 152, 159, 222, 226
Solanum lycopersicum (tomato), 96, 298
Sorghum bicolor (sorghum), 3, 96, 170, 196, 198, 278
Spartina, 158, 159
5S RNA, 81, 254, 255, 267
SSAP. *See* Sequence-specific amplified polymorphism (SSAP)
Stem-loop sequence, 194
Stowaway, 9, 114, 116–118, 120
Stress
  defense responses, 275, 276
  structural impact, 280–281

T

Target site duplication (TSD), 22, 24, 116, 117, 195, 197
Tat, 103, 105, 184, 186
TE-anchored PCR strategy, 155
TE-associated structural variant (TEASV), 60, 63, 64, 66–68
TEASV. *See* TE-associated structural variants (TEASV)
Telomerase, 84, 85
Terminal inverted repeat (TIR), 6, 22, 23, 84, 114, 116, 197, 211
  FindMite, 22
  MAK, 22, 23, 116
  Must, 22
  Transpo, 22
Terminal repeat retrotransposon in miniature (TRIM), 73, 81, 158, 159
TGS. *See* Transcriptional gene silencing (TGS)
TIR. *See* Terminal inverted repeats (TIR)
Tissue culture (cell culture, callus, in vitro), 62, 65, 134, 180, 275, 277, 278, 313, 316–319
Tnt1, 156, 158, 162, 276, 277, 280, 286, 288, 289, 302
Tos17, 61, 62, 65, 276, 277, 287, 302

Tourist, 114–118, 120
Transcriptional gene silencing (TGS), 61, 132–135, 137
Transcription factors, 74, 75, 134, 138, 156, 229, 238, 302–306, 319
Transduplication, 207
  coevolution, 84, 241
  gene creation, 208
  gene-fragment capture, 204
  Helitrons, 206–208
  pack-MULES, 60
Transposase, 7, 25, 79, 84, 90, 114, 117–119, 121, 126, 127, 194, 197, 199–201, 203, 222–224, 226, 228, 232, 233, 237–239, 241
TRIM. *See* Terminal repeat retrotransposon in miniature (TRIM)
Triticum (wheat), 3, 96, 156–160, 170, 203, 277, 278
tRNA, 2, 76, 77, 82, 84, 132, 254–256, 258–260, 263, 264, 267, 303, 319, 320
Tsc1/Tcs2, 277, 289
TSD. *See* Target site duplication (TSD)
Tto1, 128, 156, 275–277
Ty1/Copia, 6, 9, 73, 76, 85, 90, 94, 95, 99, 104, 106
Ty3/Gypsy, 6, 73, 76, 85, 90, 94, 96, 101–104
  envelope glycoprotein, 93, 106

V

Virus-like particle (VLP), 74, 79, 85
VLP. *See* Virus-like particle (VLP)
V-SINE, 264, 265

W

Wis2, 156, 158, 160, 288

Z

Zea (maize), 3, 44–47, 50, 51, 56, 170, 175, 178, 270, 277, 278